Service Industries and As
Cities

T0221286

Contributed by distinguished scholars from all over the world, this volume convincingly critiques conventional thinking about service industries and richly documents urban services across the Asia-Pacific region from not only economic but social, political and geographic perspectives.

Professor Cindy Fan, *Department of Geography, University of California, Los Angeles*

A timely contribution to our understanding of the dynamism of the region's economic landscapes, bringing together a number of fascinating studies in an agenda-setting volume.

Niall Majury, *School of Geography, Queen's University Belfast*

During the second half of the twentieth century, development in the Asia-Pacific region has been dominated by industrialisation. However, at the beginning of the twenty-first century, services, in particular finance, information and creative services, have become deeply embedded in the processes of urban growth.

In the Asia-Pacific the rise of service industries has led to national modernisation programmes and globalisation strategies. Services are also driving change in the internal form of city regions and are being actively deployed as instruments of metropolitan reconfiguration and land use changes. These changes have created problems such as social polarisation and the displacement of traditional industries and residential districts. Also, there are tensions between local and global processes in the development of service industries, and between the imperatives of competitive advantage and sustainable development.

Service Industries and Asia-Pacific Cities brings together a multi-disciplinary team of experts to explore and illustrate the theoretical, conceptual and practical issues arising from the transformation of Asia-Pacific cities by service industries.

P. W. Daniels is Professor of Geography at the University of Birmingham. **K. C. Ho** is Associate Professor of Sociology at the National University of Singapore. **T. A. Hutton** is a Faculty Associate in the Centre for Human Settlements at the University of British Columbia in Vancouver.

Routledge studies in the growth economies of Asia

Service Industries and Asia-Pacific Cities

New development trajectories

Edited by P. W. Daniels, K. C. Ho and T. A. Hutton

Routledge
Taylor & Francis Group

LONDON AND NEW YORK

First published 2005
by Routledge
2 Park Square, Milton Park, Abingdon, Oxfordshire OX14 4RN

Simultaneously published in the USA and Canada
by Routledge
711 Third Avenue, New York, NY 10017

Routledge is an imprint of the Taylor & Francis Group

First issued in paperback 2011

© 2005, editorial matter and selection, P. W. Daniels, K. C. Ho and
T. A. Hutton; individual chapters, the contributors

Typeset in Baskerville by Wearset Ltd, Boldon, Tyne and Wear

All rights reserved. No part of this book may be reprinted or
reproduced or utilised in any form or by any electronic, mechanical,
or other means, now known or hereafter invented, including
photocopying and recording, or in any information storage or
retrieval system, without permission in writing from the publishers.

British Library Cataloguing in Publication Data
A catalogue record for this book is available from the British Library

Library of Congress Cataloging in Publication Data
A catalog record for this book has been requested

ISBN 978-0-415-32749-7 (hbk)
ISBN 978-0-415-51395-1 (pbk)

Contents

Figures

Tables

Contributors

W. B. Beyers is Professor of Geography at the University of Washington. He has undertaken research and published papers on a variety of topics related to service industries over the past several decades. This work has ranged from national-level analyses of trends in employment to case studies based on survey data addressing basic research needs on services, particularly producer services. Key contributions have focused on trade in services, bases of competitive advantage and demand, and the importance of externalisation processes on growth. He has also written in recent years about cultural industries, the new economy and structural change in input–output models.

John Bowen is an Associate Professor of Geography at the University of Wisconsin Oshkosh. His research has focused on air transport, particularly in the context of Southeast Asia where he has examined airline liberalisation, the role of air transport in economic development and the use of air cargo services in global production networks.

Ji-Sun Choi is Associate Research Fellow at the Science and Technology Policy Institute, Seoul, Korea. She is interested in understanding the digital economy and the knowledge-based economy from a spatial perspective. A journal paper on this topic entitled 'Spatial analysis of transactions that use e-catalogs in public business-to-business electronic marketplaces by business model in Korea' is published in a special issue of *Urban Geography* (2004). In addition, as an economic geographer, she has also carried out several research projects on innovative clusters and regional innovations systems (RIS) in Korea.

P. W. Daniels is Professor of Geography and Deputy Dean of Physical Sciences and Engineering, University of Birmingham. He has undertaken research and published several articles and books on the location and development of office activities and on the role of service industries, especially producer services, as key drivers of metropolitan and regional restructuring at the national and international scale. He is the author of *Service Industries: A Geographical Appraisal* (1985), *Service Industries in* *the World Economy* (1993), *Service Worlds: People, Organisations, Technologies* (with J. R. Bryson and B. Warf) (2004).

David W. Edgington is Associate Professor of Geography and Director, Centre for Japanese Research, University of British Columbia. His research interests focus on trade and investment in the Pacific Rim, the geography of Japanese multinational corporations, as well as urban and regional change in Japan. He is editor of *Japan at the Millennium: Joining Past and Future* (2003); and co-editor of *Australian and Canadian Approaches to Asia in an Era of Unstable Globalization* (with Terry G. McGee) (2004).

R. Hayter is Professor of Geography at the Simon Fraser University. His main research interests have focused on British Columbia's forest economy and on the economic geography of Pacific Rim industrialisation, especially with respect to Japanese production systems. He is the sole author of *The Dynamics of Industrial Location: The Factory, The Firm and the Production System* (1997) and *Flexible Crossroads: The Restructuring of British Columbia's Forest Economy* (2000).

K. C. Ho is Associate Professor of Sociology at the National University of Singapore. An urban sociologist by training, Dr Ho's research interests include the political economy of cities, economic restructuring and sub-regional development, issues of information society, youth and leisure. He is co-author of *City-States in the Global Economy: Industrial Restructuring in Hong Kong and Singapore* (Westview, 1997) and co-editor of *Culture and the City in East Asia* (Oxford, 1997), *Critical Reflections on Cities in Southeast Asia* (Brill, 2002) and *Asia Encounters the Internet* (2003).

Thomas A. Hutton is a Faculty Associate in the Centre for Human Settlements (affiliated with the School of Community and Regional Planning), University of British Columbia in Vancouver. He has been published widely in the geographical, urban studies and planning literatures on the role of service industries in processes of urban change. His current research interests include services and urban transformation within the Asia-Pacific region, the 'new economy' of the inner city and the role of planning and public policy in reshaping the space-economy, social structure and land use patterns in the central city. Tom Hutton's books and monographs include *The Transformation of Canada's Pacific Metropolis: a Study of Vancouver* (1998) and *Service Industries, Globalization, and Urban Restructuring within the Asia-Pacific* (2004).

Michael Leaf is Director of the Centre for Southeast Asia Research (CSEAR), University of British Columbia (UBC), Vancouver, Canada, an Associate Professor, UBC School of Community and Regional Planning (SCARP) and a Research Associate of the UBC Centre for Human Settlements (CHS). The focus of his research and teaching has been on urbanisation and planning in cities of developing countries, with particular interest in Asian cities. Since his original doctoral research (PhD Berkeley, 1992) on land development in Jakarta, Indonesia, Dr Leaf has been extensively involved in urbanisation research and capacity building projects in Indonesia, Vietnam, Thailand, China and Sri Lanka.

Thomas R. Leinbach is Professor of Geography at the University of Kentucky, Lexington. He is editor of *Growth and Change: A Journal of Urban and Regional Policy* published by Blackwell. His research interests include the interlocking roles of transportation and development especially in the context of Southeast Asia. The impact of information, technology and especially e-commerce, at a variety of scales in the globalisation process is also a focus of his research. He is the co-author of *Southeast Asian Transport: Issues in Development* (1989), co-editor of *Worlds of E-Commerce: Economic, Geographical and Social Dimensions* (2001) and, most recently, *The Indonesian Rural Economy: Mobility, Work and Enterprise* (2004).

David Ley holds the Canada Research Chair in Geography at the University of British Columbia in Vancouver. His research examines social and economic restructuring of downtown and inner city districts, including studies of housing and labour markets, neighbourhood organisations and landscapes of consumption. From 1996–2003 he was the UBC Director of the Metropolis Project, an interdisciplinary and international study of immigration and urbanisation. Author of *The New Middle Class and the Remaking of the Central City* (1996), he is presently writing a book on the recent Chinese diaspora in Canada.

George C. S. Lin is Associate Professor of Geography at the University of Hong Kong, is the author of *Red Capitalism in South China: Growth and Development of the Pearl River Delta* (University of British Columbia Press, Vancouver, 1997) and has published a large number of articles on Chinese urban and regional development.

Larissa Muller is Research Scholar at the Center for Information Technology Research in the Interest of Society (CITRIS) at the University of California, Berkeley. A PhD in city and regional planning, she has consulted extensively on urban economic and environmental policy issues in Southeast Asia and China over the last ten years. Her research interests include the development dynamics of knowledge industries, and comparative urbanization dynamics in extended metropolitan regions in East Asia and North America. Dr Muller is currently researching the impact of American investment in the high-tech sector in Southeast Asia.

Kevin O'Connor is Professor of Urban Planning, Faculty of Architecture, Building and Planning at the University of Melbourne. His teaching and research explores the links between the economic system (particularly manufacturing and services) and the growth and internal structure of cities. A book entitled *The New Economic Geography of Australia: A*

Society Dividing (Oxford University Press, 2001) outlines the results of Australian applications of this research. His international focus is illustrated in the editorship of a special issue of *Asia Pacific Viewpoint* which presented the first stage results of a study into producer services and city development in the Asia-Pacific region.

Sam Ock Park is Professor of Economic Geography at Seoul National University. He is presently the President of Korean Geographical Society and the Chair of the International Geographical Union Commission on the Dynamics of Economic Spaces. He has undertaken research and published many articles and books on the locational dynamics of economic activities and regional development, especially on high tech industries and regional innovation systems. He is the author of *Modern Economic Geography* (1999) and co-editor of *The Asia Pacific Rim and Globalization* (with Richard Le Heron) (1995).

Kris Olds is Associate Professor of Geography at the University of Wisconsin-Madison. His research focuses on the geographical organisation of global networks in relation to contemporary urban transformations. Recent publications include (as co-editor) *Globalization and the Asia-Pacific: Contested Territories* (1999), *The Globalization of Chinese Business Firms* (2000) and (as author) *Globalization and Urban Change: Capital, Culture, and Pacific Rim Mega-Projects* (2001).

Nigel Thrift is Head of the Division of Life and Environmental Sciences, Professor of Geography and a Student of Christ Church at the University of Oxford. he is also an Emeritus Professor of Geography at the University of Bristol. His main research interests are in cultural economy, cities, information ecologies and the history of time. His most recent publications include *Cities* (with Ash Amin, 2002), *The Cultural Economy Reader* (co-edited with Ash Amin, 2003), *Patterned Ground* (co-edited with Stephan Harrison and Steve Pile, 2004) and *Knowing Capitalism* (2004).

Anthony G. O. Yeh is Chair Professor in the Centre of Urban Planning and Environmental Management and Chairman of the Task Force on Hong Kong–Pearl River Delta Development of the University of Hong Kong. He is currently researching the development of producer services in the region. He specialises in urban planning and development in Hong Kong and China and the applications of GIS in urban and regional planning. He is an Academician of the Chinese Academy of Sciences and Fellow of a number of professional planning institutes. He is at present Secretary-General of the Asian Planning Schools Association and Asia GIS Association. He has served in a number of consultative committees of the Hong Kong Government, including the Pearl River Delta Panel of the Central Policy Unit. He has published over 30 books and monographs and over 140 journal articles and book chapters.

Preface

Following an inaugural meeting in Sydney in 1996 of a small interdisciplinary group of geographers, sociologists and planners interested in the global shift to service industries and its implications for urban growth and change within the Asia-Pacific, a broad consensus emerged for an ongoing dialogue. The group met again during a session of the North American Regional Science Association Meeting in Hawaii in 1998 and, finally, through the good offices of Tom Hutton and the Peter Wall Institute for Advanced Studies, University of British Columbia, was again able to come together for a two-day workshop in March 2002. These meetings have brought together international service scholars, leading Asian urban theorists and important Asian urban/regional specialists to provide a rich blend of theoretical and empirical work that is at the frontier of social science research and which will also inform the development of an ongoing research agenda. Their invaluable additional insights, along with the accumulation of ideas and research by the core group, suggested that it would be timely and appropriate to assemble the research papers as an edited collection.

Central to this conviction is the premise that the experience of service industry growth within the Asia-Pacific has not followed the model based on the patterns of the old 'Atlantic core'. Industrialisation has been the dominant development paradigm for Asia-Pacific nations and regions for much of the second half of the twentieth century. Nowhere is this more evident than in the emergence of world scale industrial metropolises such as Tokyo, Los Angeles, Hong Kong, Seoul and Shanghai; they are prominent symbols of this process. However, during the last decade of the twentieth, and at the beginning of the twenty-first, century services, and more specifically advanced services (finance, business or 'producer services', information, and creative services), have become more deeply embedded in the processes of urban growth and change in the region.

These influences are manifest in at least three ways. First, rapid tertiarisation is fundamentally modifying urban functions or 'vocations', as reflected in the recomposition of metropolitan industrial structure, export base and gross domestic product. In the Asia-Pacific the growth of advanced services is a concomitant of advanced production systems, national modernisation programmes and the globalisation strategies of central and regional governments. Second, services are drivers of change in the internal form of the regions' cities and mega-urban regions and are being actively deployed as instruments of metropolitan reconfiguration and land use change. Examples include the massive Pudong international financial project in Shanghai, the reclamation of land for central business districts in Singapore and Hong Kong or the Multimedia Supercorridor south of Kuala Lumpur. Third, the rapid growth of service industries and related occupations is leading the re-formation of social class within Asia-Pacific cities, including the emergence of a powerful 'service elite' of senior managers, professionals and entrepreneurs. There is also a fast growing contingent of lower-paid service workers, shaping a distinctly bifurcated and polarised service work force.

In addition to these, and other, *issues* associated with the rise of service industries in Asia-Pacific economies, there are also corresponding sets of *problems* that underscore the seminal significance of tertiarisation within the region. There are at least three problems. First, there are tensions between local and global processes in the development of service industries, and between the imperatives of competitive advantage and sustainable development. Second, there is a tendency for advanced, higher-order services to displace or to encroach upon other urban land uses such as 'informal services', traditional industries and residential districts. Third, the problems of social polarisation may be exacerbated by the growth of advanced service occupations, as well as by the 'new social divisions of service labour' exemplified by the rise of cultural and technology-based service industries within inner city districts.

This collection of essays by some of the leading scholars in the field therefore has four main aims:

- To consolidate current thinking about the relationship between the rise of the service economy, global city formation and development trajectories in the Asia-Pacific.
- To highlight the impact of the development of service industries in the Asia-Pacific on the economic, social and spatial reconfiguration of cities in the Asia-Pacific.
- To inform understanding of the role of intra-regional and extra-regional processes on the transformation of cities in the Asia-Pacific using case studies of selected service sectors and individual cities.
- To explore the role of service industries in shaping strategic policy choices for urban governments within the Asia-Pacific region.

We believe that the papers presented in this volume represent innovative and scholarly responses to research questions raised by the emergence of service industries as influential features of development within the Asia-

Pacific. At the same time, we readily acknowledge the complexity, dynamism and differentiation of tertiarisation experiences within the region, and therefore look forward to new collaborative ventures in this important research domain.

P. W. Daniels
K. C. Ho
Thomas A. Hutton
Birmingham, Singapore and Vancouver
27 July 2004

Acknowledgements

The editors would like to acknowledge the support of the Peter Wall Institute for Advanced Studies (PWIAS), University of British Columbia, which provided funding to bring together scholars from a number of different disciplines at an Exploratory Workshop, held in Vancouver in March 2002. PWIAS also made available its excellent conference and residential facilities on the UBC campus for the duration of the Workshop. The Institute supports fundamental, innovative research in the humanities and social and natural sciences. We would like especially to thank Dr Kenneth MacCrimmon, (former) Director of PWIAS, who assiduously encouraged Tom Hutton to put together an application for funding to investigate the critical relationships between services growth and urban change within the Asia-Pacific region. We hope the willingness of the participants in the Workshop to allow their contributions to be brought together in this edited volume will provide the basis for an ongoing relationship with colleagues at PWIAS as we seek to move the research forward in this salient social science domain.

P. W. Daniels
K. C. Ho
Thomas A. Hutton

1 Service industries and Asia-Pacific cities

Introduction and overview

P. W. Daniels, K. C. Ho and T. A. Hutton

The context

The industrialization of Asia-Pacific national and regional economies over the post-war period, accelerated in many cases by assertive state policies and programmes, is one of the most remarkable experiences of economic development in the modern era. There has of course been considerable variation in the nature of the industrialization experience from place to place, contingent upon historical factors, the nature of political control and issues of scale, among other variables. In just 30 years or so many Asia-Pacific nations have navigated the progression from essentially agrarian and basic industrial economies, to export-driven manufacturing growth, and on to advanced industrial status. The most advanced economies in the Asia-Pacific region are home to manufacturing corporations that utilize the very high levels of capital and technology inputs that enable them to compete effectively in international markets against the most sophisticated European and American competitors across a diverse range of product lines. The process has been so rapid that descriptors such as 'newly-industrializing countries' (NICs) or 'newly-industrializing economies' (NIEs) are still occasionally deployed for societies such as Taiwan, Hong Kong and Singapore, although such usage in this context is clearly anachronistic.

Industrialization is of course a complex and dynamic process, rather than an 'end-state', so that Asia-Pacific nations continually generate new episodes in the industrial development story. These include the well-known 'hollowing out' of traditional manufacturing industry in parts of the region that industrialized early, notably Japan, Hong Kong and Singapore, with the attendant relocations (and dislocations) of industrial capacity and labour; the formation of 'industrial technopoles' such as the well-known example of the Hsinchu Science Park in southern Taiwan; and the continuing growth of industrial production in the 'near NICs' (Malaysia and Thailand). We can also cite more tentative industrialization programmes in places such as Indonesia and the Philippines, or the 'incipient' industrialization experiments in lagging regions such as Laos and Myanmar. In Vietnam, an ideological commitment to market-driven industrialization under the so-called 'doi moi' (economic renovation) rubric underpins efforts to attract foreign direct investment (FDI) for exigent investments in capital plant and equipment. But in many ways, notwithstanding the services focus of this volume, the biggest 'story' in the Asia-Pacific industrialization experience is the quite remarkable growth of manufacturing, for export as well as for domestic consumption, in China. The scale and growth rate of Chinese manufacturing is in some respects *sui generis* in the history of economic development since the Industrial Revolution of the eighteenth and nineteenth centuries. China is emerging as the 'world's factory' for the complete range of consumer (and, increasingly, capital) goods.

Manufacturing growth has not occurred evenly across the region or within individual countries. There are important spatial expressions of industrialization within the Asia-Pacific, represented in particular by a number of world-scale industrial metropoles that emerged within the broadly-defined region over the last quarter of the twentieth century. These metropoles, defined both by scale and industrial specialization, include Tokyo, Nagoya, Seoul, Shanghai, Hong Kong, Taipei, Bangkok and Singapore, within East and Southeast Asia, and Los Angeles, San Francisco and Seattle on the American Pacific Rim. This urbanization–industrialization nexus was of course observed earlier in the development of more mature economic systems within the Atlantic realm. But the development of Asian industrial conurbations reflects an even stronger influence of primacy and hierarchy within national urban systems, driven by agglomeration, labour market conditions, scale considerations and policy factors. In addition, rapid industrialization was a central feature of growth and development among large Asia-Pacific cities at a time coinciding with catastrophic deindustrialization among many of the old industrial societies of Europe and eastern North America.

Against this backdrop of industrialization as represented by the phenomenon of the 'factory world' and affiliated labour markets, social classes and communities we can discern the rapid expansion and development of service industries that has been taking place within the Asia-Pacific (Daniels, 1998). This (relative and absolute) growth in service industries, labour and output (or *tertiarization*) has incorporated large cohorts of informal sector workers located in many countries within the region, and services-type employment ancillary to industrial production. Increasingly, however, the tertiarization process has included expansion of the 'quaternary' sector (after Jean Gottmann, 1970) or higher-order services occupations and industries: highly-skilled professionals, managers, administrators and entrepreneurs, situated within the full range of advanced service sectors and industry groups: banking and finance; intermediate or 'producer' services; the 'New Economy' driven by information and communications technology (ICT); higher education; and the public service.

Given the relatively recent emergence of services as key actors in the region's economic firmament, what are the development implications for its nations, regions and societies? There has of course been a debate over the 'meaning' of services growth since Daniel Bell published his forecast of redefining economic and social change, *The Coming of Post-Industrial Society* (1973). Constructed largely, although not exclusively, along ideological lines, the debate included critics on the political left decrying the class and cultural implications of Bell's post-industrial scenario, while economists, geographers and others were simultaneously engaged in refuting the idea of an 'autonomous', free-standing services economy supplanting manufacturing as the basis of productive enterprise and labour (see for example Gershuny, 1978). Some well-known scholars still prefer to interpret the growth of services in metropolitan cities as essentially another 'evolving discourse' of industrial urbanism (Soja, 2000), rather than a development trajectory in its own right. Nevertheless a consensus has been reached among less polemically-oriented scholars that service industries and manufacturing constitute co-dependent elements of advanced production systems (for example Coffey, 1994).

The intricacies of this debate aside, it seems clear that the rapid growth of services represents a distinctive (though not autonomous) phase of economic development within many Asia-Pacific societies. It merits a concerted programme of basic and applied research. At the national level, the expansion of service industries is reflected in shifts in the sectoral composition of investment, industries and enterprises, labour, output and trade, although there is significant variation within the region, reflecting contrasts in stages of tertiarization, as well as the relative balance of services and manufacturing capacity within national economies. In the 'leading edge' service economies within the region, such as the US (California, Washington State), Australia, Japan, South Korea, Taiwan and Singapore, the shift towards services is more pronounced, although manufacturing maintains a stronger position within each economy than is the case in most of the old Atlantic core countries, with the notable exception of Germany.

While the expansion of service industries and employment is expressed at the national level within the Asia-Pacific, many of the economic and social impacts are manifest among major cities and city-regions. Just as the preceding industrialization phase was accompanied by the formation of propulsive industrial metropoles, so has the rapid expansion of services created a new round of urban development based on specializations in banking and finance, producer services, knowledge-based industries and 'gateway' functions. Moreover, the increasing influence of global processes, with their attendant privileging of competitive advantage, tends to reinforce functional specialization of services among Asia-Pacific cities. Globalization has also exacerbated inter-city competition for high-value, higher-order services, while imparting a greater degree of volatility or destabilization to local labour and property markets (Daniels, 1993). The mutually reinforcing tendencies of globalization and tertiarization have in some cases increased the dominance of large metropolitan cities within national urban systems, as in the example of Australia among the advanced economies of the region (O'Connor *et al.*, 2001) and in the case of China among transitional economies (Lin, 2002).

In cities such as Tokyo, Hong Kong, Singapore, Sydney, San Francisco and Vancouver, services are now the dominant influences on urban growth and change, the sectoral composition of the regional economy, new divisions of labour, demand for housing and patterns of everyday life. Strategic-level impacts of rapid service industry expansion within the Asia-Pacific region include the following:

- Rapid tertiarization is fundamentally modifying urban functions or 'vocations', as reflected in the recomposition of the metropolitan industrial structure, export base and regional domestic product. In the Asia-Pacific, the growth of higher-order services is a concomitant of advanced production systems, national modernization pro-grammes, and the globalization strategies of central and regional governments (Sassen, 1991; Edgington, 1991; Nanjiang, 1996).
- Services are increasingly drivers of change in the internal structure and form of Asia-Pacific city-regions and are being actively deployed as instruments of metropolitan reconfiguration and land use change. Examples include the massive Pudong international financial project in Shanghai, the reclamation of land for central business districts in Hong Kong and Singapore, and the Multi-Media Super Corridor in Malaysia (Wu and Yeh, 1999; Olds, 2001; Bunnell, 2002).
- The rapid growth of service industries and related occupations is leading the reformation of social class within Asia-Pacific cities, including the emergence of a powerful 'service elite' of senior managers, professionals and entrepreneurs. There is also an expanding contingent of lower-paid service workers, shaping a distinctly bifurcated and polarized service work force (Ho, 1997).

There are also important implications of rapid growth in specialized services for 'extended' mega-urban regions in the Asia-Pacific (see McGee and Robinson, 1995), expressed in the emergence of multi-centred service clusters or corridors. These extended formations encompass global-scale concentrations of specialized service functions, and are exemplified by the Pearl Delta region in southern China, the Tokyo-Kanto Plain region of central Honshu, and the San Francisco Bay Area in northern California (Hutton, 2004a).

In addition to these processes, there are also corresponding sets of more problematic issues that underscore the growing significance of tertiarization within the region. These problems include:

- Tensions between global imperatives and local interests in the development of service industries, and between the values of competitive advantage and sustainable development and competitive advantage (Thornley, 1999; Ho, 2002). These conflicts are manifest clearly among the region's 1st and 2nd-order global cities, but, are also evident within medium-size cities, underscoring the pervasive effects of inter-city competition associated with the pressures of globalization (Douglass, 2002).

- The tendency for advanced, higher-order services and tourism-related activities to displace or to encroach upon proximate land uses, including informal communities and traditional industries (Leaf, 1996a). In the Ancient Quarter (district of Hoan Kiem) of Hanoi, the imposition of a new trading and tourism economy has supplanted a vibrant, central city, craft production industrial regime dating back to the thirteenth century, severing in the process long-established linkages between the city and the countryside (Turner, 2004).

- Problems of social polarization which may be exacerbated by the expansion of advanced service occupations (Leaf, 1996b).

It is also the case that key service industries within the Asia-Pacific region are particularly vulnerable to exogenous and localized shocks, as in the case of the currency crisis and broader economic and political fallout of 1997 (which impacted banking, finance and associated producer services, as well as public services dependent on buoyant taxation revenues), the effects of 9/11 and the Bali nightclub bombing (air travel, tourism and conventions, FDI), and the recurrent SARS crisis (tourism, retail and personal services).

At a theoretical level, the search for more robust models of urban growth and change within the Asia-Pacific of the kind proposed by Lin (1994) must therefore incorporate the impacts of tertiarization, both as a key process and as the provenance of social, economic and environmental problems. The expansion of service industries and employment is an important entrée into the development of urban theory in the region (O'Connor and Daniels, 2001). Examples of specific theoretical domains subject to the increasing centrality of services growth include theories of urban transition and transformation (Sassen, 1991; Nanjiang, 1996), services and urban spatial models (McGee and Lin, 1993; Wu and Yeh, 1999), and service occupations and social change (Ho, 1997).

There are some early examples of research documenting the contours and implications of tertiarization within the Asia-Pacific, such as the seminal study of service exports from the Seattle-Puget Sound Region (Beyers *et al.*, 1985), work on the role of services in Vancouver's industrial restructuring (Davis and Hutton, 1981) and Asia-Pacific market orientation (Davis and Hutton, 1991, 1994), or research on services and global vocations for Hong Kong (Taylor and Kwok, 1989). But these are the exceptions rather than the rule; in practice relatively little has been published on the role of services in urban development in the Asia-Pacific. In view of its growth trajectory, developmental impacts and normative implications, it therefore remains one of the most exciting and potentially fruitful domains of urban inquiry in the early years of the twenty-first century.

Points of departure and purpose of the book

This book, which we have the privilege to co-edit, represents a collective intellectual enterprise, with an overarching purpose of addressing important *developmental* issues associated with the expansion of service industries and employment within the Pacific realm. In the selection of themes and issues, we have been guided by three salient principles or assumptions. First, we contend that the expansion of service industries and labour within the Asia-Pacific, while not an 'autonomous' sectoral orientation as claimed by some of the early post-industrialists, nonetheless constitutes an important new development trajectory within the region. In this respect we cannot accept the position of some scholars among the industrial urbanism contingent that the growth of services industries can be simply subsumed within the broader industrialization process and associated literatures and discourses.

Second, we acknowledge at the outset that the tertiarization process and experience within the Asia-Pacific does follow in some respects the broad outline of the trajectories shown by the mature economies of the Atlantic core, but that there are some significant contrasts. These include, for example, the juxtaposition of advanced services within large informal tertiary sector cohorts within many Asia-Pacific societies, the linkages between services and buoyant manufacturing sectors and the stronger influence of state developmental (as opposed to regulatory) policies in the growth of services, especially advanced, higher-order intermediate service industries. In addition to these salient contrasts, the rate of growth in services in some regions of the Asia-Pacific has been demonstrably greater than has ever been the experience throughout much of Europe and the old industrial (or post-industrial) regions of eastern North America.

Third, it seems very clear to us that an investigation of the experience of services growth in cities, complemented by studies of key service industry groups in the region, offers the richest field of inquiry for social science and normative scholarship. While (as noted) industrialization is closely linked to urbanization within the Asia-Pacific, it can be argued that the interdependence between cities and tertiarization is even more intimate, especially for advanced services. The concentration of services within Asia-Pacific cities contrasts both with the spatial dispersion of manufacturing in some Asian regions (notably in China), and with the marked tendency towards the decentralization of many service industries in North America. Certainly the urban setting offers the most promising opportunities for investigating nuanced features of services growth, including implications for the reformation of social class, as well as for exploring critical processes of agglomeration and clustering which constitute such fundamental aspects of modern service industry development.

The broad purpose of *Service Industries and Asia-Pacific Cities: New Development Trajectories* is to explicate the increasingly central role and influence of service industries on the transformation of cities and city-regions within the broadly-defined region. The book engages (and critiques) the extant theoretical literature, and contributes innovative models of services development in this critical new landscape of globalization, industrial restructuring and urban change. We believe that this book collectively contributes to a scholarly appreciation of the increasing centrality of service industries to many facets of urban change, as well as to strategic shifts in the trajectories of city-regions within the Asia-Pacific. At the same time we acknowledge that this collection marks an opening statement rather than a definitive position on the dynamic, complex and consequential processes under investigation. We therefore look forward to a continuing debate and discourse on the implications of tertiarization in the region, both with the authors represented in this volume, as well as with other colleagues and prospective collaborators.

Thematic overview: a pathway through the book

The structure of the book embodies the interdisciplinary quality of contemporary urban scholarship and discourse, consistent with the conduct of this collaborative research project. It is through such interdisciplinary research models that the synergies and spillover effects between industrial, social, cultural and spatial processes come into analytical focus. We have also endorsed the progressive intellectual posture that multi-perspective research on complex, dynamic problems such as those addressed in this project can yield particularly rich insights and new knowledge. The contributions to the volume are grouped into three major, inter-related themes and these are now summarized on a chapter-by-chapter basis.

I. Services and urban development in the Asia-Pacific: theory and context

Globalization and the persistence of exceptionalism

Industrial restructuring has been a defining feature of the 'industrialization paradigm' within the Asia-Pacific, driven both by exogenous and endogenous factors. The former include the increasingly pervasive influence of global forces, including the familiar contours of liberalizing financial and property markets, international migration and increasingly mobile labour both in elite and low-wage occupations, and the Asia-Pacific's leading role in the 'new international division of labour' (see for example Ho, 1994). However widespread the forces of change, the experience of globalization and restructuring in the region, far from conforming to a generic 'template' of trends, has been in important respects highly differentiated from place to place (as well as distinct from the earlier processes of urban tertiarization in Europe and North America). In the field of services trade, for example, the commitment of Asia-Pacific nations has been somewhat qualified, as a number of states endeavour to provide some regulatory bulwarks against the potentially damaging effects of unbounded trade in services, while at the same time striving to build regional and international competitive advantage.

Immediately following this Introduction, Peter Daniels ('Services, globalization and the Asia-Pacific region', Chapter 2) provides an empirical overview and conceptualization of processes and trends of services growth among the region's constituent economies. This chapter seeks to place the development of the Asia-Pacific's service industries (and more particularly its impact on the role and function of the region's cities) within the wider context of trends in the global services economy. To a considerable (and growing) extent the economies of the Asia-Pacific are integrated within global systems of trade and investment, but much of the region's economy actually operates on an intra-regional basis, with (significant and expanding) inter-regional flows. Within the Asia-Pacific there are important variations in the 'openness' of economic systems for services trade, as evidenced by the very different policy and regulatory regimes administered by (for example) China, South Korea and Japan. Technological innovation, including production technologies as well as information and communications technologies (ICT), is enabling a number of Asia-Pacific economies to 'leap-frog' development in services, although ICT can have a centralizing (or reconcentrating) effect as well as dispersion possibilities. The chapter concludes with a case study of the goods–services gap in Japan as a way of demonstrating the continuing difficulties of trying to unravel the role performed by an apparently limited business and professional services sector in a major regional industrial economy that has competed successfully within the region, as well as globally.

One of the key variables in the differentiated trajectories of services growth within the Asia-Pacific is that of public policy, including local and regional planning policy and development strategy, as well as expressions of state policy. Thomas Hutton's contribution ('Services and urban development in the Asia-Pacific region: institutional responses and policy innovation', Chapter 3) includes a discussion of the role of multi-level public policy interventions in the formation of urban service industry complexes, and in the (incipient) shaping of extended specialized service clusters or 'corridors', the latter a response in part to the pressures and opportunities of globalization. Chapter 3 underscores the critical role of development policy (as opposed to more stringent regulatory regimes characteristic of the mature economies of the 'Atlantic core') among the nations and city-regions of the Asia-Pacific, and presents as a conceptual heuristic a framework of strategic policy influences in the region since the 1980s. This approach recognizes benchmark developments in the progression of policy models for services, while acknowledging the pervasiveness of exceptionalism derived from the very different development histories, policy cultures and governance structures among states and cities in the region.

Labour markets, occupational change and social class reformation

As experienced earlier in Europe and the mature regions of eastern North America, the expansion of service industries is associated with industrial restructuring, entailing new divisions of production labour, occupational change and the reformation of urban social class structures (Ley, 1996). There are, however, significant contrasts with the experience in the older economies of the 'Atlantic core', including the coincidental growth of employment both in manufacturing and services in many Asia-Pacific societies (although not in some of the more advanced cases, such as Japan, Hong Kong and Singapore). The social and spatial juxtaposition of higher-order services with large informal sector cohorts among many Asian cities also stands as a significant element of contrast set against the record of economies within the Atlantic realm. This makes it very clear that an incisive research effort is justified.

David Ley opens this section with an investigation of the socio-spatial structures that accompany the development of the urban service economy ('The social geography of the service economy in global cities'). His analysis engages literatures that include the post-industrial city, the world city and the gateway city, as key informing concepts. As a positioning gesture Ley revisits the axial principles of Daniel Bell's contested (but in many ways prescient) theory of post-industrial society, including a judgement that in many salient areas Bell's forecast of social and cultural change has been demonstrated to be essentially correct. His research questions include the exigency of reconciling contradictory accounts of polarization and professionalization, the reconstitution of family forms and gender associated with industrial restructuring, and the connections between restructured urban labour markets and housing markets, drawing on the experiences of Toronto, Vancouver and London. Ley's contribution concludes with conjecture on the broad reconfiguration of social class in Asia-Pacific cities associated with the expansion of advanced services.

Next, K. C. Ho addresses the 'sociology of the service economy' (after Saskia Sassen) in 'Service industries and occupational change: implications for identity, citizenship and politics (Chapter 5). Ho is fundamentally concerned with the socio-political consequences of the concentration of producer (or intermediate) services in certain Pacific Asian cities. He is interested in creating a more precise understanding of consumption behaviour by investigating the interdependencies between the way new forms of work are organized, new social practices, and innovative lifestyles in the Asian city. In examining the links between occupational shifts and consumption outcomes, the study's focus is to read the social into what are essentially economic spatial patterns, exploring implications of services development in terms of income inequality, polarization and accompanying politics. By examining consumption practices surrounding service work, Ho discusses the implications of service development on identity and lifestyle. Finally, Ho argues for a broader understanding of urbanization economies to incorporate not only production practices but its links to consumption practices. Ho's analysis is drawn from the important Singapore example, but includes implications for other rapidly tertiarizing cities within Pacific Asia.

The final contribution to this section is Michael Leaf's investigation of problems associated with the 'imposition' of advanced services trajectories upon cities in transition (Chapter 6: 'The bazaar and the normal: informalization and tertiarization in urban Asia'). Here Leaf brings together critical perspectives on two sets of issues traditionally treated in isolation: first, the role of services in 'step-wise modernization' and contemporary urban development; and, second, long-standing (and as yet unresolved) debates concerning the developmental meaning of the informal sector in transitional cities of the Asia-Pacific. This multiperspectival approach yields fresh insights on the contradictions posed by the modernization aspirations of the state, and the persistence of the informal sector. As Leaf observes, a modernization trajectory characterized in part as a shift to advanced services is likely to imply a reinterpretation of informal sector activities as 'lower order services', constituting a significant constraint upon state aspirations of 'transition' and 'progress'.

II. Services and urban development in the Asia-Pacific: sectoral perspectives

Service industries and city-regions in the Asia-Pacific

Just as the region's industrialization experience has produced both giant manufacturing concerns (for example Mitsubishi, Sony, Samsung, San Miguel and Siam Cement) and the aforementioned industrial metropoles (Tokyo, Seoul, Shanghai and Bangkok among others), the rise of the Asia-Pacific's services economy has underpinned the emergence of important service sector corporations and increasingly tertiarized city-regions. These include for the purposes of illustration, national air carriers (JAL, Cathay

Pacific, Singapore Airlines and Qantas), some of which are world leaders, hotel companies (Tokyu, the Shangri-La group), banking and financial corporations (Dai-ichi Kangyo, Hong Kong and Shanghai Banking Corporation, United Overseas Bank), property development and management corporations (Cheung Kong Holdings, Hutchison Whampoa), and telecommunications companies (Singtel). There are also major state-owned (or directed) service corporations within the region, such as the Bank of China, as well as state projects such as the Malaysian Putrajaya and Cyberjaya ICT 'corridors' (Corey, 2000).

In addition to these 'mainstream' or established services corporations and institutions, there is an expanding roster of firms in the entertainment, creative and cultural services, and educational and knowledge sectors. These industries typify the growth of the new 'urban cultural economy' (Scott, 2000), the expansion of the associated 'creative class' (Florida, 2002), and the centrality of 'knowledge industries' in regional production and trade (Beyers, 2000). As a spatial expression of innovation we can acknowledge the idea of the urban 'innovative milieu' as a 'crucible' or incubator for creative, knowledge-intensive industries (Mugerauer, 2000). This latest phase of the tertiarization experience within the Asia-Pacific includes the role of Los Angeles and Tokyo as world-scale centres of cultural production, as well as the emergence of new territorial forms of creative activity (including 'New Economy' sites and creative services precincts) within the core areas of cities such as Singapore, Seoul, Tokyo, Shanghai, Seattle, San Francisco and Vancouver (Hutton, 2004b). The growth of the service sector among states and regions of the Asia-Pacific now includes new service 'products' (such as 'cultural products' and higher education) as well as emergent production regimes, industry groups and occupations.

As experienced in the manufacturing sector, globalization has forced changes in the market scope and operating practices of many of the region's service corporations. The partial liberalization of some national markets for banking, finance, property development and other services has increased pressure on indigenous service firms, while at the same time opening up new opportunities in international markets. Postures toward globalization in the service sector are to a large extent contingent upon national development aspirations and local circumstances. Vietnam has encouraged the inflow of FDI to stimulate service industries (investments in hotels for tourism and in offices for producer services) crucial to the national 'doi moi' programme, while in the well-known Shanghai case the municipal government has assiduously sought the participation of international 'superstar' architects as well as foreign investors in support of globalization visions for the Pudong urban mega-project (Olds, 2001). But in many other cases (as Peter Daniels outlines in Chapter 2 of this volume) states are concerned about increasing penetration of domestic services markets by global corporations from the US and Europe, and are attempting to 'manage' the globalization of services process by regulatory and other means.

The pressures (and opportunities) of globalization notwithstanding, there is a distinctive regional quality to the growth of service industries and development of services firms and enterprise in the Asia-Pacific, related in part to indigenous development culture, to state or local policies, or to (natural or induced) regional competitive advantage. This section of the volume presents illustrative features of the regional and global contours of service industry development within selected jurisdictions by means of incisive case studies and analysis.

Larissa Muller examines foreign direct investment (FDI) in professional business services, and more especially advertising and marketing, with regard to the positionality of transnational corporations (TNCs) and local firms, factors of development and impacts on developing metropolitan economies of Southeast Asia ('Localizing international business services investment: the advertising industry in Southeast Asia', Chapter 7). Muller challenges the idea that because no developing country has produced transnational firms in the advanced business services, they can be relegated to the status of 'passive recipients' of foreign investment. Focusing on the case of advertising firms in Bangkok (with comparative references to firms based in Manila and Singapore), she constructs a profile of a very lively and distinctive regional advertising culture which effectively synthesizes western ideas and technology with local talent. The Bangkok advertising sector is characterized by successful innovation and adaptation, which Muller interprets as the hallmark of a 'creative milieu', rather than as a 'passive periphery'. This effective hybridization process suggests the broader generative possibilities of Southeast Asian cities as sites for development in advanced service industries.

While services, notably advanced, contact-intensive services, are closely associated with specific urban locations, reflecting agglomeration and labour market access considerations among other locational determinants, there is (as Hutton's Chapter 3 proposes) a significant and growing *regional* dimension to tertiarization processes within the Asia-Pacific. The Pearl (Zhujiang) Delta region of southern China presents an especially interesting and instructive case study for examining regional dynamics of advanced services development in the Asia-Pacific context, as demonstrated by Anthony Gar-On Yeh ('Producer services and industrial linkages in the Hong Kong–Pearl River Delta region', Chapter 8). Yeh depicts a dynamic profile of signifying changes in the structure and pattern of intra-region production linkages. An initial stage typified by a 'front shops back factories' relationship, with factories dispersed from Hong Kong to other locations within the Pearl Delta, was supplanted by a phase in which Taiwan (among other sources) emerged as a source of manufacturing in the region, and in which Hong Kong became instead a supplier of producer services for firms situated in the Delta. But in the most recent period the relative significance of Hong Kong as supplier of intermediate services for the Pearl Delta has changed (and in some respects declined), raising questions concerning both the emerging pattern of producer service networks in the region, as well as the increasingly extra-regional and global orientation of Hong Kong's producer services sector.

As observed, the dynamics of advanced services location and production in the Asia-Pacific involves not so much a total commitment to globalization, but rather to a strategic repositioning of operations, decision-making and expertise within the broader region. These experiences of service firm branch operations in business centres within the Asia-Pacific can disclose insights into both the spatial preferences and behaviour of service corporations in the region, as well as the impacts upon host cities and city-regions, as suggested in David Edgington and Roger Hayter's study of Japanese logistics firms in Hong Kong (Chapter 9). As Edgington and Hayter demonstrate, Hong Kong's status as global city and regional control centre is underpinned by important elements of competitive advantage, but may be compromised by a set of new forces including China's accession to the WTO, the rising power of Shanghai, and the possible convertability of the renminbi.

The latest phase of tertiarization within the advanced economies of the Asia-Pacific region includes greater emphasis on cultural, creative, educational and knowledge-based industries, and on new rounds of policy innovation and experimentation. Both of these defining features of advanced urban services development in the region are addressed in Kris Olds and Nigel Thrift's treatment of the globalization of higher education in relation to global city formation processes in Singapore ('Assembling the "global schoolhouse" in Pacific Asia', Chapter 10). The study examines the articulation of the 'cultural circuit of capitalism' with state development policy in Singapore, seen by the authors as a 'rapidly evolving laboratory for the corporate interests' of both capital and the state. The more strategic purpose involves the re-engineering of Singapore as a 'global education hub' to complement the city-state's specialized service roles and discursive identity. This strategy, if successful, has the potential to 'lift' Singapore out of the increasingly limited Southeast Asian ambit to a new positionality in which 'the region can be the globe itself', but as Olds and Thrift make clear there are inherent risks to this venture as well.

As we have observed, a defining feature of contemporary service industry development within many states of the Asia-Pacific is the interface between services and still-expanding manufacturing sectors. In the final chapter of this section, Thomas Leinbach and John Bowen address a crucial feature of this interdependency: air cargo services and related logistics as aspects of competitive advantage for manufacturing firms ('Air cargo services, global production networks and competitive advantage in Asian city-regions', Chapter 11). These air cargo and logistics services are depicted here not merely as ancillary activities to production, but rather as advanced, knowledge- and information-intensive producer services which are increasingly key to the success of manufacturing enterprises in competitive export trade markets. Case studies informed by interviews both with users and providers of air cargo services based in Malaysia, the Philippines and Singapore are deployed to assess the changing structure of the air cargo industry in Asia, and also to offer a perspective on implications for the performance of Asian city-regions in the global economy.

III. Services and urban development in the Asia-Pacific: city case studies

Service industries and urban change: retheorizing the Asia-Pacific city

In Part II we deployed a set of service industry case studies to generate insights on the changing spatiality of tertiarization in the Asia-Pacific, as a means of understanding the key role of cities in the growth of specialized service industries within the region. Here we are directly concerned with investigations into services as agents of structural change within Asia-Pacific cities and city-regions. In the early years of the twenty-first century service industries are implicated in multiscalar processes of change in cities, from the globalization of the CBD, and the emergence of new service production spaces in the old industrial (and in some cases residential) districts of the inner city, to new advanced-technology services production complexes in the suburbs and on the metropolitan periphery. Furthermore, new rounds of industrial restructuring driven increasingly by service-led growth, including the new urban cultural economy and the so-called New Economy, profoundly influence the social morphology, class structure and patterns of everyday life of Asia-Pacific cities.

To some extent, these processes are widely-experienced within the region, imparting a measure of commonality to patterns of urban transformation. But consistent with our position acknowledging the persistence of differentiation and exceptionalism in experiences of urban change within the Asia-Pacific, our case studies demonstrate as well aspects of contingency, including contrasts in urban scale, location and spatial context, regional development culture and governance and policy system, among other factors. To this end Part III presents case studies of services and urban development drawn from four Asia-Pacific cities: Melbourne, Seattle, Guangzhou and Seoul. Melbourne and Seattle are medium-size, 'Western' cities with large and dynamic service sectors, as well as still-substantial manufacturing sectors, while Seoul and Guangzhou exemplify specific and instructive tertiarization experiences within Pacific Asia. While there are limits to the universalizing possibilities of any specific sample of cities, these case studies in important respects demonstrate the breadth of experiences and implications of service-led trajectories among major metropolitan cities of the Asia-Pacific.

As observed above, service industry growth has a powerful influence on the recomposition of metropolitan employment structure and labour markets, with associated impacts for the reshaping of urban housing markets, especially in the central and inner city. Kevin O'Connor's examination of the Melbourne example ('Understanding mega city-region development: a case study of Melbourne', Chapter 12) focuses on changing linkage patterns between the inner city and suburbs. Analysis of changes in the journey-to-work movements of specific occupations and industrial employment over the period 1986–1996 can generate insight into producer service employment/residential linkages, as well as associated labour in industries such as courier services and accommodation. This analysis is designed to evaluate changes in the function of the inner city within the broader Melbourne metropolitan area, an issue of normative as well as scholarly significance.

Seattle represents an instructive case study of industrial development among medium-size cities, given both the presence of world-scale industrial corporations (notably Boeing, Weyerhaeuser and Microsoft), as well as an established record of services-led growth in the metropolitan employment and economic (export) base. William Beyers' treatment of the Seattle case study addresses changes in the employment structure of the Seattle metropolitan region, emphasizing the role of service industries in these processes ('Service industries and development trajectories in the Seattle metropolitan region', Chapter 13). Beyers' analysis includes both times series data of employment change in the Seattle metropolitan region over the last several decades, as well as references to trends in the US as a whole, to provide context for the Seattle experience. The empirical analysis is deployed to generate insights on the structural forces at work in the growth of Seattle's service economy, as well as on implications for theory and further research.

One of the most exigent research needs in the domain of urban tertiarization in the Asia-Pacific is that of developing deeper theoretical positions to link trends in service industry growth to broader processes of urban transformation, including globalization. George C. S. Lin critically explore conceptual aspects of these generative urban processes using Guangzhou as a case study ('Service industries and transformation of city-regions in globalizing China: new testing ground for theoretical reconstruction', Chapter 14). As a point of reference, Lin rehearses the position that urban services growth in China is seen as a consequence both of China's internal change and its continuing rearticulation with the global economy. What distinguishes the Chinese experience is a trajectory comprising rapid growth both in services and manufacturing industries, requiring a robust theoretical response to the previous industrial-deterministic urban development discourse. Lin concludes that there is an urgent need to incorporate the growth of services (which includes lower-as well as high-value service industries) in new discourses to account for recent and divergent trajectories of urban transformation in countries of the Socialist Third World undergoing globalization.

As is well known, the so-called Asian currency and financial crisis of 1997 negatively impacted major service industries within the region, while the crash of the 'dot.coms' and other ITC sector industries in 2000 checked at least temporarily the expansion of the advanced-technology sector. Seoul, South Korea's global city, encompasses important advanced-technology industrial complexes and capacity, as well as world-scale banking and finance, producer services and advanced manufacturing industries. Sam Ock Park's contribution to this volume takes the form of a case study of high-technology clusters in Seoul's Kangnam district ('IT service industries and the transformation of Seoul', Chapter 15). The implications of the new, post-1997 high-tech services and industrial clusters transcend localized effects to suggest far-reaching changes in Seoul's development trajectory and space-economy, as well as the transformation of Kangnam into a 'learning region' or district with the metropolitan area. Park's analysis of inter-firm interactions within Kangnam's high-tech service cluster underscores the development of a new phase of industrialization within Pacific Asia, as well as presenting the logical platform for new urban theory in the region.

References

Bell, D. (1973) *The Coming of Postindustrial Society*. New York: Basic Books.

Beyers, W. B. (2000) 'Cyberspace or human space: wither cities in the age of communication?', in Wheeler, J. O., Aoyama, Y. and Warf, B. (eds), *Cities in the Telecommunications Age: the Fracturing of Geographies*. London and New York: Routledge, 161–80.

Beyers, W., Alvine, M. J. and Johnsen, E. G. (1985) *The Service Economy: Export of Services in the Puget Sound Region*. Seattle: Central Puget Sound Economic Development District.

Bunnell, T. (2002) 'Multimedia utopia? A geographical critique of high tech development in Malaysia's multimedia supercorridor', *Antipode*, 34(2), 265–95.

Coffey, W. J. (1994) *The Evolution of Canada's Metropolitan Economies*. Montréal: The Institute for Research on Public Policy.

Corey, K. E. (2000) 'Intelligent corridors: outcomes of electronic space policies', *Journal of Urban Technology*, 7(2), 1–22.

Daniels, P. W. (1993) *Services in the World Economy*. Oxford: Blackwell.

Daniels, P. W. (1998) 'Economic development and producer services growth: the APEC experience', *Asia-Pacific Viewpoint*, 39(2), 145–59.

Davis, H. C. and Hutton, T. A. (1981) 'Some planning implications of the expansion of the urban service sector', *Plan Canada*, 21, 15–23.

Davis, H. C. and Hutton, T. A. (1991) 'An empirical analysis of producer service exports from the Vancouver metropolitan region', *Canadian Journal of Regional Science*, XIV(3), 371–89.

Davis, H. C. and Hutton, T. A. (1994) 'The marketing of Vancouver's producer services to the Asia-Pacific', *The Canadian Geographer*, 38(1), 18–28.

Douglass, M. (2002) 'Globalization, intercity competition and the rise of civil society: towards livable cities in Pacific Asia', *Asian Journal of Social Science*, 31(1), 131–52.

Edgington, D. W. (1991) 'Economic restructuring in Yokohama: from gateway port to international core city', *Asian Geographer*, 10, 62–78.

Florida, R. (2002) *The Rise of the Creative Class: and How its Transforming Work, Leisure, and Everyday Life*. New York: Basic Books.

Gershuny, J. (1978) *After Industrial Society?* London: MacMillan.

Gottmann, J. (1970) 'Urban centrality and the interweaving of quaternary activities', *Ekistics*, 29, 322–31.

Ho, K. C. (1994) 'Industrial restructuring, the Singapore city-state, and the regional division of labour', *Environment and Planning A*, 26, 33–51.

Ho, K. C. (1997) 'The global economy and urban society in Pacific Asia', *International Sociology*, 12(3), 275–93.

Ho, K. C. (2002) 'Globalization and Southeast Asian urban futures', *Asian Journal of Social Science*, 30(1), 1–7.

Hutton, T. A. (2004a) 'Service industries, globalization, and urban restructuring within the Asia-Pacific: new development trajectories and planning responses', *Progress in Planning*, 61(1), 1–74.

Hutton, T. A. (2004b) 'The New Economy of the Inner City', *Cities* 21(2), 89–108.

Leaf, M. (1996a) 'Building the road for the BMW: culture, vision and the extended metropolitan region of Jakarta', *Environment and Planning A*, 28, 1617–35.

Leaf, M. (1996b) *Redevelopment before Development: Asian Cities Prepare for the Pacific Century*. Vancouver: Centre for Human Settlements, University of British Columbia.

Ley, D. F. (1996) *The New Middle Class and the Remaking of the Central City*. Oxford: Oxford Geographical and Environmental Studies, at the University Press.

Lin, G. C. S. (1994) 'Changing theoretical perspectives on urbanization in Asian developing countries', *Third World Planning Review*, 16, 1–23.

Lin, G. C. S. (2002) 'The growth and structural change of Chinese cities: a contextual and geographic analysis', *Cities*, 19(5), 299–316.

McGee, T. G. and Lin, G. C. S. (1993) 'Footprints in space: spatial restructuring in the East Asian NICs 1950–1990', in Dixon, C. and Drakakis-Smith, D. (eds), *Economic and Social Development of Pacific Asia*. London: Routledge.

McGee, T. G. and Robinson, I. M. (1995) *The Mega-Urban Regions of Southeast Asia*. Vancouver: UBC Press.

Mugerauer, R. (2000) 'Milieu preferences among high-technology companies', in Wheeler, J. O., Aoyama, Y. and Warf, B. (eds), *Cities in the Telecommunications Age*. London: Routledge, 219–27.

Nanjiang, O. (1996) The Development and Transformation of Guangzhou. Centre for Urban and Regional Studies, Zhongshan University, Guangzhou, mimeo.

O'Connor, K. and Daniels, P. W. (2001) 'The geography of international trade in services', *Environment and Planning A*, 33, 281–96.

O'Connor, K., Stimson, R. and Daly, M. (2001) *Australia's Changing Economic Geography: A Society Dividing*. Melbourne: Oxford University Press.

Olds, K. (2001) *Globalization and Urban Change: Capital, Culture and Pacific Rim*

Mega-Projects. Oxford: Oxford Geographical and Environmental Studies, at the University Press.

Sassen, S. (1991) *The Global City: New York, London, Tokyo.* Princeton: Princeton University Press.

Scott, A. J. (2000) *The Cultural Economy of Cities.* London: Sage.

Soja, E. (2000) *Postmetropolis: Critical Studies of Cities and Regions.* Oxford: Blackwell.

Taylor, B. and Kwok, R. Y.-W. (1989) 'From export center to world city: planning for the transformation of Hong Kong', *Journal of the American Planning Association,* 55(3), 309–22.

Thornley, A. (1999) 'Urban planning and competitive advantage: London, Sydney and Singapore'. Metropolitan and Urban Research Discussion Paper No. 2, London School of Economics and Political Science.

Turner, S. (2004) 'The paradox of Hanoi's ancient quarter: a landscape's heritage called into question'. Paper presented to the 100th annual meeting of the Association of American Geographers, 15–19 March: Philadelphia.

Wu, F. and Yeh, A. G.-O. (1999) 'Urban spatial structure in a transitional economy: the case of Guangzhou, China', *Journal of the American Planning Association,* 65(4), 377–94.

Part I

Services and urban development in the Asia-Pacific

Theory and context

2 Services, globalization and the Asia-Pacific region

P. W. Daniels

Introduction

The objective of this chapter is to provide a survey of economic trends within the Asia-Pacific region, with particular references to services, and where possible to place these in their global context. A brief case study of some aspects of the production and consumption of producer services in Japan is also included as an illustration of some of the reasons why it is sometimes difficult to compare the level of service activities in the Asia-Pacific economies with those of Western Europe or North America.

The Asia-Pacific region is notable for its rich diversity of demographic, cultural, social and economic environments. It has an abundance of human resources, accounting for almost one-third of the world's population in 2000 (Table 2.1). Although the economic crises of the late 1990s put a brake on growth, the region has some of the fastest growing economies in the world; China has achieved and is expected to continue achieving rates in excess of 10 per cent per annum until 2010 and beyond. In many respects it is still the engine of the world economy even though, on an intra-regional scale, it incorporates the full spectrum from low-income countries (Bangladesh, Cambodia, China, India, Pakistan, Vietnam), middle-income countries (Indonesia, Philippines), high income countries (Malaysia, Thailand), and highly industrialized countries (Australia, Japan, New Zealand, Republic of Korea, Singapore).[1] In 1991, the region's combined GDP was approximately US\$3.5 trillion and accounted for about one-sixth of the combined global GDP. By 2000 this had risen to US\$7.8 trillion and around 25 per cent of global GDP (see Table 2.1). In many ways the Asia-Pacific is proof of the notion first expounded by Adam Smith in the *Wealth of Nations* that globalization would be accompanied by an acceleration of economic growth (Sachs, 2000).

In broad economic terms it is still the case that the Asia-Pacific is not yet punching its full weight; Western Europe accounts for 6 per cent of world population and 26 per cent of GDP while North America with just one-twentieth of world population generates one-third of world GDP. But

Table 2.1 Asia-Pacific: population and gross domestic product, 2000

Country/region	Population ('000s)	GDP (US$m)	GDP per capita (US$)
Australia	19,138	388,461	20,298
Cambodia	13,104	3,183	243
China	1,253,392	1,086,156	867
Hong Kong	6,860	162,642	23,709
Indonesia	212,092	153,255	723
Japan	127,096	4,765,313	37,494
Korea, Rep. of	46,740	457,219	9,782
Laos	5,279	1,709	324
Malaysia	22,218	89,659	4,035
New Zealand	3,778	50,781	13,441
Philippines	75,653	74,733	988
Singapore	4,018	92,252	22,959
Taiwan	22,185	309,377	13,946
Thailand	62,806	122,166	1,945
Vietnam	78,137	31,344	401
Asia-Pacific[1]	**1,952,496**	**7,788,250**	**3,989**
North America[2]	313,987	10,510,772	33,475
Western Europe[3]	389,282	8,272,980	21,252
World total	6,056,715	31,362,624	5,178

Source: Extracted from data in the UNCTAD Handbook of Statistics, accessed from http://stats.unctad.org/restricted/eng/TableViewer/wdsview.

Notes
1 No data for Brunei Darussalam and Myanmar.
2 US and Canada.
3 Includes Scandinavian countries.

whereas the average real GDP growth rate for the world as a whole was of the order of 2.6 per cent in 2001, International Monetary Fund estimates put the corresponding rate for the Asia-Pacific economies at 4.6 per cent (it was 7.6 per cent in 2000) (World Trade Organization, 2002). Thus, it is anticipated that Asia-Pacific GDP will be larger than that of Western Europe and equal with that of North and South America by 2025. From an industry sector perspective, the growth of the region's output between 1980 and 2000 has depended largely on a stronger performance by manufacturing than by services (Table 2.2) although there are inter-country differences. In the case of Singapore, for example, output growth by services exceeded manufacturing in both decades shown in Table 2.2 whereas in China, Thailand and Malaysia annual output growth by manufacturing has always exceeded that for service industries. The diversity of the Asia-Pacific economies is underlined by Sachs (2000) who groups countries worldwide by growth mechanism or forms of development (a product of each country's resource endowment, economic policies and geography).

Table 2.2 Asia-Pacific economies: growth of output by sector (% per annum), 1980–2000[1]

Country	Total GDP		Agriculture		Manufacturing[2]		Services	
	1980–1990	1990–2000	1980–1990	1990–2000	1980–1990	1990–2000	1980–1990	1990–2000
Australia	3.5	4.1	3.4	3.1	1.9	2.4	4.0	4.5
Cambodia	–	4.8	–	1.9	–	8.2	–	6.9
China	10.1	10.3	5.9	4.1	11.1	13.4	13.5	9.0
Hong Kong	6.9	4.0	–	–	–	–	–	4.0
Indonesia	6.1	4.2	3.6	2.1	12.8	6.7	6.5	4.0
Japan	4.1	1.3	1.3	-3.2	–	0.5	4.2	2.5
Korea, Rep. of	8.9	5.7	3.0	2.0	12.1	7.5	8.4	5.7
Laos	3.7	6.5	3.5	4.9	8.9	11.7	3.3	6.5
Malaysia	5.3	7.0	3.4	0.3	9.3	9.8	4.9	7.2
Myanmar	0.6	6.6	0.5	5.3	-0.2	7.0	0.8	6.8
New Zealand	1.9	3.0	3.8	2.7	–	–	2.7	3.7
Philippines	1.0	3.3	1.0	1.6	0.2	3.0	2.8	4.1
Singapore	6.7	7.8	-5.3	-1.6	6.6	7.1	7.6	7.8
Thailand	7.6	4.2	3.9	2.1	9.5	6.4	7.3	3.7
Vietnam	4.6	7.9	4.3	4.8	–	–	–	7.7

Source: World Bank: 2002 World Development Indicators, accessed from www.worldbank.org/data/wdi2002/economy, August 2003.

Notes
1 The World Bank data does not include Taiwan.
2 Excluding mining, construction, electricity, water, and gas.

Several of the countries in the region are therefore grouped with the US, France, Germany or the UK as economies driven by endogenous growth (Australia, Hong Kong, Japan, Korea, New Zealand, Singapore, Taiwan). Yet others are 'catching up' (China, Indonesia, Malaysia, Philippines, Thailand, Vietnam) and in the case of two other countries (Cambodia, Laos) the growth mechanism is Malthusian.

Much of the recent economic growth in the Asia-Pacific reflects increasing intra-regional interdependence in the form, for example, of increased trade in goods and services. This has been complemented with the widespread adoption of trade liberalization and the relaxation of restrictions on inward flows of foreign capital from outside the region. This has helped to sustain the export-oriented growth (intra- and inter-regional) that has made the Asia-Pacific economies so successful (Lee, 2002). As noted elsewhere (Daniels, 1998) this export-led growth has been largely dominated by merchandise rather than service transactions, even amongst those economies, such as Japan or Korea, that are classed as highly industrialized. But, in common with the global trend, the Asia-Pacific economies are also undergoing a sectoral shift towards service industries and their expansion in this direction is expected to continue.

China, for example, is likely to be in the forefront of this trend. Service industries are forecast to become a major sector of the Chinese economy in the next 15 years, accounting for over 50 per cent of the gross domestic product (GDP) by 2020 (*Peoples Daily*, 2004). The State Development and Reform Commission (SDRC) places great importance on the leading role of the service industry in economic growth and recognizes the potential of this sector as China opens up to private and foreign investment and as living standards improve. Between 1978 and 2002, service industry value added increased by 380 per cent and its share of GDP from 21.4 per cent to 33.7 per cent. During the same period, the number of employees in the service industry surged from 48.9 million to 210.9 million with the net growth of employees in the service industries about twice that in the manufacturing and mining industries.

According to government estimates, the service industry's share of GDP and total urban employment will grow to 36 per cent and 33 per cent respectively by 2005 and by 2020 the service industries will provide about 40 per cent of the total jobs in cities and towns. This is still a long way behind equivalent figures for developed economies such as the UK, the US or Japan but the scale of the absolute numbers of service workers and consumers that even the lower percentages represent will fundamentally change the balance within the global service economy.

Services and cities in the Asia-Pacific

Apart from the impacts of these global level trends on employment, trade, and investment in the region, amongst the major beneficiaries are the cities that have been the focus for many of the changes, if only because of their inexorable population growth, the associated demand for employment, and their function as nodes in the transport and telecommunications systems that support intra-regional economic relations as well those with the rest of the world (Lo and Yeung, 1996; Haila, 2000; Sassen, 2000). The dynamics of the Asia-Pacific economies are intimately intertwined with their cities, increasingly so as recent and longer-term structural changes, regionally and globally, are associated with the rise of services, especially knowledge- and information-intensive services (see Hutton, 2004, for a detailed and in-depth exposition). It has been noted that the mega-cities of the Asia-Pacific 'are at once the products and instruments of the system of economic internationalization and regional integration' King (1990: 54) while the 'urban corridors ... will be the locales for furthering the globalization of production and services in Pacific Asia' (Lo and Yeung, 1996).

The task of measuring the size and attributes of service industries in the Asia-Pacific economies and how they interact with the global service economy continues to be fraught with difficulties. In common with studies of the service sector in other parts of the world, some of the answers await better data, longer and consistent time series and perhaps a more disaggregated approach in which individual components of the service-producing sector are examined. While the details are elusive, there will be agreement that the cities within the Asia-Pacific are its principal points of contact with the global service economy and key actors such as service TNCs (Grant and Nijman, 2002). Cities are the national focus for the growth of service activities and employment, especially business and professional services (BPS) and internationally traded services including tourism, some aspects of education, health and cultural services.

Most of the chapters in this book examine from different perspectives the way in which service industries, including their growing role in globalization processes, have shaped the planning and development of Asia-Pacific cities and mega-urban regions (Forbes, 1997). They add to what is already a substantial concern about the sustainability of these regions in the face of pressure on social and environmental conditions generated by very rapid growth and competition between cities within the region for the symbols of global success (Douglass, 2000, 2002). Ironically, many of these symbols are linked with services: gleaming office buildings, modern shopping malls, high technology transportation systems, state-of-the-art ports and airports, and high capacity, high speed ICT networks. Acquiring these artefacts of global success is not just about links and flows involving other world regions, they also demonstrate growing interconnections among the metropolitan areas within the Asia-Pacific. But the trajectories of these cities are diverse although they have been usefully categorized into post-industrial capital exporting cities (Tokyo, Seoul, Taipei); border-less cities (Hong Kong and Singapore); cities with very high growth rates driven by globalization (Shanghai, Jabotabek [Jakarta urban mega-region], Bangkok); and amenity cities (Sydney and Vancouver) (Lo and Marcotullio, 2000, 2001). Other detailed typologies are also suggested by Hutton (2004: 29–42).

Although home to a diverse set of national economies and a wide variety of urban environments, much of the Asia-Pacific region is a relatively new player in those dimensions of globalization that can be linked with the rise of service industries during the last thirty years (Daniels, 1998). This chapter seeks to place the development of the region's service industries and, in particular, its impact on the role and function of the region's cities within the wider context of trends in the global service economy (see for example Aharoni and Nachum, 2000; Stern, 2001; Cuadrado-Roura *et al.*, 2002). Although globalization remains a contested concept (Olds *et al.*, 1999), there is little doubt that the world's major regions are better connected than ever, whether viewed as a space of flows or as a product of the connections between the discrete centres of corporate command and control (global/world cities). Yet much of the 'global' economy actually operates on an intra-regional regional basis with some inter-regional flows; witness the highly regionalized flows of commercial services trade revealed by the OECD statistics published for the first time in 2002 (OECD, 2002). There continues to be evidence for the dependence of producers in the Asia-Pacific on higher-order services supplied from Europe and North America. Will this change as the balance of the global economy shifts as a consequence, for example, of China's increasing involvement and what will it mean for the sourcing and distribution of services within the Asia-Pacific, between the region and other parts of the world, and for the contribution of advanced services to the transformation of cities and regions (see for example Morshidi, 2000; Abdul Fatah Che *et al.*, 2001, on Kuala Lumpur)?

Some dimensions of services growth in the Asia-Pacific

Growth of services in total employment

As noted in an earlier paper (Daniels, 1998), trends in employment in services are a useful starting point for any discussion of the extent of economic restructuring in Asia-Pacific economies. Since this book is focusing on the growth trajectories of the region's major cities, it is useful to concentrate on those services that show a disproportionate tendency to cluster in urban areas. The performance of finance, insurance, real estate and business services (FIRB) is therefore compared with total employment. It is actually quite difficult to compile good comparative time series employment data across the full range of economies within the region but the relatively small sample shown in Table 2.3 is likely typical of what has been happening more generally. It is clear that for both overall employ-

Table 2.3 Asia-Pacific: change (%) in FIRB[1] and total employment ('000s), selected countries, 1984–2001

Country	Sector etc.	1984	1993	2001	% change 1984–1993	% change 1993–2001
Hong Kong	*FIRB*	133	289	478	162.1	65.4
	Total	2,505	2,800	3,249	18.6	16.0
	%	5.3	10.3	14.7	–	–
Indonesia	*FIRB*	–	565	1,128	–	99.6
	Total	–	79,201	90,807	–	14.7
	%	–	0.7	1.2	–	–
Japan	*FIRB*	3,830	5,470	6,290	44.9	15.0
	Total	57,660	64,500	64,120	12.0	-0.6
	%	6.6	8.5	9.8	–	–
Korea, Rep. of	*FIRB*	501	1,362	2,085[2]	226.5	53.0
	Total	14,429	19,328	21,061[2]	41.2	9.0
	%	3.5	7.0	9.9	–	–
Malaysia	*FIRB*	–	330	462[2]	–	40.0
	Total	–	7,383	9,322[2]	–	26.3
	%	–	4.5	5.0	–	–
Philippines	*FIRB*	370	496	678[2]	48.9	36.7
	Total	19,368	24,443	27,775[2]	32.7	13.6
	%	1.9	2.0	2.4	–	–
Singapore	*FIRB*	109	173	352	132.5	103.0
	Total	1,269	1,592	2,047	34.0	28.6
	%	8.6	10.9	17.2	–	–

Source: Extracted from ILO, *Yearbook of Labour Statistics*, 1994, and ILO Bureau of Statistics, accessed from http://laborsta.ilo.org/, September 2003.

Notes

1 Financing, insurance, real estate and business services (Major Division 8, International Standard Industrial Classification (ISIC-Rev.2, 1968)).

2 Year 2000.

ment and FIRB employment the rate of change has slowed down between 1984 and 2001, especially in Japan where total employment increased by 12 per cent (1984–1993) only to decrease by 0.6 per cent (1993–2001). Elsewhere the slowdowns are less dramatic, no doubt reflecting the varying exposures of individual economies to the Asian financial crisis during the second half of the 1990s. Some economies are now recovering although Japan has consistently struggled to recover from problems, such as a weak financial sector, limited competition, over-regulation and movement of production to other locations in Southeast Asia, that have lasted for more than a decade.

Employment in FIRB has not been exempt from similar influences but in all the economies shown in Table 2.3 it has been expanding its share of total employment and growing in absolute terms at a significantly higher rate than total employment. In all cases the baseline share of FIRB in 1984 was low by global standards; only Japan, Hong Kong and Singapore had a share exceeding 5 per cent. This had doubled by 1993 and increased again by 2001 so that more than one in six of Singaporeans were employed in FIRB, closely followed by Hong Kong. This reflects significant absolute growth which has not been the case in Japan which only reported one in ten employed in FIRB in 2001. The Republic of South Korea started behind Japan but by 2001 had caught up with its larger neighbour, the proportion of FIRB in total employment also being marginally higher (9.9 per cent). But there are also less developed economies such as the Philippines (2.4 per cent) and Indonesia (1.2 per cent), and to some extent Malaysia, where FIRB has yet to make an impact. Overall, it remains the case that, despite the recent expansion of FIRB in the Asia-Pacific economies, the majority still lag behind many Western European and North American economies on this measure. This was noted by the OECD (1999) which observed that more open economies with relatively fewer regulatory barriers (such as Australia) showed strong growth in services compared with more closed economies such as Japan or Korea. We will return to this, using Japan as a case example, later in this chapter.

Telecommunications indicators

Information and communications technology have played a large part in shaping the depth and reach of globalization (Graham, 1999; Shin and Timberlake, 2000). The Asia-Pacific economies have in many ways been one of the most significant beneficiaries in the sense that with some exceptions, as comparatively late entrants, they have leapfrogged the stages of earlier telecommunications infrastructure encountered by competitor economies in Europe or North America. In more recent years, economic growth, coupled with greater consumer spending power has driven the growth in the supply and demand for telecommunication services in the Asia-Pacific region (International Telecommunications Union (ITU),

2002). Economies such as the Republic of Korea, Hong Kong, Australia, New Zealand, Taiwan and Singapore compare very favourably and, on some indices, are ahead of the US and the UK (Table 2.4). Again, however, there are wide intra-regional variations in telecommunications use and access and it is not difficult to see the link between the indices in Table 2.4 and the shares of FIRB in total employment (see Table 2.3).

Worldwide, seven of the ten most profitable public telecommunication operators originate in the Pacific region which is distinguished not just by the size of its market for telecommunications services but also by the increases in teledensity and technology deployment, and the innovative and flexible nature of its telecommunications policies. More than one new telephone user was added every second between 1992 and 2002. Japan and the Republic of Korea lead the world in commercial deployment of 3G networks while deregulation has encouraged more competition;

Table 2.4 Asia-Pacific telecommunications indicators, 2001

Country	Main telephone lines	Mobile phone subscribers	Personal computers	Internet users	Broadband (% of internet users)[3]
Australia	520	578	517	372	–
Brunei Darussalam	245[1]	289[1]	75	105	–
Cambodia	20	17	1	1	–
China[2]	138	113	19	26	–
Hong Kong	581	819	374	445	68(2)
Indonesia	37	25	11	19	–
Korea, Rep. of	476	617	255	518	83(1)
Laos	9	5	3	2	–
Malaysia	199	315	133	252	–
Myanmar	6	0	1	0	–
New Zealand	471	635	394	287	–
Philippines	40	137	22	26	–
Singapore	472	696	511	365	35(14)
Taiwan	573	967	224	338	59(4)
Thailand	94	119	27	56	–
Vietnam	38	16	10	5	–
US	665	444	623	500	19(11)
UK	578	790	369	403	–

Sources: Extracted from the UNCTAD Handbook of Statistics, accessed from http://stats.unctad.org/restricted/eng/TableViewer/wdsview/, September 2003 and International Telecommunications Union http://www.itu.int/ITU-D/ict/statistics/at_glance/top15_broad.html, October 2003.

Notes
1 Data for 2000.
2 Not including Macao SAR.
3 Rank in top 15 countries for broadband users, 2002.

Singapore, Hong Kong and China, for example, have withdrawn the exclusivity of their national telecommunications operators' licenses for the supply of international services and found ways of encouraging inward investment into their telecommunications markets.

Although overall mobile telephone penetration remains low, the Asia-Pacific region has overtaken Europe as the world's largest market. On the basis of the number of mobile phones per capita, two of the top three 'mobile' economies worldwide (Taiwan and Hong Kong) are in the region. Mobile technologies have also enabled teledensities to increase in developing economies that might otherwise have been relatively excluded from the telecommunications revolution; only Myanmar had a teledensity of less than one by mid-2002 (ITU, 2002). At that time the region accounted for one-third (160 million) of the total internet users in the world with innovative projects for community access, such as Indonesia's *warung internets* or *warnets* and prepaid user cards, expected to continue boosting internet usage. Broadband internet access in some of the Asia-Pacific economies also compares very favourably with other regions; measured by penetration rates, it has five of the top 12 economies worldwide. The capacity on internet bandwidth is crucial for many businesses and this has also expanded quickly from 8 Gbit/s in 1999 to 65 Gbit/s at the end of 2001 (ITU, 2002). By late 2003 broadband internet connections had spread to 5 per cent of households (compared with 3 per cent one year earlier) with a very high penetration rate of 70 per cent in South Korea (*The Economist*, 2004). The equivalent statistic for Japan was 30 per cent with China, despite the advances by its economy, remaining a digital laggard by international standards. There remain a wide range of policy issues and strategic opportunities for the deployment of ICTs within the Asia-Pacific, such as its prospects in the global digital economy, further development of its information infrastructure, or how to move towards a more inclusive information economy and society (Organization for Economic Cooperation and Development, 2003a).

Trade in services

An expansion of service industries/employment and major advances in access to ICTs have provided a platform for the Asia-Pacific economies to trade in services as well as goods. The volume (by value) and share of international trade in goods is much larger than for services but as manufacturing and services increasingly converge (see for example Daniels and Bryson, 2002; Bryson *et al.*, 2004) the significance of the smaller services share will be diminished. It could also be argued that the tacit and codified knowledge provided by some traded services (such as many FIRB activities) creates benefits, such as improved host country innovative capacity or service business formation rates, that are far more significant than their share of world trade by value. Most of the APEC countries, for example, have achieved a considerable expansion of services trade (exports), often of course from a low baseline, during the 1990s, and by 2000 they accounted for almost 20 per cent of the world total. This was not far behind the US and Canada (21 per cent) but well behind Western Europe (45 per cent) although relative to the region's share of world population its share of service exports is less than expected. But the balance is clearly changing as Western Europe's dominance of commercial service exports has declined considerably since 1980 (59 per cent), North America's share has grown only slowly since 1990, and the Asia-Pacific has expanded its share at a higher rate (Table 2.5). Most of the countries shown in Table 2.5 have achieved an increased share of world service exports in line with the increased regional share of world trade; the exception is Japan with a regional share which is lower in 2000 than in 1980. China, on the other hand is making big strides from a position of virtually no involvement in 1980 to more than 2 per cent of world exports

Table 2.5 Asia-Pacific: value and share of total trade in commercial services

Country/region	1980		1990		2000		Average annual change 1980–2000
	US$m	%	US$m	%	US$m	%	%
Australia	3,660	1.05	9,833	1.26	17,908	1.22	24.5
China	379	0.01	7,221	0.93	32,735	2.24	431.9
Hong Kong	5,763	1.65	18,128	2.32	41,548	2.84	36.0
Indonesia	–	–	2,488	0.32	5,060	0.35	–
Japan	18,760	5.37	41,384	5.31	68,303	4.67	18.2
Korea, Rep. of	2,402	0.69	9,155	1.17	29,746	2.03	61.9
Malaysia	1,046	0.3	3,769	0.48	13,649	0.93	65.2
Myanmar	48	0.01	93	0.01	509	0.03	53.0
New Zealand	950	0.27	2,415	0.31	4,270	0.29	22.5
Philippines	1,214	0.35	2,897	0.37	3,935	0.27	16.2
Singapore	4,774	1.37	12,719	1.63	26,960	1.84	28.2
Taiwan	1,944	0.56	6,937	0.89	25,766	1.76	66.3
Thailand	1,366	0.39	6,292	0.81	13,785	0.94	50.5
Asia-Pacific[1]	42,306	12.12	123,331	15.81	284,174	19.44	33.6
North America[2]	45,225	12.95	151,230	19.39	308,560	21.1	34.1
Western Europe[3]	207,135	59.32	411,934	52.81	652,973	44.66	15.8
World total	349,159	100	779,971	100	1,462,006	100	20.9

Source: Extracted from the UNCTAD Handbook of Statistics, accessed from http://stats.unctad.org/restricted/eng/TableView/wdsview, August 2003.

Notes
1 No data for Brunei Darussalam, Cambodia, Laos or Vietnam.
2 US and Canada.
3 Includes Scandinavian countries.

in 2000. Its accession to the WTO in 2001 will probably ensure further expansion of its involvement over the next decade. Further gains are likely if the Asia-Pacific economies adopt a more open stance towards the services trade although, since the majority of the economies in the region are emerging or newly industrializing, they have yet to be persuaded that their inefficient and non-competitive domestic service providers will not lose out to advanced economy service providers (Organization for Economic Cooperation and Development, 2003b).

Such reservations are important but they have not prevented some of the Asia-Pacific economies from establishing a significant presence in a number of services, including construction (a service according to international classifications), data processing, port and shipping services and audiovisual services. Many of these are labour-intensive services for which the Asia-Pacific economies have clear comparative advantage. The outstanding example, among a number of existing and potential cases, is India which has capitalized upon the availability of well-educated labour force, low wage rates and access to ICTs to export computer and related services and outsourced call centre work worldwide (OECD, 2003b). A leading example is Tata Consultancy Services, an Indian software company with headquarters in Mumbai, employing 25,000 across operations in 55 countries. In 2003 it had several Fortune 500 companies amongst its 600 clients worldwide and its export transactions were approximately US$184 million. Three data processing companies based in the Philippines are amongst the leading data processing companies worldwide with clients in Europe and North America. Software producers in countries on the Pacific Rim such as Chile, Colombia and Peru have formed an association (Latin American Association of Companies of Technology and Information) to promote and encourage the development of their software industries (OECD, 2003b). Numerous international manufacturing and service corporations have outsourced their business and IT service requirements to Indian companies (in Bangalore and Delhi for example), China and Indonesia are also entering the market, and the Philippines is becoming a hub for outsourced financial and accounting services (such as payroll management systems).

In the case of audiovisual services, countries such as Thailand provide production and post-production film, music, video and documentary services for international companies such as Warner Brothers; the Indian film industry, now the largest in the world, has a rapidly growing export market that includes significant international audiovisual producers in their own right such as the UK and the US/Canada. The Asia-Pacific has significant specialization in port and shipping services with four of the top five container terminals in the world (Hong Kong, Korea, Taiwan, Singapore) and some leading international corporate operators such as International Container Services Inc. which operates several container ports in the Philippines as well as six terminals overseas or PSA Corporation of Singapore. Finally, Asia-Pacific construction services with major shares of their revenue obtained from outside their home markets are prominently represented in the international arena; China is the base for 24 of the top 50 worldwide, along with Korea (5), and Chinese Taipei (2).

As elsewhere in the world, the Asia-Pacific economies have developed their service export capacity after they have established a strong presence in their domestic markets. This is usually followed by the development of intra-regional exports, involving partners with whom individual countries may have historical links (Poon, 1997, 2000; Murphy, 1995). Geography is important for shaping the regional structure of services trade with the exception of tourism (in terms of market reach) and services delivered electronically using ICTs. In some instances export activity can be boosted by imports, especially where these involve direct representation of foreign firm branches or subsidiaries. In financial services, for example, the presence of foreign banks can trigger spillover benefits in the form of improving local business practices (as a condition for underwriting business loans, for example) and raising skill levels. Unfortunately, even the detailed breakdowns of partner country service exports and imports published for the first time by the OECD in 2002 make it difficult to closely examine the degree to which Asia-Pacific trade is regionally oriented and/or characterized by an imbalance towards imports (Organization for Economic Co-operation and Development, 2002). For Japan and Korea the two main partners account for approximately 50 per cent of their trade and 40 per cent for Australia. A large proportion of Australia, Japan and Korea's trade is with China, the other 'dynamic' Asian countries, and New Zealand (for Australia). There is partial data for Australia, Japan and Korea (all members of the OECD) showing the top seven trade in services partners in 2000. Australian services exports and imports are more or less balanced but the EU15 and the United States account for 37 per cent of exports but 45 per cent of imports. Intra-regional service exports to Japan (12 per cent), New Zealand, Singapore, Hong Kong China and Indonesia amount to just over 31 per cent compared with less than 27 per cent for imports. Trade imbalance in the case of Japan is much clearer with imports exceeding exports by a ratio of almost two to one (1.7) but the imbalance is not a function of EU15 and United States transactions (51 per cent of exports and 51 per cent if imports) but with other parts of the world. Its other five trading partners are all within the region, accounting for 25.5 per cent of exports and 20 per cent of imports. Korea reveals a more balanced distribution of service exports and imports but relies more on the EU15 and the United States for imports (51 per cent) than for exports (42 per cent). Japan is a major trading partner and accounts for 23 per cent of Korean service exports but only 16 per cent of imports.

After growth of 6 per cent in 2000 world commercial services trade stagnated in 2001, underlining the difficulty of generalizing about wider trends and the role of the Asia-Pacific (World Trade Organization

(WTO), 2002). The Asia-Pacific's commercial services imports declined due to a fall in imports of transport and travel services which was only partly offset by a moderate increase in imports of other commercial services. The WTO notes that the region's exports of commercial services between 1990 and 2001 (by category) differed significantly from its imports; travel receipts continued to increase while other services exports (including advanced services) stagnated. There are also inter-country contrasts in export performance, especially between the two largest commercial services traders; Japan and China. The latter increased commercial services exports and imports by 9 per cent; the latter contracted by 7 per cent. Several other countries in the region also experienced a decline in service exports in excess of 5 per cent and this was usually accompanied by a decline in imports (WTO, 2002). For countries like India which were experiencing vigorous economic expansion and GDP growth export and import growth rates exceeded 10 per cent.

FDI and services

With TNCs responsible for a significant share of global export and import activity that is dependent in many cases upon foreign direct investment, especially by those classified as services, it provides another indicator of the role of the Asia-Pacific. The data in Table 2.6 are not divided between manufacturing and services but it is certainly the case that the services share of FDI has grown steadily since 1970 to reach levels in excess of 50 per cent of total stock in 1995 (Mullampally and Zimny, 2000). Between 1998 and 2000 the Asia-Pacific's share of world FDI flows (all industries) declined from a peak of 13 per cent at the beginning of the period to less than 10 per cent. It had, however, recovered to the 1998 level in 2001 (13 per cent). Although services now lead the way in FDI, the level of transnationalization of the service sectors of host countries, whether measured using employment or value-added, remains low by comparison with manufacturing (Mullampally and Zimny, 2000). Service foreign affiliates in China accounted for 0.5 per cent of total value added in 1996 (19.1 per cent by manufacturing foreign affiliates); equivalent figures for India in 1991 were 0.5 per cent and 12.2 per cent respectively (the latter has no doubt changed in more recent years). On the basis of employment the shares were 14.3 per cent and 16.0 per cent respectively for Hong Kong China in 1994, 7.6 per cent and 21.1 per cent for Taiwan (1995), 1.5 per cent and 14 per cent for Vietnam (1995), and 0.4 per cent and 4.7 per cent for Indonesia (1996). However, a substantial part of services FDI is actually undertaken by manufacturing TNCs in the form of trading and financial operations to support their global value chains, rather than 'pure' service TNCs setting up affiliates in foreign markets. The dominance of manufacturing in total FDI outflows in 1994 from countries within the region has also been noted by Lo and Marcotullio (2000). They

Table 2.6 Inward foreign direct investment (UD$m), Asia-Pacific, 1998–2001

Country/region	1998	Share (%)	1999	Share (%)	2000	Share (%)	2001	Share (%)
Australia	6,112	6.8	5,686	5.5	11,957	8.4	4,090	4.3
Brunei Darussalam	573	0.6	596	0.6	600	0.4	244	0.3
Cambodia	230	0.3	214	0.2	179	0.1	113	0.1
China[1]	43,733	48.6	40,328	38.8	40,765	28.5	46,841	48.8
Hong Kong	14,770	16.4	24,596	23.7	61,938	43.3	22,834	23.8
Indonesia	−356	−0.4	−2,745	−2.6	−4,550	−3.2	−3,277	−3.4
Korea, Rep. of	5,412	6.0	9,333	9.0	9,283	6.5	3,198	3.3
Laos	45	0.1	52	0.1	34	0.0	24	0.0
Malaysia	2,714	3.0	3,895	3.7	3,788	2.6	554	0.6
Myanmar	315	0.4	253	0.2	255	0.2	123	0.1
New Zealand	1,191	1.3	1,412	1.4	3,209	2.2	1,699	1.8
Philippines	1,752	1.9	578	0.6	1,241	0.9	1,792	1.9
Singapore	6,389	7.1	11,803	11.4	5,407	3.8	8,609	9.0
Taiwan	222	0.2	2,926	2.8	4,928	3.4	4,109	4.3
Thailand	5,143	5.7	3,561	3.4	2,813	2.0	3,759	3.9
Vietnam	1,700	1.9	1,484	1.4	1,289	0.9	1,300	1.4
Asia-Pacific	89,945	13.0	103,972	9.6	143,136	9.6	96,012	13.1
North America[2]	197,243	28.4	307,811	28.3	367,529	24.6	151,900	20.7
Western Europe[3]	274,739	39.6	507,222	46.6	832,067	55.8	336,210	45.7
World total	694,457	80.9[4]	1,088,263	84.4	1,491,934	90.0	735,146	79.5

Source: Extracted from UNCTAD World Investment Report 2002, accessed from http://www.unctad.org/Templates/WebFlyer.asp.

Notes
1 Including Macao SAR.
2 US and Canada.
3 Including Scandinavian countries.
4 Sum of Asia-Pacific, North America, Western Europe as share of world total.

also observe that Asia-Pacific FDI is largely regionally-based (see also Edg-ington and Hara, 1998) with Japan's early domination declining as a result of its own economic problems and as other countries in the region (espe-cially the NIEs) have introduced their own initiatives.

Declining share of service corporate headquarters

Although there are limitations attached to the use of the Fortune Global 500 as a basis for assessing the relative importance of the Asia-Pacific as a service-producing region, a comparison of the situation in 1995 with that in 2002 is quite revealing (Table 2.7). The first point of note is that the number of service headquarters in the top 500 has declined from 299 in 1995 to 291 in 2002 but the number of employees in the worldwide opera-tions of the service corporations increased by almost twice (41 per cent) the figure for manufacturing corporations (23 per cent). However, the contraction in the overall share of service headquarters in the top 500 comprises expansion in some trade blocs and contraction in others. By far the largest reduction has occurred in the Asia-Pacific; it was the location of 99 service headquarters in 1995 but this had declined to just 57 (20 per cent) in 2002. Meanwhile, Western Europe showed a very small decline (−2) while North American moved ahead of Western Europe with 43 per cent of the service headquarters in the top 500 (up from 29 per cent in 1995). This shift is mirrored by manufacturing headquarters which have become even more concentrated in North America (40 per cent in 2002). Even though there were fewer service headquarters in the Asia-Pacific in 2002, the region's share of total employees has remained constant although the change in the absolute number (34 per cent) was the lowest of all the trade blocs (Table 2.7). The picture is very different for manu-facturing with Asia-Pacific based corporations whose share of total employ-ment grew more quickly than any other trade bloc to almost match the share of Western Europe which declined from 40 per cent in 1995 to 32 per cent in 2002. It seems, then, that from the perspective of corporate control of service production the relative position of the Asia-Pacific is declining even though the average size of the corporations still headquar-tered there is growing, albeit at a slower rate than in North America or Western Europe (see Table 2.7).

There are other ways of demonstrating the relatively weak position of the Asia-Pacific as a location for major service-producing corporations. It is the location of only five of the 52 service firms in the top 100 (19 of the 48 manufacturing firms in the top 100) in 2002. Further down the ranked list the picture across the trade blocs is rather more complicated but at each level the Asia-Pacific manufacturing corporations are more numer-ous than their service counterparts while the reverse is the case in Western Europe and North America. Almost all the employees of the region's man-ufacturing corporations are employed by firms in the top 100 (over

Table 2.7 Share of top 500 corporate headquarters, major world trading blocs, 1995 and 2002

Trade bloc	Year	Manufacturing HQs				Service HQs			
		No	%	Total employees[3]	%	No	%	Total employees	%
Asia-Pacific[1]	2002	64	30.6	5,995,422	30.2	57	19.6	3,835,496	14.4
	1995	62	30.8	3,593,272	22.2	99	33.1	2,852,945	15.1
Change		2		66.8		-42		34.4	
North America[2]	2002	84	40.2	7,213,495	36.4	124	42.6	11,899,551	44.7
	1995	70	34.8	5,789,525	35.8	88	29.4	8,475,556	44.8
Change		14		24.6		36		40.4	
Western Europe	2002	57	27.3	6,299,785	31.8	106	36.4	10,395,999	39
	1995	65	32.3	6,537,269	40.4	108	36.1	7,296,467	38.6
Change		-8		-3.6		-2		42.5	
Other regions	2002	4	1.9	334,552	1.7	4	1.4	518,360	1.9
	1995	4	1.9	262,750	1.6	4	1.3	298,074	1.6
Change		nc		27.3		nc		73.9	
Totals	2002	209	100	19,843,254	100	291	100	26,649,406	100
	1995	201	100	16,182,816	100	299	100	18,923,042	100
Change		8		22.6		-8		40.8	

Source: Compiled from data in *Fortune*, 5 August 1996, 21 July 2003.

Notes
1 Excludes US and Canada (1995); US, Canada and Mexico (2002).
2 US and Canada (1995); US, Canada and Mexico (2002).
3 Employees in all affiliates, subsidiaries etc., world-wide.

4 million) while the reverse is true for services (some 3 million employed by firms ranked lower than 200). This does not compare favourably with North America and Western Europe (service corporations in the top 200 globally account for more than half of total employment). The *Fortune* data also shows (Table 2.8) that the region specializes in selected services, notably trading companies such as Mitsubishi, Mitsui and Sumitomo (Japan), Samsung, LG International and Hyundai (South Korea), and Sinochem and Cofco (China). It is also strong in engineering and construction services with six of the ten corporations in the top 500 such as Kajima, Shimizu and Tokyo (Japan). More than 60 per cent of the mutual life and health insurance firms are headquartered in the region. It has a smaller share than expected for most other services, especially those that can be broadly defined as producer services. Commercial and savings banks, computer services and software, diversified financials, securities or diversified outsourcing services are all under-represented. With the excep-

Table 2.8 Service corporations in Fortune top 500: Asia-Pacific share, by activity, 2003

Activity	No. in global 500	No. in Asia-Pacific	Share (%)
Banks: commercial and savings	62	11	17.7
Computer services and software	4	0	0.0
Diversified financials	6	0	0.0
Diversified outsourcing services	2	0	0.0
Engineering, construction	10	6	60.0
Entertainment	6	1	16.6
Food services	3	0	0.0
Food and drug stores	23	4	17.4
General merchandisers	12	1	8.3
Health care	12	0	0.0
Insurance: life, health (mutual)	13	8	61.5
Insurance: life, health (stock)	19	3	15.7
Insurance: P&C (mutual)	2	0	0.0
Insurance: P&C stock	15	3	20.0
Mail, package, freight delivery	8	1	12.5
Railroads	6	3	50.0
Securities	4	0	0.0
Speciality retailers	13	0	0.0
Telecommunications	24	6	25.0
Trading	16	15	93.8
Wholesalers: electronics, office	2	0	0.0
Wholesalers: food and grocery	4	0	0.0
Wholesalers: health care	6	1	16.6
Miscellaneous	10	2	20.0
Total	282	65	23.1

Source: Extracted from *Fortune*, 21 July 2003.

tion of the trading companies, many of the services in which the region is relatively specialized are oriented towards national/regional markets rather than global markets. A recent study of producer services in Hong Kong, for example, shows that local clients dominate (76 per cent) with only 11 per cent engaged in international transactions (Yeh, 2003). This contrasts with much earlier research (Beyers and Alvine, 1985; Beyers, 1986; Davis and Hutton, 1981) on the other side of the Pacific Rim that revealed the high export propensity of business service firms in the Puget Sound region (Seattle) and in Vancouver.

Redistribution of foreign bank branches and representative offices

A measure of the internationalization of the Asia-Pacific economy is the presence of the offices of foreign banks. These are set up to serve the needs of existing corporate clients (and may also service retail clients), and in anticipation of the banking requirements of overseas and domestic corporations looking to set up business operations in the region for the first time or, conversely, to expand from the region into other markets. Depending on the scale and status of the foreign bank operation, they also perform an important strategic research role for the parent bank, with information distributed amongst its constituent activities to enable better client service or identification of emerging business opportunities. Foreign banks also need to be familiar with local and sub-regional market characteristics (opportunities and weaknesses) as well as with the finer points of undertaking business transactions within the diverse social and cultural contexts characteristic of the region. When European and North American bank representation (i.e. excluding Asia-Pacific banks with branches in other countries within the region) was first looked at for 1996 (Daniels, 1998) the numbers were increasing at a time of strong growth rates and increasing openness to businesses from other parts of the world. The region's most vibrant financial centres with significant stock exchanges and strong foreign exchange transactions, Hong Kong and Singapore, hosted at least one bank from almost all of the EU15 countries, especially the UK, France, Italy, Germany and the Netherlands. US banks also had a prominent presence (15 in Hong Kong, 12 in Singapore). A combination of historical factors and competition for the position of leading regional corporate control centre underpinned the dominant position of these two cities. It was also linked to the fortunes of a small number of major banks with longstanding links to the region, notably the HSBC group and Standard Chartered Bank. Not only did a lion's share of both banks come from their operations in the Asia-Pacific, they were also actively opening new or additional offices or branches in Australia, China, Vietnam, the Philippines and elsewhere; in 1996 Standard Chartered operated in every Asia-Pacific country except North Korea and the Asian part of Russia (Warner, 1996). Tokyo, despite its much hyped position as a member of the global triad, fell some way behind in terms of both the absolute number of foreign banks and the range of countries represented.

On the basis of the 1996 data it was noted that 'it seems apparent that European and North American banks have been attracted to the region in greater numbers than would have been expected' (Daniels, 1998: 155) and this was an indication that further growth of foreign bank presence was likely. In the event the number of foreign banks represented in ten Asia-Pacific countries[2] marginally declined (92 to 89) and the range of home country representation has remained the same with minor additions from Cyprus, Denmark and Luxembourg. Of the leading countries represented in 1996, only German banks had increased their representation by 2001. It seems that the financial crises that beset the region during the late 1990s and the collapse of Japan's bubble economy at the beginning of that decade have discouraged its closer integration with the international banking system via foreign bank representation. Restoring domestic banking systems that were hit hard by previous bad loans and weak banking sectors has undermined the confidence of foreign banks even though their participation could actually help the recovery process. However, ongoing inadequacies in local regulatory and legal structures, especially in China, as well as political sensitivities about allowing foreign banks into domestic markets are still an obstacle (Blanden, 2001). The geopolitical uncertainties created by the return of Hong Kong to Chinese jurisdiction also seems to have unsettled foreign banks in that the number represented there fell from 70 to 63, allowing Singapore to slightly strengthen its position as the principal location for foreign bank offices in the Asia-Pacific. Japan's position has weakened further (only 43 foreign banks in 2001) with Australia, Thailand and Malaysia experiencing significant increases in foreign bank representation. China was not included in the 1996 analysis but its much more open stance towards foreign investment has encouraged foreign banks from the leading home countries such as France and Germany; it already had 27 branches and subsidiaries by 2001 and the number will certainly have increased since then, even though there are ongoing uncertainties about the robustness of its financial infrastructure (Bremner *et al.*, 2004).

Japan: a case example

In view of the indisputable role of a number of economies in the Asia-Pacific region (such as Japan, South Korea and Taiwan) as global suppliers of manufactured goods (Dicken, 1991; Eli, 1990; Stevens, 1996) it is reasonable to anticipate parallel expansion of business and professional services (BPS).[3] However, for reasons outlined elsewhere, but not substantiated in any detail here, there seems to be a gap in countries such as Japan between the rise of manufacturing production and the parallel development of BPS (Daniels, 1998, 2001). Certainly, the prevailing structural conditions suggest that the demand for services should be substantial; it seems unlikely that the scale of industrial development and exports, in many cases associated with global market leadership in things such as sophisticated electronic products, has been achieved without considerable inputs of the knowledge and expertise provided by a full range of producer services. It could be that manufacturing firms based in the Asia-Pacific find it easier to obtain such services, and of appropriate quality, from suppliers in North America and Europe.

A global regional division of labour whereby countries such as the UK export services while countries such as Japan specialize in the export of goods may make this situation sustainable. The UK has a positive balance of trade in services with Japan, except for transportation and travel but, with exception of construction and to a much lesser degree financial services, Japan imports more services from the UK than it exports. However, investment in Japan by UK firms is quite modest (£4,477 million in 2000) relative to the size of the economy compared with similar investment in Hong Kong (£4,681 million) and Singapore (£6,300 million). But there has been a discernible shift in the UK's net international investment position in Japan with respect to services; financial services accounted for a much larger share than real estate and business services in 1997 and 1998 but the pattern reversed in favour of the latter between 1999 and 2000 (Office for National Statistics, 2001). Outward foreign direct investment from Japan in 2000–2001, even though the national economy has been depressed by comparison with the boom decade of the 1980s, continued to be dominated by manufacturing (£68 billion), transportation (£127 billion) and finance and insurance-related services (£49 billion) (Ministry of Finance, 2001) with knowledge and information intensive services, as part of a general category 'services', minor players in foreign direct investment (just over £10 billion in 2000–2001). There were no net investment flows from Japan into the UK for real estate and business services between 1997 and 2000; the net investment flows were dominated (for services) by finance and insurance.

From the perspective of overseas BPS firms looking to invest in new business opportunities in countries such as Japan and South Korea there are a number of issues. First, the differences in culture and markets; second, the intensity of the competitive environment that requires suppliers of producer services to simultaneously combine quality, flexibility, efficiency and agility; third, the subtle rules of doing business that are very different from those of Europe or North America (Lassere and Redding, 1995; Yeung and Li, 2000; Yeung, 2002) and, fourth, the regulatory obstacles to developing commercial opportunities. The difficulties confronted by overseas service firms seeking direct representation in the Japanese market are well documented (House of Commons, 1998). Deregulation of the insurance market, where the concept of insurance broking for example is still relatively novel, is now underway. London-based firms now have opportunities for Japanese asset management previously denied to overseas firms but foreign lawyers and legal services firms are still subject to significant restrictions such as tight limits on forming partnerships with Japanese lawyers. This has the effect of encouraging Japanese corporations, many of whom are familiar with the quality of the producer services available in North America and Europe, to source their requirements outside Japan.

At first glance the structure of employment in Japan is broadly similar to that of the UK (Table 2.9).[4] Service industries comprised just over 68 per cent of Japanese employment in 1999 compared with some 77 per cent in the UK. More than one in five employees in Japan is in the manufacturing sector while in the UK the ratio is slightly higher than one in six. Employment in the construction sector in Japan was more than twice as large as that in the UK. Within the service sector the shares of employment in transport and related services, financial intermediation and in real estate are very similar; the major discrepancies occur with respect to wholesale and retail trade (services related to final consumption) and the catch-all category 'services' which includes activities related to intermediate demand such as BPS (see Table 2.9). In this case the difference between the UK and Japan is especially marked; other 'services' comprise almost 42 per cent of all employment in the UK compared with just over 25 per cent in Japan.

Approximately 30 per cent of the establishments and 42 per cent of the employment in 'services' can be classified as producer services (*Census of Service Industries of Japan*, 1999).[5] More telling perhaps is the data in Table 2.10 which shows that while there are three times as many business establishments in Japan as in the UK (which is partly a function of the difference in the size of the total labour force, see Table 2.9) the total number of selected producer service establishments is very similar. Because the average size of these establishments (number of employees) may be different, some caution is necessary when interpreting such figures. But even if this is a consideration it is probably insufficient to explain away the large difference in the share of producer service establishments in the two economies; just over 2 per cent of the total for Japan compared with almost 8.5 per cent for the UK in 1999. Some of the major differences in the share of establishments are found amongst those services that have been most prone in the UK to the demand generated by externalization, namely software consultancy and supply, advertising, legal services, labour recruitment and accounting and auditing services. While the data in Tables 2.9 and 2.10 are far from comprehensive, they do suggest that there are some significant structural differences in the sectoral composition of employment in Japan and the UK. Further, the major source of these differences is the share of producer services, whether measured by employment share or by share of total establishments.

In view of the major role of Japanese manufacturing in the domestic

Table 2.9 Employment structure, Japan and UK, 1999

Sector	Japan ('000s)	UK ('000s)	Japan (%)	UK (%)
Agriculture, hunting, forestry and fishing	216	302	0.40	1.24
Mining and quarrying, electricity, gas and water supply	269	199	0.50	0.82
Manufacturing	11,452	3,944	21.28	16.18
Construction	5,090	1,080	9.46	4.43
Manufacturing and others	17,027	5,525	31.64	22.67
Wholesale and retail trade, repairs,[1] hotels and restaurants	17,245	5,846	32.05	23.98
Transport, storage, post and telecommunications	3,254	1,449	6.05	5.94
Financial intermediation	1,723	1,060	3.20	4.35
Real estate	869	311	1.62	1.28
Services	13,687	10,190	25.44	41.80
All services	36,778	18,856	68.36	77.35
Total	53,807	24,380	100.0	100.0

Source: Census of Enterprises of Japan, 1999; Labour Market Trends, Monthly Digest of Statistics, 2000, UK.

Note
1 Japanese data exclude repairs.

Table 2.10 Number of establishments, selected producer services, Japan and UK, 1999

Industry	SIC No.		No. of establishments		Share in all industries (%)	
	JP	UK	JP	UK	JP	UK
Renting of machinery and equipment without operator and of personal and household goods	79	71	31,612	12,430	0.51	0.78
Software consultancy and supply	821	72.2	14,136	55,895	0.23	3.50
Data base activities	82B	72.4	1,264	455	0.02	0.03
Data processing	82A	72.3	4,449	860	0.07	0.05
Advertising	83	74.4	10,779	10,020	0.17	0.63
Legal activities	841,842	74.11	24,542	22,685	0.40	1.42
Accounting, book-keeping and auditing activities; tax consultancy	843	74.12	33,115	19,675	0.53	1.23
Labour recruitment and provision of personnel	865	74.5	4,500	10,665	0.07	0.67
Research and development	92	73	2,723	2,250	0.04	0.14
Total: selected producer services			127,120	123,935	2.04	8.45
All industries			6,203,249	1,595,705	100.0	100.0

Source: Census of Service Industries of Japan, 1999; Business Monitor, Service Analysis of UK Business, 1999.

and, in particular, the global economy, the apparently limited representation of producer services is surprising. There are a number of possible explanations. First, there is a fundamental difference between Japan and the UK with reference to the value attached to 'service'. In Japanese culture something that is intangible such as a service is not regarded as possessing value; 'to serve' is to provide something that is 'free'. For example, UK firms accept, if only implicitly, the idea that the higher the fee the better the service; much less so their Japanese counterparts. Second, it could be hypothesized that Japanese firms do not need to use the expertise provided by knowledge and information intensive producer services and, third, it is possibly the case that the 'needs-demand gap' is wider in Japan than in the UK. Finally, since Japanese firms regard services as an expense rather than a cost, managers are inclined to bundle the supply of producer services in-house. This has the effect of limiting the demand for externally-provided producer services which, in turn, contributes to market underdevelopment for producer services. It also makes them more expensive because of the limited competition.

Importance of large corporations for the supply of producer services

The market underdevelopment of Japanese producer services can also be understood in relation to the tendency for very large corporate organizations, such as the banks, to provide a comprehensive portfolio of services for their clients that includes producer services that in markets such as those of the UK are provided by freestanding, often small, specialist firms. For example, the Tokyo-Mitsubishi Bank (turnover ¥3.6 trillion) comprises 50 group companies (stock holdings) in Japan of which some 40 per cent are only loosely related to the provision of the bank's core business in financial services. This is typical of all banks in Japan. Although relatively small by domestic standards, this is illustrated by a local bank such as the Bank of Fukuoka (turnover ¥212 million) that reveals a diversity of service provision which incorporates a considerable range of non-financial services in its portfolio (Table 2.11). It diversified into computer services as early as 1979 (Fukuoka Computer Service) which, together with the Joint Data Service established in 2000, employs more than any other of the Group's companies which also include personnel placement (established in 1984), real estate maintenance (1988), leasing (1981), and a printing and bookbinding service (1999).

The Japanese general trading companies (*sogo shosha*) also perform an important role in the supply of producer services. Trading companies, such as Mitsui & Co which has 240 group companies, supply goods *and* producer services both to clients within Japan and to the establishments of major Japanese corporations located elsewhere in the world. Although held in stock by the parent company the group companies are independent so that their transactions are classified by their activity etc. in the Japanese census, not by the activity of the parent company. If the latter was the case, it could be argued that the level of producer services employment in total employment in Japan is a function of the internalization of transactions amongst the companies incorporated within *sogo shosha* or the *keiretsu*. But this is not the case and the explanation for the low level of producer service activity must therefore rest with the preference of Japanese firms to source most of their producer services in-house. Alongside this they may also, where necessary, undertake these services transactions primarily with other companies within their own group of companies.

The preference for internalization and bundling of services

A tendency to bundle services as much as possible accounts for the Japanese preference for internalization of producer services. This is partly a legacy of traditional Japanese business practice but also enables firms to maintain a high level of confidentiality about their activities. Furthermore, externalization of the consumption of producer services is considered more expensive than internal production. This is accentuated by the

Table 2.11 The group companies of Fukuoka Bank, 2001

Company	Year of foundation	Share of stock holdings (%)	Employees	Turnover (¥ million)	Main business
Joint data service	2000	5	269	–	System development
Fukuoka Bank guarantee	1978	25	48	2,480	Guarantee of debts
Fukuoka card	1983	25	51	1,829	Credit card business
Fukuoka computer service	1979	5	232	3,260	Information processing, software development
Fukuoka property management	1995	100	6	163	Real estate
Fukuoka Bank office service	1984	100	119	1,882	Personnel placement
Fukuoka Bank deposit service	1980	100	20	324	Deposit of slips, account book and documents
Fukuoka Bank system service	1999	100	170	432	System maintenance
Fukuoka Bank office work service	1999	100	178	548	Printing and bookbinding of documents
Fukuoka Bank investment counselling	1986	25	9	145	Consultancy for investment
Fukuoka Bank business service	1976	100	176	1,730	Cash adjustment
Fukuoka Bank real estate management	1988	100	16	284	Maintenance of real estate
Fukuoka Bank real estate investigation	2000	100	4	–	Estimation of collateral real estate
Fukuoka Bank leasing	1981	25	111	33,399	Leasing
Fukuoka Bank loan service	1999	100	86	463	Concentrated dealing of loan

Source: Japanese Group Companies. Toyokeizai.

difficulty of accessing the kind of BPS advice that meets, for example, corporate expectations on price and quality. The result is a vicious circle: demand does not stimulate supply and the supply of producer services does not improve and diversify because of limited demand.

It seems that as far as firms in Japan are concerned, the importance of acquiring high quality specialized knowledge at whatever the cost has yet to develop. Japanese firms do not therefore generate the significant external demand for knowledge intensive or specialized services that are considered essential by similar firms in the UK. Even if they do consider this to be a priority, they tend to internalize these services. It is the case, however, that labour intensive, routine and non-confidential services *may* be externalized on cost-saving grounds. Because they take it for granted that they as individuals should have the appropriate knowledge about all aspects of management, Japanese managers believe that it is a waste of resources to purchase knowledge from external sources. Even if they do wish to purchase such services, they lack confidence in the willingness of or the ability of the supplier to undertake the amount of work that they have commissioned. Such market underdevelopment makes Japanese producer services expensive and low-quality; this reinforces the passive management attitude towards externalization.

Although service production is by no means insignificant, the preference for 'service bundling' continues to have a strong influence in Japanese business culture. It not only disguises the 'real share' of producer service activities in the economy but also helps to sustain an immature market for such services. Market segmentation and a drive towards higher quality producer service knowledges will be discouraged as long as a significant number of client firms want a one-stop, comprehensive service from their suppliers.

Conclusion

Tertiarization of the Asia-Pacific economies and its impact on urban development opportunities and challenges has undoubtedly been an important feature of the region's development trajectory over the last 20–30 years. Such is the economic, social and cultural diversity of the region that generalization is difficult, but as a general rule the tertiarization process has been dominated by intra-regional service transactions and trade flows, although there are of course some significant extra-regional linkages. Many of the latter, especially those embedded in financial, business and professional services, are mediated through the region's global city-regions. The really interesting question, which is explored at different levels from the macro- to the micro- by the contributions later in this volume, is whether, and how, the region from a services perspective will become more fully integrated into the global economy and whether this will be reflected in the organization of urban hierarchy or the form and structure of individual cities (Sassen, 2000).

The prospect for the future has been summarized by Hutton (2004: 66) as the:

> extent to which full globalization processes may override established regional service markets, for example, in finance, property and real estate ... *and whether* ... the specialised service centres and corridors described earlier may emerge as platforms for the globalization of services produced within the Asia-Pacific, as well as engines of development within the region.

In contrast to the influence of market forces on the way in which services have adopted a significant role in the development of the economies and cities of Europe and North America (the 'Atlantic core' countries), the prospects for the Asia-Pacific may depend much more on the role of institutions and the state. They have already been instrumental in creating a strong emphasis on promotional and developmental strategies designed to encourage inward service investment based on, for example, rapid modernization of urban infrastructures and human resources. By comparison with the Atlantic core, in many cases such policies have been underwritten by softer regulatory environments in activities such as city and regional planning. There are, therefore, numerous policy and development issues that confront the cities and urban regions at the leading edge of the tertiarization of the Asia-Pacific economies. These are explored in much greater detail by Hutton in the chapter that follows.

Acknowledgements

I am indebted to Dr David Shaw, Honorary Research Fellow, Service Sector Research Unit, University of Birmingham, for his invaluable assistance with the not inconsiderable task of compiling most of the statistical material for this chapter. The case study of producer services in Japan arises from exploratory research undertaken in collaboration with Dr Tetsuji Ishimaru, Lecturer in Human Geography, Faculty of Humanities and Social Studies, Fukuoka University of Education, Fukuoka, Japan. I am grateful to him for permission to reproduce some of the findings here.

Notes

1 At the global level all the countries in the region, except the highly industrialized countries, are classed as developing or newly industrialized countries.
2 Australia, Hong Kong, China, Indonesia, Japan, South Korea, Malaysia, Singapore, Taiwan, Thailand, Vietnam.
3 Also referred to as producer services.
4 The figures reproduced here are estimates based on matching the relevant classes of activity listed in the *Census of Enterprises in Japan* (1999) and *Labour Market Trends* (2000).

5 Defined as: professional services (SIC84), other producer services (86), information services (82), Leasing (79), research institute (92), advertising (83).

References

Abdul Fatah Che, H., Morshidi, S. and Nooriah, Y. (2001) *Producer Services in Cities of the Asia Pacific Economic Region: The Case of Kuala Lumpur, Malaysia.* Huntington: Nova Science Publishers.

Aharoni, Y. and Nachum, L. (eds) (2000) *Globalization of Services: Some Implications for Theory and Practice.* London: Routledge.

Beyers, W. B. (1986) *The Service Economy: Understanding the Growth of Producer Services in the Central Puget Sound Region.* Seattle: Central Puget Sound Economic Development District.

Beyers, W. B. and Alvine, M. J. (1985) 'Export services in post-industrial society', *Papers of the Regional Science Association,* 57, 33–45.

Blanden, M. (2001) 'Healing old wounds', *The Banker,* May, 71–6.

Bremner, B., Roberts, D. and Balfour, F. (2004) 'Headed for a crisis?', *Business Week,* 3 May.

Bryson, J. R., Daniels, P. W. and Warf, B. (2004) *Service Worlds: People, Organisations, Technologies.* London: Routledge.

Cuadrado-Roura, J. R., Rubalcaba-Bermejo, L. and Bryson, J. R. (2002) *Trading Services in the Global Economy.* Cheltenham: Edward Elgar.

Daniels, P. W. (1998) 'Economic development and producer services growth: the APEC experience', *Asia Pacific Viewpoint,* 39, 145–59.

Daniels, P. W. (2001) 'Globalization, producer services and the city: is Asia a special case?', in Stern, R. M. (ed.), *Services in the International Economy.* Ann Arbor: University of Michigan Press, 213–30.

Daniels, P. W. and Bryson, J. R. (2002) 'Manufacturing services or services manufacturing?: new forms of production in advanced capitalist economies', *Urban Studies,* 39, 177–99.

Davis, H. C. and Hutton, T. A. (1981) 'Some planning implications of the expansion of the urban service sector', *Plan Canada,* 21, 15–23.

Dicken, P. (1991) 'The changing geography of Japanese investment in manufacturing industry: a global perspective', in Morris, J. (ed.), *Japan and the Global Economy: Issues and Trends in the 1990s.* London: Routledge, 14–44.

Douglass, M. (2000) 'Mega-urban regions and world city formation: globalization, the economic crisis and urban policy issues in Pacific Asia', *Urban Studies,* 37, 2315–35.

Douglass, M. (2002) 'From global intercity competition to cooperation for liveable cities and economic resilience in Pacific Asia', *Environment and Urbanisation,* 14, 53–68.

The Economist (2004), 19–25 June.

Edgington, D. W. and Hara, H. (1998) 'Japanese service sector multinationals and the hierarchy of Pacific Rim cities', *Asia Pacific Viewpoint,* 39, 161–78.

Eli, M. (1990) Japan Inc.: *Global Strategies of Japanese Trading Corporations.* London: McGraw Hill.

Forbes, D. (1997) 'Metropolis and megaurban region in Pacific Asia', *Tijdschrift voor Economische en Sociale Geografie,* 88(5), 457–68.

Graham, S. (1999) 'Global grids of glass: on global cities, telecommunications and planetary urban networks', *Urban Studies*, 36(5–6), 929–49.

Grant, R. and Nijman, J. (2002) 'Globalization and the corporate geography of cities in the less-developed world', *Annals, Association of American Geographers*, 92(2), 320–40.

House of Commons (1998) *Select Committee on Trade and Industry, Eleventh Report: Industrial and Trade Relations with Japan.* Cmnd 569, London: Her Majesty's Stationery Office.

Hutton, T. A. (2004) 'Service industries, globalization, and urban restructuring within the Asia-Pacific: new development trajectories and planning responses', *Progress in Planning*, 61, 1–74.

International Telecommunications Union (2002) *ITU Asia-Pacific Telecommunications Indicators.* Geneva: ITU.

King, A. D. (1990) *Global Cities: Post Imperialism and the Internationalization of London.* London and New York: Routledge.

Lassere, P. and Redding, G. (1995) 'Corporate strategies for the Asia-Pacific', *Long Range Planning*, 28, 11–69.

Lee, K. T. (ed.) (2002) *Globalization and the Asia Pacific Economy.* New York: Routledge.

Lo, F.-C. and Marcotullio, P. J. (2000) 'Globalization and urban transformations in the Asia-Pacific region: a review', *Urban Studies*, 37(1), 77–111.

Lo, F.-C. and Marcotullio, P. (eds) (2001) *Globalization and the Sustainability of Cities in the Asia Pacific Region.* Tokyo: UNU Press.

Lo, F.-C. and Yeung, Y. M. (1996) *Emerging World Cities in Pacific Asia.* Tokyo and New York: United Nations University Press.

Ministry of Finance (2001) Foreign Direct Investment, 1986–2001. Available at http://www.mof.go.jp (accessed 10 April 2004).

Morshidi, S. (2000) 'Globalising Kuala Lumpur and the strategic role of the producer services sector', *Urban Studies*, 37(12), 2217–40.

Mullampally, P. and Zimny, Z. (2000) 'Foreign direct investment in services: trends and patterns', in Ahorini, Y. and Nachum, L. (eds), *Globalization of Services: Some Implications for Theory and Practice.* London: Routledge, 25–51.

Murphy, A. B. (1995) 'Economic regionalisation and Pacific Asia', *Geographical Review*, 85, 127–40.

Office for National Statistics (2001) *Foreign Direct Investment (MAG).* London: Office for National Statistics.

Olds, K., Dicken, P., Kelly, P. F., Kong, L. and Yeung, H. W.-C. (1999) *Globalization and the Asia-Pacific: Contested Territories.* London: Routledge.

Organization for Economic Cooperation and Development (1999) *The Service Economy* (Business and Industry Policy Forum Series). Paris: OECD.

Organization for Economic Cooperation and Development (2002) *OECD Statistics on International Trade in Services: Partner Country Data and Summary Analysis, 1999–2000.* Paris: OECD.

Organization for Economic Cooperation and Development (2003a) *OPEC-APEC Global Forum: Policy Frameworks for the Digital Economy: Summary Report.* Paris: OECD.

Organization for Economic Cooperation and Development (2003b) *Opening up Trade in Services: Opportunities and Gains for Developing Countries.* Paris: OECD.

People's Daily (2004) 'Service industry to grow in share of economy', at http://english peopledaily.com.cn (accessed 26 April 2004).

Poon, J. P. H. (1997) 'Inter-country trade patterns in Europe and the Asia-Pacific: regional structure and extra-regional trends', *Geografiska Annaler*, 79B, 41–55.

Poon, J. P. H. (2000) 'Myth of the triad? The geography of trade and investment "blocs"', *Transactions of the Institute of British Geographers*, NS25(4), 427–44.

Sachs, J. D. (2000) 'Globalization and patterns of economic development', *Weltwirtschaftliches Archiv*, 136, 579–600.

Sassen, S. (2000) *Cities in a World Economy*. Thousand Oaks: Pine Forge Press.

Shin, K.-H. and Timberlake, M. (2000) 'World cities in Asia: cliques, centrality and connectedness', *Urban Studies*, 37(12), 2257–85.

Stern, R. M. (ed.) (2001) *Services in the International Economy*. Michigan: University of Michigan.

Stevens, R. (1996) *Japan and the New World Orders Global Investments, Trade and Finance*. Basingstoke: Macmillan.

Warner, A. (1996) 'Follow the herd: foreign banks in Asia', *The Banker*, May, 40–3.

World Trade Organization (2002) *World Trade Developments in 2001 and Prospects for 2002*. Geneva: WTO.

World Trade Organization (2002) *Statistics on Globalization, 2001*. Geneva: WTO.

Yeh, A. (2003) 'Hong Kong's producer services linkages with the Pearl River Delta', in *The Hong Kong Servicing Economy*. Hong Kong: Hong Kong Coalition of Service Industries, 2–3.

Yeung, H. W. (2002) *Entrepreneurship and the Internationalisation of Asian Firms*. Cheltenham: Elgar.

Yeung, Y.-M. and Li, X. (2000) 'Transnational corporations and local embeddedness: company case studies from Shanghai, China', *The Professional Geographer*, 52(4), 624–35.

3 Services and urban development in the Asia-Pacific region

Institutional responses and policy innovation

T. A. Hutton

The last quarter-century has seen the rapid growth of service industries as a feature of development within the broadly-defined Asia-Pacific region.[1] This secular growth in services (or *tertiarization*) has largely been associated with market forces, demographic trends and rising incomes. But tertiarization within the region can also be attributed to generally increasing policy commitments at national and local government levels, notwithstanding the paramountcy of the industrialization paradigm (in the categorical sense of manufacturing-led growth and industrial labour formation as preeminent policy aspirations). Further, we can identify distinct phases of service policy approaches within the region, in response to economic shocks and experiences of industrial restructuring, changing development values and goals, and, increasingly, the pressures of globalization. This chapter will offer a framework for delineating benchmarks in the evolution of services industry policy within the Asia-Pacific, and will include examples of illustrative strategies, as well as comparative references to policy approaches among 'Western' jurisdictions. Given the scale and complexity of the region, and the sharp contrasts in forms of political control among constituent states, the narrative will inevitably over-simplify processes and outcomes. The (necessarily limited) intention, then, is to stimulate further discussion in this critical sphere of inquiry, and to provide a measure of context for the substantive chapters to follow.

Services at the leading edge of urban change

Over the last half-century, the development record within the Asia-Pacific region has emphasized the remarkable growth in industrial production and trade among the leading economies. Manufacturing has underpinned the growth of every dynamic national economy in the region, as seen in the post-war recovery and ascendancy of Japan, the development of the four Newly-Industrialized Economies (NIEs), the so-called 'little tigers' or dragons (South Korea, Taiwan, Hong Kong and Singapore), and of course China, which has emerged as the chief factory for the world's consumer markets. Industrialization is also a principal development instrument of the 'near-NICs', notably Malaysia and Thailand, and represents an aspiration of many lagging nations and regions within the Asia-Pacific.

Much of the industrial production within Asia-Pacific nations is located in urban areas, as the industrial infrastructure is concentrated within primate cities, with the exception of China, where a major share of national manufacturing capacity is situated within exurban or rural areas. But the industrialization experience of nations within the Asia-Pacific also featured the emergence of truly world-scale industrial metropoles over the 1960s and 1970s, including Tokyo, Nagoya, Shanghai, Beijing, Seoul, Hong Kong, Taipei and Singapore, among others. While the precise mix of sectors varied from place to place, in each of these city-regions the centrepiece of the urban industrial economy comprised an ensemble of manufacturing engaged in the production of capital goods and consumer products, systems of light industry, craft-scale production and workshops, and ancillary warehousing and distribution activities. Over a relatively compressed period, the manufacturing sectors within these industrial cities shifted production from basic, lower-end goods to higher value-added products, involving more sophisticated production technologies and skilled labour. Concomitant social features included new divisions of production labour, the rise of an industrial class comprised of manufacturing and allied occupations, and high-density working class neighbourhoods and communities.

As is well known, these rapid industrialization trajectories within the Asia-Pacific were achieved in large measure with the support of national (and complementary local) government policies, in the form of central planning directives, investments in industrial production capacity and infrastructure, tariff barriers and other import controls, fiscal incentives, land use policies, and housing for industrial workers and communities, among other instruments. That said, these industrialization programmes exhibited significant variation from place to place, reflecting sharply contrasting political ideologies, cultures and ideals. To illustrate, we can reference the command economy and highly centralized industrial management of the People's Republic of China (PRC) (Lin 1994), the *dirigisme* style of Japanese industrial sector policies (Vogel 1979), and the 'developmental state' model of post-colonial Singapore (Perry *et al.* 1997). Even in Hong Kong, to many the last bastion of freebooting, mercantilist capitalism, the colonial administration intervened in support of industrial development, as seen in major infrastructure investments, industrial new town planning, and overseas trade marketing (Taylor and Kwok 1989).

Even at the height of this manufacturing-led national and regional development programme within the region, service industries performed significant roles, for example in ancillary distribution industries and trade, recalling long-established entrepôt functions; in the marketing and sale of manufactured goods; and in the financing of industrial production and exports. Asia-Pacific urban economies also incorporated rich and complex patterns of traditional (or 'informal') services, including food-stalls and street-vendors, often juxtaposed amid high-rise towers domiciling professional, financial and commercial services (Leaf 1996).

But since the 1980s, services – and more particularly specialized, knowledge-based and intermediate (or producer) service industries – have become more central to (rather than merely supportive of) the sustained development of Asia-Pacific city-regions and national economies (Yeung and Lin 2003). The rapid growth of service industries among many Asia-Pacific states and city-regions represents not simply another phase in the industrialization experience, but rather, a defining feature of new developmental trajectories, with strategic linkages to larger processes of change. To illustrate:

1   *producer services* are critical to the operation of more advanced, 'flexible' post-Fordist production regimes (Coffey and Bailly 1992), and to the formation of the 'new middle class', comprised principally of professional and managerial workers (Ley 1996);
2   *information technology* (IT) firms are essential to international communications and 'connectivity' (Rimmer 1996), and to the development of the 'knowledge economy';
3   *banking and financial* institutions are key intermediaries for savings, investment and the flows of capital that support high-growth economies (Daniels 1993);
4   *creative and applied design* services constitute the basic industrial underpinnings of the new 'cultural economy' of the city (Scott 1997); and
5   *international airports* and allied service industries represent the critical infrastructural elements and connecting systems of global 'gateway' cities.

Within the public domain, service sector institutions can act as key agencies of growth and change: universities and colleges are engaged in education, training and the enhancement of human capital, while efficient, forward-looking governments are increasingly acknowledged as developmental (as well as administrative) institutions. While manufacturing and industrial production sectors remain major elements of dynamic economies within the Asia-Pacific, then, positioning of cities within national and global urban hierarchies is increasingly influenced (and defined) by clusters of strategic service sectors and industries (Hutton 2001).

Contours of urban services development within the Asia-Pacific

In some respects, the tertiarization process within Asia-Pacific states and cities follows the earlier pattern of services growth within the Atlantic (or Western) realm, but there are some important distinctions to observe.

First, there is only a rough correlation between urban scale and degree of specialization in advanced services. The economies of very large city-regions such as Manila and Jakarta incorporate relatively small advanced services sectors, while medium-size cities such as Singapore, Seattle and Melbourne are highly tertiarized with respect to industrial and employment structure. We can compare these circumstances with the situation in the Atlantic sphere, where the largest cities, such as London, Paris and New York, are all characterized by high levels of specialization in services, and by employment in services exceeding 80 or even 85 per cent of the total metropolitan labour force. Second, in contrast to the generally linear sequence of sectoral development within the Atlantic realm (crudely, from agrarianism to manufacturing, followed by tertiarization), many cities within the Asia-Pacific are experiencing significant growth both in manufacturing and services, including cities within the Pearl Delta (Lin 2001) and the Shanghai-Lower Yangzi urban region (Marton 1996). These cities are experiencing rapid tertiarization, but are by no means 'post-industrial'. Finally, in contrast to the mature (and in many cases stagnant) economies of Atlantic core regions, urban economic structures and systems within the Asia-Pacific are experiencing high levels of growth and change, as well as exposure to external shocks (such as the recent SARS epidemic and the effects of the Bali bombing), imparting a measure of dynamism (or volatility) to the urban order within the region.

With these important distinctions in mind, Table 3.1 demonstrates that growth in advanced services is strongly associated with global city formation, incorporating notions of hierarchy, and with complex and differentiated patterns of urban development. Tokyo and Los Angeles are conventionally situated at the peak of the Asia-Pacific urban hierarchy, a status defined by global-scale concentrations of banking, finance and corporate control. Consistent with the first-level world cities within the Atlantic sphere, both Los Angeles and Tokyo had embarked upon a specialized services trajectory by mid-century. Machimura observes that 'while Tokyo retained its position as one of the major productive industrial centres, its economic base had already changed from manufacturing to management and services' by the end of the 1950s (Machimura 1997: 162), reflected in a central area office boom phenomenon also experienced in London and New York. Each also contains nationally- and internationally-significant cultural, creative and educational activity, as well as leading-edge IT capacity, reinforcing the links between advanced tertiarization and global city status. But the economies of the Los Angeles and Tokyo regions also incorporate giant industrial complexes, on a larger scale than most higher-order global cities within the Atlantic domain. Many second- and third-tier cities within the Asia-Pacific also encompass both fast-growing service sectors and substantial industrial production sectors, although an increasing number are experiencing losses in some lower-end Fordist manufacturing activity and employment. At the same time, the classification also includes more narrowly-specialized port/industrial and travel/tourism urban centres, as well as mega-urban regions in which tertiarization is sharply exacerbating developmental dualism and social polarization at levels not generally observed within the mature Atlantic sphere.

Apart from the nodal forms of services-led urban development described above, the most recent period has also seen the emergence of more spatially-extended, multinuclear urban regions within the Asia-Pacific, reflecting in large part the reorganization of economic space at the regional level to respond to the pressures and opportunities of globalization. Here we can distinguish between two forms of spatial development (Table 3.1). First, we can discern contiguous (or semi-contiguous) metropolitan clusters which include world-scale platforms of specialized service industries, installations and institutions (including corporate control and banking, strategic gateway functions, cultural assets, higher education and other suites of service activity). Examples in this category include the Kanto Plain-Tokyo Bay area, the Pearl Delta Cluster, and the northern California Bay Area centred on San Francisco.

The second category of multinuclear service regions takes the form of more spatially-extended, linear, and generally non-contiguous corridors, which exhibit aspects of both competition and complementarity. To illustrate, the Pacific Northwest Corridor includes major airports and seaports which compete with each other, but also perform specialized functions, or cater to different markets. Bulk cargo exports and cruise ship travel comprise the principal functions of the Port of Vancouver, while Seattle and Tacoma are major container ports. Each is of course endeavouring to attract business from its regional rivals, but at the same time they collectively offer a more diverse and flexible array of transportation services than any could individually, enabling the corridor to compete against other gateway regions (such as the San Francisco Bay area, in this instance), in an era of relentless seaborne trade route consolidation. Similarly, the combined service infrastructures and advanced service industries of the cities of the Southeastern Australian corridor (Sydney, Canberra and Melbourne) constitute Australia's 'window' on the world and access to global trading and investment systems, overcoming to some extent the relative isolation of Australia within the Asia-Pacific (O'Connor *et al.* 2001).

Contrasts in service industry policy approaches

Policies for service industries within the Asia-Pacific have been imbued with a strongly developmental character, associated with modernization, industrial restructuring and globalization strategies. In marked contrast, the record of institutional and policy responses to growth in services within most Atlantic jurisdictions since the 1960s has tended to emphasize

Table 3.1 Spatial forms of urban service industry development within the Asia-Pacific

Urban classification and hierarchy	Exemplary centres, clusters and corridors
I Metropolitan-urban service industry centres	
A Global cities	
1 1st-order global cities: primate Asia-Pacific financial and corporate centres	Tokyo, Los Angeles
2 2nd-order global cities: major Asia-Pacific financial and business centres	Hong Kong, Shanghai, Seoul, Singapore, Osaka, Sydney, San Francisco
3 2nd–3rd-order global cities: important Asia-Pacific business and industrial centres	Taipei, Nagoya, Beijing, Guangzhou, Seattle, Melbourne, Yokohama, Bangkok, Kuala Lumpur
4 Transnational cities: advanced, highly-tertiarized, medium-size cities	Vancouver, San Diego, Portland, Honolulu, Fukuoka, Brisbane, Adelaide, Auckland
B Southeast Asia mega-cities: tertiarization and developmental dualism	Jakarta, Manila, Ho Chi Minh City
C Asia-Pacific port/industrial cities	Pusan, Kaohsiung, Dalien, Vladivostok, Kitakyushu, Oakland, Long Beach, Tacoma, Haiphong, Surabaya
D Asia-Pacific travel/tourism centres	Sapporo, Macao, Kona, Denpasar, Santa Barbara, Gold Coast, Victoria, Whistler
II Extended multinuclear regions (after Gottmann, 1961; McGee and Robinson, 1995)	
A Multinucleated metropolitan service clusters (contiguous or semi-contiguous urban development patterns): world-scale service industry platforms	Kanto Plain-Tokyo Bay: Tokyo, Chiba, Kawasaki, Yokohama Pearl Delta: Hong Kong, Guangzhou, Macao, Shenzen Bay Area: San Francisco, Oakland, San Jose Southern Malay Peninsula: Kuala Lumpur, Malacca, Johor, Singapore
B Extended metropolitan service corridors (extended linear, non-contiguous urban development systems): aspects of competition and complementarity	Southeast Australian corridor: Sydney, Canberra, Melbourne Pacific Northwest corridor: Vancouver, Seattle, Tacoma, Portland

regulatory functions and development control, derived from a preoccupation with negative externalities generated by high-growth office-based services, such as the growth of long-distance commuting (Daniels 1975). At the regional scale, the national dominance of office complexes in primate cities such as London, Paris and Amsterdam was perceived as inimical to the fortunes of services-deficient peripheral regions and cities. In response, stringent development control regimes were established to suppress office development and specialized services in the largest centres, augmented in some cases by special agencies and institutions mandated to encourage the decentralization of offices from 'core' cities to the lagging periphery (Goddard 1975). The developmental potential of services in some areas of the Atlantic sphere was also inhibited by perceptions of the inferiority of services vis-à-vis manufacturing in terms of regional growth potential, as disclosed in debates conducted in academic journals in the 1970s and 1980s.[2] This debate was eventually resolved by a broad expert consensus that both services and manufacturing were essentially co-dependent components of advanced industrial systems. Programmes for attracting or nurturing IT industries, creative and cultural services and other advanced service industries are now elements of local policy and institutional structures within European and North American cities (see, for example, Evans 2001). Generally, however, progressive policies for services lagged the rising importance of tertiary and quaternary industries for urban and regional development (Coffey 1994).

The process of tertiarization has a more recent provenance within most Asia-Pacific jurisdictions than in the Atlantic sphere, so there has been an opportunity to learn from the experience of European and North American urban and national governments. This may account in part for the clearly more developmental thrust of policies for services within the Asia-Pacific. It has also been the case that governments within much of the Asia-Pacific region have been prepared to accept relatively high environmental and social costs as the price of accelerated growth and transformation. That said, while policies for industrialization were at the heart of development strategies from the early post-war years in many Asian states, in some cases associated with quite fundamental reformations of the state and society, developmental programmes for services have been in most cases established *in response* to trends and events over the last quarter century.[3] What follows is a narrative of service industry policy experiences among selected Asia-Pacific states and cities, situated within three distinct phases since the late 1970s.

Policy responses to economic and political shocks

The decade spanning the late 1970s and 1980s constituted in some respects a high point in the accelerated industrialization of the Asia-Pacific, marked by the power of Japanese and South Korean industrial conglomerates, by the expansion of increasingly advanced industrial production in Taiwan, Hong Kong and Singapore, and by the achievement of 'take-off' stages of industrial development in Southeast Asian economies, notably Thailand and Malaysia. Robust growth in advanced-technology industrial production was also recorded in Pacific North America, notably in the Los Angeles-Orange County region, in Silicon Valley, and in the Seattle-Puget Sound region (and to a lesser extent Vancouver and Portland), driven by propulsive aerospace, electronics and communications industries and corporations.

But paradoxically, perhaps, this period also saw a number of quite significant policy innovations for service industries, both with respect to direct (sector-based) policies and more general development strategies, in response to a sequence of economic shocks and political shifts. In general, growing interest in the possibility of service industries assuming more central economic roles reflected concerns about the sustainability of Fordist manufacturing and industrial labour among advanced societies, as well as new appreciations of the potential developmental roles of specialized service industries. An example here is the incipient concern of central and prefectural governments with the socioeconomic implications of the 'hollowing out' of the Japanese manufacturing sector, and the clear need to assign greater importance to advanced technologies and allied services in the interests of improving industrial productivity (Edgington 1989).

Over the late 1970s and mid-1980s, services were also seen as a means of ameliorating the swings of business cycles and recessionary downturns in a number of states and cities within the Asia-Pacific. In the well-known case of Singapore, a sudden and sharp recession in 1985, following two decades of sustained high growth, stimulated a comprehensive reappraisal of the city-state's development strategy and industrial policies, reflecting concerns about the stability of Singapore's entrepôt roles and basic industries. Consistent with the orientation of an exemplary developmental state, the Singapore government had in 1979 signalled an intention to shift investment and labour to high-technology industry, but the 'shock value' of the 1985 recession led to a more far-reaching set of reforms, including new policies to encourage a more specialized service vocation (Clad 1986). Research supported by the Economic Development Board (EDB) disclosed that Singapore possessed a strong competitive advantage in higher-order services, including business or 'producer' and professional services, with Hong Kong and Tokyo seen as chief regional rivals (Krause *et al.* 1987). Labour productivity data demonstrated that output values were higher among higher-order service workers than for industrial workers, reinforcing the EDB's commitment to fostering a new service industry vocation, incorporating a more highly-skilled, professionalized labour force (Government of Singapore 1986). The mid-1980s reforms in Singapore therefore included redefining occupational and social class

implications, as well as a repositioning of development strategy and programmes to favour high-value services.

The case of Vancouver on the eastern littoral of the Pacific represents
an analogous case to that of Singapore in important ways, as here again a
sharp recession, precipitated by downturns in established production
sectors and in the city's hinterland markets, stimulated new policies for
advanced services as 'lead' (rather than supporting) sectors. Severe price
shocks experienced in British Columbia's staple regions were transmitted
to Vancouver via the tight linkages and connectivity characteristic of a
core-periphery staple economy, including flows of capital, resource products, information and management services, leading to unemployment
levels in metropolitan Vancouver reaching as high as 14 per cent in the
deep recession of 1982–1984 (Ley and Hutton 1987). The city embarked
upon a development policy process which led to the approval by the
Mayor and Council of a new strategy in 1983 in which advanced services
(including producer services, banking and finance, higher education,
international transportation and communities) were assigned leading
roles (City of Vancouver 1983). Significantly, Vancouver's linkages with
the markets, cities and societies of the Asia-Pacific (rather than with the
traditional resource hinterland of British Columbia) were emphasized, a
comprehensive reorientation to be supported by trade in services,
tourism, travel and educational exchange. Given the differences in autonomy and powers between a sovereign city-state and a city situated within
a federal state (and subject to provincial jurisdiction), Vancouver's economic strategy could not match the programmatic or institutional depth
of Singapore's new policy direction, and there are after all few true counterparts anywhere to Singapore's EDB. But the 1983 policy statement
served to underscore Vancouver's commitment to pursue a new vocation
with specialized services at the vanguard of growth and change.

In a number of cases, notably that of the PRC, new possibilities for
service industries were stimulated by quite radical shifts in state ideology
and overarching developmental strategy. The death of Mao Zedong in
1976 marked the end of China's calamitous Cultural Revolution and
enabled the accession of Deng Xiaoping as Chairman of the Communist
Party of the PRC. Upon consolidating power, Deng embarked upon a well-
known series of economic reforms, including the approval of Special
Enterprise Zones in 1979 to encourage foreign investment in export-
oriented industrial production, and the designation of fourteen 'open
coastal cities' in 1984. The reforms initiated a new era of 'marketization
and globalization' of the Chinese economy, aimed at accelerating the
modernization of the PRC's primitive manufacturing sector, but they also
afforded opportunities for the expansion of urban service industries. In
part these possibilities for encouraging services reflected the acknowledged need for tertiary industries (transportation, sales and marketing)
to facilitate industrial production and trade, and for quaternary activity

(higher education, market intelligence, R & D) as a means of enhancing human capital and industrial productivity.

But in important ways, the real stimulus for tertiarization in China ensued from the designation of the coastal cities as the spatial foci for economic transformation (emphasizing urban gateway and trading roles, financial and investment functions of key cities and higher education, among other service industries), and in the partial devolution of development and planning from central to local agencies. The relinquishing of a measure of central control over economic development, seen by Deng and his allies as crucial to the freeing up of development potentials within key regions and urban areas, enabled some cities to pursue new vocations which encompassed larger roles for services. In particular, cities were at least tacitly allowed to abandon the old model of the 'producer city' as designated in the five-year plans commencing in 1953, which insisted on the primacy of factories (and more especially heavy industry) for each city. New possibilities beyond ideologically-prescribed industrialization trajectories were envisaged for Beijing, the PRC's capital and preeminent cultural centre, and Shanghai, until 1949 China's chief banking and financial centre. But perhaps the most immediate effects were felt in Guangdong and Fujian provinces in southern China, in which traditional entrepreneurial and mercantile energies were released by the reforms. High growth in these southern provinces, stimulated in large measure by Deng's reforms (and capital inflows; see Hsing 1995), imparted a distinctive new spatiality to regional development in China, and underpinned the resurrection of service industries in cities such as Guangzhou and Xiamen which had been suppressed during Mao's regime.

Policy responses to globalization and restructuring

The 1980s proved to be a highly significant period in the evolution of policies for service industries within the Asia-Pacific, not only with respect to the important experimentation described above, but also in the shaping of a distinctly new developmental context. In this regard, the 1980s saw a dramatic acceleration of inter-related processes including industrial restructuring among advanced and transitional economies, privatization and de-regulation, and globalization, typified by larger flows of capital, finished goods and commodities, information and international migrants. These processes set the stage for a far more wide-ranging set of development policy innovations over a decade, spanning the late 1980s and 1990s, as an increasing number of states and local governments within the Asia-Pacific attempted to deploy service industries as instruments of globalization, modernization and industrial restructuring. While the specific policy models and programme mix naturally varied significantly from place to place, again reflecting highly differentiated forms of political control, governance structures and ideologies, we can characterize this period as one in which policies for service industries assumed a more 'mainstream' (as opposed to marginal) status within developmental strategies among Asia-Pacific states and city-regions. What follows is a necessarily selective overview of some instructive reference cases, designed to underscore the depth and range of service industry policy experiences in the region.

Service industries and responses to restructuring

The many successes of state-level industrialization strategies notwithstanding, by the late 1980s processes of industrial restructuring (in the sense of major sectoral shifts and new divisions of production labour) were well-advanced among leading economies within the Asia-Pacific. Although cities within the region were not exactly in a post-industrial trajectory along the lines experienced in Western or Atlantic societies (i.e. a rapid, catastrophic and permanent collapse of basic manufacturing capacity and labour), some traditional production sectors and industries in Japan, South Korea, Hong Kong and Singapore were clearly in decline, in relative or even absolute terms. These experiences of industrial decline, consequences of corporate outsourcing strategies, product cycle factors, accelerating technological innovation and displacement, represented a longer-run challenge than the recessionary conditions that stimulated the policy initiatives for Singapore and Vancouver described above, and thus policy and institutional responses assumed in some cases a more structural character. There was, however, significant variation in policy postures at the state level, as well as a general trend toward more engagement with service policies at local (as opposed to central) government levels.

This trend toward a greater service industry policy commitment at the local level was part of the story in China, where the 1984 open cities policy and subsequent reforms led to new opportunities for growth in financial, corporate and commercial activity. The marketization of coastal cities in particular included new local planning and development policies for services, and an aggressive courting of both domestic and foreign entrepreneurs prepared to invest in office development, hotels and retail centres. A new generation of municipal and city-region development plans for major Chinese cities, notably Shanghai, Beijing, Guangzhou and Xiamen, among others, specifically identified services (and more categorically advanced, intermediate service industries) as part of the 'next wave' of sectoral/industrial development, signalling urban futures less reliant on manufacturing and industrial production labour formation. Instruments deployed to promote services in Chinese cities exhibited considerable local variation, in keeping with the 'localization' of planning and development cultures, but typically included more permissive policies for urban structure and land use, investments in transportation and communications infrastructure, marketing and information services. Here we can cite the well-known case of Shanghai, where corporate, financial and producer service functions were promoted as part of a 'back to the future' scenario, in which the city's pre-1949 role as China's banking and business centre was to be re-established (Wu 1999).

In Hong Kong and Singapore, an industrial restructuring trend comprising in part the continued relocation of lower-end, Fordist production to lower-cost neighbouring regions was acknowledged by government and public development agencies, accompanied by new programmes to support specialized services industries. In the case of Singapore, a new vocation emphasizing higher-order services was formally articulated in policy statements and complementary programmes, including the redirection of public investment, infrastructure planning, and educational support policy (Ho 1994), consistent with (and supportive of) the ascendancy of a markedly 'professionalized' workforce and society (Baum 1999). In Hong Kong, the policy response of the colonial government was of course more pragmatic (and in some ways implicit), but nonetheless favoured higher-order services in urban structure and land use planning (including provisions for expanding the CBD eastward into Wanchai and Causeway Bay) (Taylor and Kwok 1989). Hong Kong and Singapore represent exceptionalist cases in the Asia-Pacific, as they correspond more closely to the Western post-industrial, professionalized urban society model of development than other cities in the region. But each is seen by other states and local authorities as a leading-edge economy and as policy models worthy of emulation in some respects.

In Japan, Edgington observes that governments (central and prefectural) 'have been disinclined to endorse a future where the service sector is dominant, and in the spirit of "manufacturing still matters", have been willing to implement policies to guard against eventual deindustrialization at a local level' (Edgington 1996). This may be viewed by some as reflective of a general inability of economic policy-makers to acknowledge the harsh realities of Japan's economic stagnation over the 1990s, and as an expression of over-caution in responding to structural industrial problems, in contrast to the far-sighted postures of the expansionist 1960s and 1970s. At the same time, national planning in Japan (including *Yonzenso*, the comprehensive development plan administered by Japan's National Land Agency) has acknowledged the need to encourage advanced services among each of three principal regions (Kanto, Kansai and Chukyo), both to support increasingly advanced industrial production, and to 'infill' some portion of the 'hollowed out' labour capacity.

But the impacts of industrial restructuring are of course most keenly felt at local levels, with highly differentiated implications for regional labour markets and urban communities within national space, so we can discern more assertive policies for service industries at the prefectural and municipal echelons within Japan. The land use and infrastructural policies for office, financial and commercial development in Tokyo, Yokohama and Osaka are well known, expressed in local development strategies and investment programmes, and evidenced in a range of urban mega-projects in each of these cities, although in the 'post-bubble' era many of these have been 'delayed, frozen, or . . . abandoned', suggesting that 'the first phase of urban restructuring has already come to an end' (Machimura 1997: 162). In the case of Chukyo, centred on Nagoya, a region specializing in car production, the current development strategy calls for a new emphasis on knowledge-intensive industries, including information technology, industrial design and engineering, each of which complements both Nagoya's car manufacturing and other heavy indus-tries, as well as a more service-intensive orientation (Edgington 1996). In Kitakyushu, an industrial port city in Kyushu, services have been desig-nated in the current development strategy as a larger part of the city's future development, in the face of steady decline in Kitakyushu's tradi-tional industries.

The emergence of an information-intensive 'knowledge economy' has offered another pathway to accelerated industrial restructuring, and here the experience of Southeast Asian states and city-regions offers dramatic and instructive exemplars. Corey has written an optimistic account of the 'digital tigers' of Singapore, Malaysia and Thailand, stressing the central role of governments and public authorities in the pursuit of grandiose strategies of 'leapfrogging' development based on investments in IT infra-structure and systems. These include, for example, Malaysia's 'Multimedia Super Corridor', which incorporates the 'intelligent' cities of Putrajaya (the administrative capital) and Cyberjaya (education, housing, research and development) (Corey 1997). As Corey himself acknowledges, however, there is a clear (if imperfectly acknowledged) need to reconcile the engineering-oriented applications of IT hardware with the social impact of new technology in these particular settings. In addition to issues of social exclusion and alienation, these states also confront tensions created by the accelerated diffusion of IT innovation and enhanced public access to electronic communications, coupled with state insistence on high levels of domestic social, cultural and political control.

Service industries and globalization strategies

There are of course important connecting points between industrial restructuring and global processes, exemplified by the emergence of a 'new international division of labour' (NIDL) shaped by the globalization of production across a spectrum of sectors and product groups (Cohen 1981; Ho 1991). But we can also identify quite specific instances of special-ized service industries deployed as agencies of globalization at the state and local levels among many Asia-Pacific jurisdictions. Here, the policy goals may include both substantive purposes (for example securing new markets for export trade, reducing balance of payments deficits in services trade, supporting key service industries and institutions), as well as more symbolic aspirations (including policies for re-imaging cities and urban communities as 'global players' in particular markets, escaping to some extent tightly-bounded regional roles and subordination to primate cities). City policies can of course entail features of both substantive and symbolic aspirations. Fukuoka's municipal strategy, ambitiously titled 'Global Reach', endeavours to reposition Fukuoka within an international urban network of some 17 designated Asia-Pacific urban centres, based upon the key service functions of commerce, transportation, exchange, sports, leisure and culture, and supported by investments in international 'connectivity', in the form of trading, cultural and city-to-city relationships (Fukuoka City 1998).

Clearly, services are central to globalization strategies, given the high export-propensity of producer and other intermediate service industries, the key functions performed by certain specialized services (banks and other financial institutions) in managing FDI and other global flows of capital, and the role of major development and real estate firms in the globalization of property markets. Creative and cultural service industries (architects, graphic designers, entertainment and leisure companies) are also implicated directly in the reconstruction of urban space and the reformation of urban identities, which can create new international or global imagery for cities. The interplay of these specialized services can be seen most spectacularly in the comprehensive re-imaging of Shanghai over the past decade, promoted by massive state and private investments in the redevelopment of the Pudong New Area. As recounted in Kris Olds' narrative of Asia-Pacific urban mega-projects, the new global status for Shanghai was to be underscored by the participation in a 'global intelligence corps' (GIC) of elite international architects and designers in the reproduction of the Lujiazui Central Financial District within the Pudong site. According to this grandiose vision, the status and profile of this new global financial centre would be confirmed by leading members of the GIC, with plans and design features 'formulated by transnational cultures active at a global scale' (Olds 2001: 229). In the end the Shanghai Municipal Government's (SMG) plan for the Lujiazui site entailed a pragmatic synthesis of Sir Richard Rogers' winning design and more prosaic features reflecting budgetary and political realities, but both the international financial centre designation and the engagement of the most prominent GIC actors signalled the global aspirations of the SMG and the Chinese government.

While the Pudong mega-project stands at the apex of Asia-Pacific urban globalization programmes, we can readily cite other examples, including the Minato Mirai (MM21) project in Yokohama, designed both to bring Yokohama out of the deep shadow of Tokyo, and also to project a more global image to the international business community, investors, and entrepreneurs.[4] The Petronas Twin Towers in Kuala Lumpur, at 451.9 metres the world's tallest buildings, project a powerful international business image, consistent with Mahathir's 2020 development strategy for the capital and Malaysia as a whole, while incorporating indigenous, Islamic design features and details. Cesar Pelli, Charles Thornton and Leonard Joseph, the elite international architects principally responsible for the design, describe the Petronas Towers as the 'preeminent examples' of a trend toward 'the tall building as cultural emblem' (Pelli *et al.* 1997: 93), and as signifiers of Kuala Lumpur's global aspirations. But a thoughtful observer has noted in this specific context that world city rankings are undertaken 'in terms of public standards of living which places a high emphasis on quality of life, public safety ... [and] culture ... and not based on the height of our buildings' (Morshidi n.d.: 24).

These mega-projects are the centrepieces of urban globalization strategies, but we can round out our profile with other important examples. In Vietnam, the national *Doimoi* ('economic renovation', or liberalization) programme initiated in 1986 included the encouragement of foreign direct investment (FDI) as key to a 'marketization with socialist orientation and government regulation' strategy. By the early 1990s inward investment had achieved significant momentum, and in the case of Hanoi tended to favour new hotels, commercial buildings and other service infrastructure in the capital (Trinh 1997: 176). These new service sector projects promised high returns on investment (ROI) for foreign interests, furthered the government's *Doimoi* strategy, and accelerated Hanoi's tertiarization process, while (on the cost side of the ledger) increasing social dislocation and polarization in Vietnam's capital city. Ho Chi Minh City (formerly Saigon) has also experienced rapid growth in services infrastructure over the last decade, but this FDI-induced tertiarization may be more congruent with Ho Chi Minh City's entrepreneurial (pre-1975) past than is the case in Hanoi.[5]

A number of cities within the Asia-Pacific have been active in supporting export-oriented services, following the dissemination of empirical studies establishing the importance of specialized intermediate services to the economic (export) base of Asia-Pacific cities (see for example the influential study of service industry export-propensity in Seattle-Central Puget Sound, by Beyers *et al.* 1985; and the Vancouver case study, Davis and Hutton 1991). Supporting local policies here can include marketing and information services, service trade missions, and improved access to central or federal government trade programmes. Public service sector institutions can also participate as partners in urban globalization strategies. Major universities within the Asia-Pacific, for example, seek to achieve global status not only through excellence in research, but also by attracting international faculty and (especially) graduate students, by establishing partnerships with international universities, or, in some cases, by establishing an institutional presence in foreign centres; these include Stanford University, the University of California at Los Angeles (UCLA), Tokyo University and the National University of Singapore (NUS), among many others.

Policy responses to crisis and the 'cultural turn'

Over the past five or six years there has been evidence to suggest that many Asia-Pacific cities have entered another phase of service industry policy, one characterized by responses to new signals, cues and influences, and by aspects of experimentation and innovation. These shifts occur at the margins, as market-oriented policy models, driven by imperatives of restructuring, globalization and modernization, are still pervasive (and likely dominant, overall) within the region. Vigorous competition among states, cities and transnational corporations remains a powerful motif within the Asia-Pacific. But other values can be introduced to development strategy processes, in response to shocks and ensuing crises, to progressive urban and regional development concepts, and to changing social and political sensibilities.

We can identify perhaps two principal sets of new service industry policy influences within the Asia-Pacific since 1997, each embodying distinct but in some respects related themes and conditions. The first, and likely most exigent, concerns state and local responses to a series of shocks over the past half decade or so, starting with the so-called 'Asian crisis' of 1997 and after, initiated by the well-known Thai currency crisis, and a subsequent pattern of fiscal, socioeconomic and political dislocations spread unevenly throughout the region thereafter (Glassman 2000). The depth and pervasiveness of the Asian crisis was implied in the title of a book by McLeod and Garnaut titled 'East Asia in Crisis: from being a miracle to needing one?' (McLeod and Garnaut 1998). This was followed by a series of abductions (and in some case murders) of foreign tourists in the southern Philippines, the bombing of two Bali nightclubs in 2002, and the recent SARS epidemic, which apparently originated in southern China, but which soon spread among populations elsewhere in Asia and in North America.

These events have complex causalities, but there are some common attributes of relevance to this present discussion. First, each 'shock' started as an essentially localized episode, but effects were rapidly transmitted throughout the region and beyond, demonstrating (among other things) the conditions of interdependency between and among states and societies. Second, each event has given rise to charges that these problems have been caused or greatly exacerbated by regulatory shortcomings (in terms of monitoring, transparency and compliance issues) among certain Asian states, notably in the domains of banking, corporate governance and public health. As some would see it, even advanced Asia-Pacific states are characterized by a 'regulatory deficit', relative to Western societies, and it can be argued that this reflects more fundamental contrasts in forms of governance and policy cultures between Asia-Pacific and Atlantic states. Finally, in each case the crisis (or shock) has especially impacted service industries in the region, providing at least a check to the growth of these industries, and potentially compromising the viability of some service sectors and major corporations. To illustrate, the so-called Asian crisis of 1997 and after severely impacted regional banks and financial institutions, as well as the revenue base of governments and public agencies, while the series of terrorist attacks and the recent SARS experience have seriously affected major regional airlines, hotels and tourism industries (as well as restaurants and other ancillary services).[6]

In response to these events, governments in the region have in some cases embarked upon regulatory reform (with mixed results to date), both to improve conditions of development and to send a positive signal to external markets and actors, while undertaking (again with varying degrees of resolution and results) a series of 'crackdowns' against insurgents seen as undermining both domestic security and the safety concerns of foreign tourists and visitors. This series of crises and shocks over the past five years or so (exacerbated by the global repercussions of the 9/11 experience in New York) has led to large-scale marketing programmes by governments, industry associations and major corporations, designed to allay international concerns about the problems of key service industries and allied security issues among Asia-Pacific states. Overall, though, it might not be unfair to characterize the state responses to crises as essentially *tactical* in nature in most cases, designed to achieve 'damage control' purposes.

A second (and potentially more strategic) set of influences on service industry policy within the Asia-Pacific can provisionally be situated within the admittedly 'fuzzy' rubric of cultural change within the region. Here we can identify at least the contours of a number of incipient policy directions, related in some fashion to the cultural influence on (or interdependency with) development in the region; to (as yet modest) shifts in local and state policy cultures which have the potential of inserting more progressive social values (in part a response to the crises outlined above); and also to what Allen Scott describes as the 'new cultural economy' of cities (Scott 2000). There are quite significant implications of this (for want of a better term) 'cultural turn' in development trajectories, social attitudes and policy preferences for service industries and for service policies. As Mike Douglass has observed, inter-city competition has underpinned the formation of urban development strategies within the Asia-Pacific (Douglass 2000), but a growing appreciation of the costs of unrestrained competition (exemplified by the proliferation of airports and seaports in the Zhujiang [Pearl] Delta), and the co-dependency of societies underscored by the recent shocks described above, has heightened interest in exploring more co-operative development models.

To illustrate the potential significance of new policy 'cultures' or frontiers in the Asia-Pacific, service industries are seen as crucial elements of a growing interest in *sustainable development*, as evidenced by both the titles and contents of a new generation of urban and city-region plans in the region. The location of service industries and clusters (for example business districts, airports and research parks) has major implications both for efficiency of land use and for transport demand in fast-growing Asian cities. Employment generation capacity of service industries will be critical for the sustainability of urban economies and labour markets, especially in contexts of contraction in traditional industries and production employment. To date, development strategies among Asia-Pacific city-regions tend to exhibit in many cases a rhetorical affiliation (rather than a deep programmatic commitment) to principles of sustainable development, and a stronger regulatory element of urban plans will be required to avoid (for example) the sprawl and dispersal of new commercial development associated with tertiarization (see Wu and Yeh 1999). Sustainability values and concepts are however increasingly part of the discourse of younger academics and planners in the region (many trained in Western universities where such ideas are in common currency), and are likely to be part of a progressive recasting of policy cultures in the region in the future.[7]

Another dimension of the 'cultural turn' in early twenty-first century urban development in the region comprises the potential of cities as centres of cultural production, interpretation, transmission and exchange. These possibilities may become increasingly important as limits to competition for corporate office development (and thresholds of risk for speculative office development) are approached. In the case of Tokyo, grandiose plans during the 'pre-Bubble' economy of the late 1980s for some 60 40-storey office towers (the so-called 'Marunouchi Manhattan plan') and the subsequent collapse of the commercial property market have encouraged a reexamination of global city assumptions (Douglass 1993), and conjecture on more imaginative visions of the capital's future. This could combine aspects of conventional world city aspirations (emphasizing corporate control and financial industries) with a 're-Japanization' of the capital, suggesting a new emphasis on 'traditional values and cultural elements', and offering 'another context for urban discourse on Tokyo' (Machimura 1997: 163). Beijing's current development plan calls both for a repudiation of the former 'industrial city' model, and an evocation of the capital's traditional cultural and administrative roles (Ke 1996), implying a significant reorientation of the capital's development trajectory. In both the Beijing and Tokyo cases, a 'cultural turn' in development orientation can include both endogenous attributes (derived from their roles as centres of national culture, tradition and memory), as well as 'global city' roles in cultural production and transmission, for example in film and video production, multimedia services and applied design, a process already well-established in Los Angeles. Further, Tokyo, Los Angeles and Seoul (among other cities) exemplify the cultural dimensions of the twenty-first century transnational city, acting as centres of diasporic networks, and connoting *inter alia* functions of inter-cultural transaction, dialogue and discourse.

As a final example, 'internal' urban policies and programmes may be deployed to support this emerging cultural orientation, in nurturing new landscapes of cultural production. In Singapore, the government's (admittedly belated, but typically thorough) conservation policies for heritage districts in parts of Chinatown immediately to the west of the CBD have secured a unique environment for 'New Economy' clusters, including creative design and cultural industries, as well as technology-based industries such as computer graphics, Internet services and software design. Local heritage policy, land use planning and amenity provision are essential to the formation of inner city New Economy clusters in Seattle (Belltown) and Vancouver (Yaletown, Gastown and Victory Square) (Hutton 2004). There are similar processes at work in new media clusters in Tokyo's Ropponggi and Shibuya districts, and creative-design industries are represented in the spatial reconfiguration of Seoul's metropolitan core (Park 1993). But in Shanghai a creative and design precinct within the Suzhou Creek district is threatened by the inexorable waves of neighbourhood demolition and comprehensive redevelopment, suggesting a more ambivalent policy posture toward this latest expression of advanced urban tertiarization within the region.

Conclusion

This chapter has presented an outline of important shifts in state and public policy deployment of service industries as instruments of urban growth and change within the Asia-Pacific over the last three decades. The approach has been to present an exemplary series of reference cases within three distinctive periods, expressed as a sequence of responses to 'economic and political shocks', 'globalization and restructuring' and 'crisis and the cultural turn' (summarized in Table 3.2). This framework and analysis has emphasized the influence of exogenous events as policy influences, especially since the 1980s, in contrast to the typically more inward-directed motivations of state industrialization programmes. This periodization model has its limitations, notably in over-simplifying trends and processes, and perhaps in conveying (unintentionally) an impression that the complex policy records of highly variegated states and local jurisdictions can be conveniently positioned within discrete eras. That said, the notion of situating illustrative policy experiences within a broader context of developmental conditions may have heuristic value, and can yield some useful insights into important features of service industry policy within the Asia-Pacific.

A first observation derived from the narrative is a confirmation of the strong association between service industries and overtly *developmental* (as opposed to regulatory) public policy goals and values throughout much of the region. Further, there was a relatively short gap between the policy experiments in the first phase (*c.*1978–1986) and the subsequent 'main-

Table 3.2 Exemplary service industry policy models within the Asia-Pacific, 1978–2003

Phase I. Policy responses to economic and political shocks: new development trajectories, c.1978–1986

A *Changes in overarching national development strategies*	Peoples Republic of China • designation of SEZs, 1979 • designation of 14 open coastal cities, 1984
B *Acute recessionary conditions:* sharp contractions in basic industries and regional entrepôt roles	1 Vancouver (recession, 1982–1984; economic strategy response, 1983) 2 Singapore (recession, 1985; 'New Directions' development strategy, 1986)

Phase II. Policy responses to globalization and restructuring: 'mainstreaming' service industry policies, c.1987–1997

A *Services and new development pathways:* post-Fordism and postindustrialism	1 Hong Kong: policies for the post-industrial metropolis 2 Nagoya: services and advanced industrial production 3 Seoul: services in an ascendant global city-region
B *Services and the 'knowledge economy':* IT and the 'urban technopole'	1 Singapore: planning for the 'intelligent city' 2 Malaysia: the 'multimedia super corridor' 3 Japan: the 'technopolis' model
C *Services and globalization strategies:* foreign direct investment, urban mega-projects, and re-imaging the global city	1 Shanghai: the Pudong 'New Area' UMP 2 Yokohama: the MM21 Project 3 Vietnam: *Doi Moi* and FDI-led tertiarization 4 Beijing: the Olympics as global marker, 2008

Phase III. Policy responses to crisis and the 'cultural turn': experimentation and innovation, c.1998–2003

A *Impacts of 'shocks' on service industries:* the 'Asian crisis', SARS, and terrorism	Tactical responses: regulatory reform, price adjustments, marketing, security 'crackdowns' and 'damage control' programmes
B *Services and the 'cultural turn':* emergent urban policy cultures and the 'new urban cultural economy'	1 Services and sustainable development: Melbourne, Seattle, Portland, Brisbane 2 The 'Sustainable Olympics': Sydney, 2000; Vancouver, 2010 3 Services and co-operative regional development: the Pearl River Delta 4 Cities as global centres of cultural production: Los Angeles, Tokyo 5 Creative services and the 'New Economy' of the inner city: Singapore, San Francisco, Seattle, Vancouver

---

streaming' of service industry development purposes toward the end of the 1980s, in response to the imperatives (or pressures) of globalization, modernization and industrial restructuring. From an implicit subordination of services to manufacturing during the hegemony of the industrialization paradigm, advanced service industries have been assigned more central roles in state and local development, although the specification of these vocations have varied considerably from place to place. To illustrate, in Singapore, Hong Kong and Vancouver, advanced services are entirely central to post-industrial trajectories, but in Japanese cities are conceived in policy terms as complements to advanced industrial production regimes, while in Shanghai tertiarization is a 'recovered' trajectory rather than development induced *de novo*.

Second, there is a fairly clear trend line comprising the transfer of policy responsibilities from state (i.e. national) to 'local' (e.g. regional, prefectural, metropolitan or municipal) authorities over the past two decades, largely as a consequence of deregulation, privatization, and liberalization programmes. National governments are in many cases concentrating policy resources in an engagement with bilateral/multilateral trade fora and processes, rather than in the micro-management of internal regional development. As in other spheres, too, globalization has tended to increase the significance of local authorities (relative to national governments) as actors in service industry development and trade, reinforced and underscored by the tight spatial association of specialized services with cities (relative to the more dispersed patterns of industrial production).

Finally, following two decades of experience in policies for services, a large and increasing number of governments within the region incorporate programmes for service industry development, although (with some exceptions) there is not yet the degree of institutional depth and complexity characteristic of mature, post-industrial jurisdictions within the Atlantic sphere. Policy initiatives within the Asia-Pacific are in most cases directed toward enhancing competitive advantage in service production, specialization and trade, and there is no present counterpart to (for example) the European Union's spatial development programmes for inter-city cooperation and collaboration (see Moulaert 2000). There is however some evidence of a new spirit of interest in progressive policies for services, inspired by innovative development paradigms and policy ideas, including sustainable development, and, perhaps, by an appreciation of the limits to (and costs of) unrestrained competition.

Notes

1 I am following here the conventional definition of the Asia-Pacific as including East and Southeast Asia, as well as Australasia, California, Oregon, Washington and British Columbia (but not the Pacific coastal states of Latin America). As well as reflecting common usage, this conception of the Asia-Pacific acknowledges the depth and complexity of linkages (trade, investment, travel and ethnicity) between Asian states and the increasingly transnational societies of Australia, New Zealand, and 'Pacific America'.

2 Although many aspects of the tertiarization experience were subject to debate and contention in this period, perhaps the most consequential concerned:

   a the extent to which the service sector constituted an autonomous 'service economy';

b the regional growth potential of services industries vis-à-vis manufacturing; and

c the social implications of tertiarization, the latter explicated in Daniel Bell's theory of the 'post-industrial' society (Bell 1973).

Much of the debate was conducted along ideologically prescribed lines, with Marxists in particular comprising the most vigorous critics of both the 'service economy' and the post-industrial society models. With the election of the Thatcher government in Britain (1979) and Ronald Reagan's election as President in the US (1980), the ideological tone to the manufacturing vs. services discourse intensified, as each was implicated by social critics and academics in the continuing decline of Fordist production and employment in their respective countries. From the vantage point of 30 years on, the significance of service industries as *leading* (rather than merely supporting or ancillary) elements of advanced economies is well established, as is the central role of service occupations and employment in reshaping urban housing and labour markets, and in the reformation of social class. In many ways, the most meaningful occupational and employment divisions within advanced societies are among certain classes of service industries, rather than between services and manufacturing. See David Ley's chapter in this present volume for a more detailed and incisive exploration of this debate.

3 Wong observes that with the introduction of China's initial Five Year Plan in 1953, 'Industrialization became a principal component of the regime's programme for transformation of the country into a socialist success story ... the First Five Year Plan aimed to develop the economic infrastructure of a strong state in charge of an independent country' (Wong 1997: 56). Thus, *industrialization* in this context concerns the institutional underpinnings of the 'new' nation and society, as well as the state's preferred development modality.

4 As is well known, Yokohama's MM 21 project suffered from unfortunate timing, coinciding with the collapse of the property market in Japan, and exacerbated by the project's emphasis on commercial (as opposed to residential) development, for which there would likely have been greater demand. In contrast, Concord Pacific's original proposal for a major office-commercial orientation for the Pacific Place urban mega-project on the north shore of Vancouver's False Creek was rejected by City Council, in favour of a land use plan privileging housing and public amenities. This decision in the end worked to Concord Pacific's advantage (and arguably to that of the City as well), as the speculative office market collapsed in the early 1990s in Vancouver, while housing demand (for owner-occupiers and investors, as well as for a large secondary rental market) has remained very robust. For a detailed analysis and critique of the Concord Pacific Place UMP, see Olds (2001).

5 To illustrate, Lisa Drummond has observed the inclusion of high-rise point towers as part of the 'approved' new imagery of Ho Chi Minh City, as depicted in government-sanctioned posters commemorating the City's historical development (Drummond 2001).

6 In April of 2003 Morgan Stanley reduced its forecast for economic growth in Asia for 2003 from 5.1 per cent to 4.5 per cent, estimating a cost to Asian business of the SARS epidemic in the order of US$15 billion. Among service industries, impacts for major Asia-Pacific airlines and hotels would be particularly severe (quoted in the *International Herald Tribune*, Tuesday, 8 April 2003: 8).

7 Rapid growth and change, coupled with sometimes dramatic shifts in overarching policy values at national levels, and the kinds of exogenous shocks (and domestic crises) described in this present narrative, place considerable pressure

on local planning systems, as seen in the Chinese experience (see Leaf 1998, and Yeh and Wu 1999). Planning reform along progressive lines will require national policy commitments and changing social attitudes, as well as a generational change in the planning profession among Asia-Pacific societies.

References

Baum, S. (1999) 'Social transformations in the global city: Singapore', *Urban Studies*, 36, 1095–117.

Bell, D. (1973) *The Coming of Postindustrial Society*. New York: Basic Books.

Beyers, W., Alvine, M. and Johnsen, E. (1985) *The Service Economy: Export of Services in the Central Puget Sound Region*. Seattle: Central Puget Sound Economic Development District.

City of Vancouver (1983) *An Economic Strategy for Vancouver in the 1980s: Proposals for Policy and Implementation*. Vancouver: Department of Finance.

Clad, J. (1986) 'Long-neglected services get new attention', *Far Eastern Economic Review* (special feature on 'Singapore reconstructs'), 27 March, 77.

Coffey, W. (1994) *The Evolution of Canada's Metropolitan Economies*. Montreal: The Institute for Research on Public Policy.

Coffey, W. and Bailly, A. (1992) 'Producer services and systems of flexible production', *Urban Studies*, 29, 857–68.

Cohen, R. (1981) 'The new international division of labor, multinational corporations, and urban hierarchy', in Dear, M. and Scott, A. (eds), *Urbanization and Urban Planning in Capitalist Society*. London: Methuen.

Corey, K. (1997) 'Creating and controlling cyber communities in Southeast Asia and the United States', paper presented at the Pacific Rim Council on Urban Development, Singapore, October 1997.

Daniels, P. (1975) *Office Location: An Urban and Regional Study*. London: Bell.

Daniels, P. (1993) *Service Industries in the World Economy*. Oxford: Blackwell.

Davis, C. and Hutton, T. (1991) 'An empirical analysis of producer service exports from the Vancouver Metropolitan Region', *Canadian Journal of Regional Science*, XIV, 371–89.

Douglass, M. (1993) 'The "new" Tokyo story: restructuring space and the struggle for place in a world city', in Fujita, K. and Hill, R. (eds), *Japanese Cities in the Global Economy: Global Restructuring and Urban-Industrial Change*. Philadelphia: Temple University Press.

Douglass, M. (2000) 'Globalization, intercity competition and civil society in Pacific Asia', keynote address presented to the Workshop on Southeast Asian Urban Futures, Singapore, July 2000.

Drummond, L. (2001) 'Socializing the modern Vietnamese family', paper presented to seminar at the Institute of Asian Research, University of British Columbia, January 1997.

Edgington, D. (1989) 'New strategies for technology development in Japanese cities and regions', *Town Planning Review*, 60, 1–27.

Edgington, D. (1996) *Planning for Industrial Restructuring in Japan: The Case of the Chukyo Region*. Vancouver: Centre for Human Settlements, University of British Columbia, Research paper AURN WP # 11.

Evans, G. (2001) *Cultural Planning: An Urban Renaissance?* London and New York: Routledge.

Fukuoka City (1998) *Global Reach: An Administrative Guide to Fukuoka City.* Fukuoka: International Relations Section, Fukuoka Municipal Government.

Glassman, J. (2000) 'Is the economic crisis in Thailand over?', in Hainsworth, G. (ed.), *Globalization and the Asian Economic Crisis: Indigenous Responses, Coping Strategies, and Governance Reform in Southeast Asia.* Vancouver: Centre for Southeast Asian Research, Institute of Asian Research, University of British Columbia.

Goddard, J. (1975) *Office Location in Urban and Regional Development.* Oxford: Oxford University Press.

Government of Singapore (1986) *The Singapore Economy: New Directions.* Singapore: Report of the Economic Committee.

Ho, K. C. (1991) *Studying the City in the New International Division of Labor.* Department of Sociology Working Paper 117, National University of Singapore.

Ho, K. C. (1994) 'Industrial restructuring, the Singapore city-state, and the regional division of labour', *Environment and Planning A*, 26, 33–51.

Hsing, Y. (1995) *Blood, Thicker than Water: Networks of Local Chinese Bureaucrats and Taiwanese Investors in Southern China.* Vancouver: Centre for Human Settlements, University of British Columbia, Research paper AURN WP #4.

Hutton, T. (2001) 'Service industries and the transformation of Asia-Pacific city-regions', Singapore: Centre for Advanced Studies, National University of Singapore, CAS Research Paper Series No. 36.

Hutton, T. (2004) 'The new economy of the inner city', *Cities*, 21(2), 89–108.

Ke, H. (1996) 'An across the century development plan for Beijing', paper presented at Towards a Sustainable Urban Future: Mega-Cities, Regional Development and Planning, Beijing: April.

Krause, L., Koh, A. and Lee, Y. (1987) *The Singapore Economy Reconsidered.* Singapore: Institute of Southeast Asian Studies.

Leaf, M. (1996) 'Building the road for the BMW: culture, vision, and the extended metropolitan region of Jakarta', *Environment and Planning A*, 28, 1617–35.

Leaf, M. (1998) 'Urban planning and urban reality under Chinese economic reforms', *Journal of Planning Education and Research*, 18, 101–9.

Ley, D. (1996) *The New Middle Class and the Remaking of the Central City.* Oxford: Oxford University Press.

Ley, D. and Hutton, T. (1987) 'Vancouver's corporate complex and producer services sector: linkages and divergence within a provincial staples economy', *Regional Studies*, 21, 413–24.

Lin, G. (1994) 'Changing theoretical perspectives on urbanization in Asian developing countries', *Third World Planning Review*, 16, 1–23.

Lin, G. (2001) 'Metropolitan development in a transitional socialist economy: spatial restructuring in the Pearl Delta, China', *Urban Studies*, 38, 383–406.

McLeod, R. and Garnaut, R. (1998) *East Asia in Crisis: From Being a Miracle to Needing One?* London: Routledge.

Machimura, T. (1997) 'Building a capital for emperor and enterprise: the changing urban meaning of Central Tokyo', in Kim, W., Douglass, M., Choe, S.-C. and Ho, K. C. (eds), *Culture and the City in East Asia.* Oxford: Oxford University Press.

Marton, A. (1996) 'Restless landscapes: spatial economic restructuring in China's Lower Yangzi Delta', unpublished thesis, University of British Columbia.

Morshidi, S. (n.d.) 'Re-imaging Kuala Lumpur for a global role: urban livability aspect examined', Penang: working paper, Geography Section, School of Humanities, Universiti Sains Malaysia.

Moulaert, F. (2000) *Globalization and Integrated Area Development in European Cities*. Oxford: Oxford University Press.

O'Connor, K., Stimson, R. and Daly, M. (2001) *Australia's Changing Economic Geography: A Society Dividing*. Melbourne: Oxford University Press.

Olds, K. (2001) *Globalization and Urban Change: Capital, Culture, and Pacific Rim Mega-Projects*. Oxford: Oxford University Press.

Park, S. (1993) 'Industrial restructuring and the spatial division of labor: the case of the Seoul Metropolitan region, the Republic of Korea', *Environment and Planning A*, 25, 81–93.

Pelli, C., Thornton, C. and Joseph, L. (1997) 'The world's tallest buildings', *Scientific American*, 277 (December), 92–101.

Perry, M., Kong, L. and Yeoh, B. (1997) *Singapore: A Developmental State*. Chichester: Wiley.

Rimmer, P. (1996) 'International transport and communications interactions between Pacific Asia's emerging world cities', in Lo, F. and Yeung, Y.-M. (eds), *Emerging World Cities in Pacific Asia*. Tokyo: United Nations University.

Scott, A. (1997) 'The cultural economy of cities', *International Journal of Urban and Regional Research*, 21, 323–33.

Scott, A. (2000) *The Cultural Economy of Cities*. London: Sage.

Taylor, B. and Kwok, R. (1989) 'From export center to world city: planning for the transformation of Hong Kong', *APA Journal*, 55, 309–22.

Trinh, D. (1997) 'Balancing market and ideology', in Kim, W., Douglass, M., Choe, S.-C. and Ho, K. C. (eds), *Culture and the City in East Asia*. Oxford: Oxford University Press.

Vogel, E. (1979) 'Nation-building in modern East Asia: Early Meiji (1868–1890) and Mao's China (1949–1971)', in Craig, A. (ed.), *Japan: A Comprehensive View*. Princeton: Princeton University Press.

Wong, R. (1997) 'Chinese understandings of economic change: from agrarian empire to industrial society', in Brook, T. and Luong, H. (eds), *Cultural Economy: The Shaping of Capitalism in Eastern Asia*. Ann Arbor: University of Michigan Press.

Wu, F. and Yeh, A. (1999) 'Urban spatial planning in a transitional economy', *APA Journal*, 65, 377–94.

Wu, W. (1999) 'City profile: Shanghai', *Cities*, 16, 207–16.

Yeh, A. and Wu, F. (1999) 'The transformation of the urban planning system in China from a centrally-planned to a transitional economy', *Progress in Planning*, 15, 167–252.

Yeung, H. and Lin, G. (2003) 'Theorizing economic geographies of Asia', *Economic Geography*, 79, 107–28.

4 The social geography of the service economy in global cities

David Ley

The concept of the service economy has permitted economic geographers to describe and account for occupational and industrial transitions in advanced societies. At the same time a parallel literature in social geography has pointed to new phenomena such as gentrification, new family structures, heightened immigration and growing social polarisation that have affected in particular the major cities of nation states. This chapter will seek to integrate these two literatures in the particular context of the global city, the service-oriented metropolis that serves as the focal point of a nation's linkages with the world economy. My primary, though not exclusive, examples will be Toronto and Vancouver, and I will be drawing upon extensive research in these cities that has moved around the interconnected questions of urban society and economy in a broadly post-industrial era. The economic transition to post-industrialism is also being experienced by Asia-Pacific global cities like Tokyo, Hong Kong and Singapore. It is perhaps a moot point as to what extent the same social transitions will accompany economic transformation in the rather different cultural contexts of East Asia. I will return to this question toward the end of this chapter.

Theoretical precedents: the post-industrial society

A decisive challenge to the modernist paradigm with its fixation upon the factory and industrial production – a paradigm shared by neo-classical and Marxist economic geographers alike – was issued by Daniel Bell's ambitious manifesto, *The Coming of Post-Industrial Society* (1973), as much a forecasting exercise as a description of existing reality. Declaring the arrival of a service society was not new, but one of the attractions of Bell's argument was its integrative ambition, as it sought to find the interconnections between economic, socio-cultural and political trends. In the economy, Bell conceptualised the emergence of a knowledge-based society, where human capital would reside in theoretical knowledge that could be extended flexibly to decision-making tasks. These characteristics defined a growing middle-class ascendancy, with the development of the

professional/managerial/technical cohorts as an emergent class. A
second societal theme was the consolidation of a watchful welfare state
dedicated to the facilitation of the public household as a metaphor for a
more collective social and political life. Third, however, there was a more
ambivalent sense of popular culture that was erring toward what Bell
called antinomianism, a permissive lifestyle that later authors would link
with a sensuous consumerism, but which Bell, writing in the waning days
of the student movement, associated with counter-cultural and artistic
excess (also Bell 1976).

Bell's argument was not unique. In France, Alaine Touraine (1971)
also developed a thesis that the holders of specialised information were a
class in ascendancy, though in the context of the highly centralised
French state he saw this in more dangerous terms as an instrumental and
inaccessible bureaucracy. There were similarities too with Jurgen Habermas' theses on advanced society (1970, 1975). Habermas, though carrying
forward Weber's thinking on the 'iron cage' of bureaucracy, agreed with
the new centrality of theoretical knowledge in society and shared the pes-
simism of other members of the Frankfurt School that it would be
deployed instrumentally, steadily invading the everyday life-world. But in
the context of the Cold War and the ideological fights of the 1970s, it was
Bell who was singled out for the most widespread criticism. His book
received hostile attention in Moscow where it was reviewed by *Pravda*, and
its economic and cultural diagnoses generated considerable intellectual
dissent. Like all academic work, it was certainly a product of its time. Its
optimistic scenario of a white-collar, middle-class economic trajectory in
the United States was influenced by the seemingly endless post-war boom,
though this was about to end just as his book was published, with the first
oil shock of 1973–1974.

While the political left may have been correct to challenge Bell's forecast they did so for the wrong reason. They argued that the dominant
foreseeable employment trend was not the swelling of the quaternary or
advanced service sector, but rather de-skilling and emiseration accompanying automation and the dispersal of jobs to lower-cost regions. The de-
skilling thesis, propounded by Braverman (1974) and others, was a vehicle
for postulating growth rather than contraction of the proletariat. But
actual employment trends supported Bell rather than his detractors. In an
act of intellectual contrition, Erik Ohlin Wright, a long-time strong critic
of post-industrialism, acknowledged that a 20-year record supported Bell's
forecasts of a growing middle-class of service workers, rather than his own
labour projections (Wright and Martin 1987). The same judgement was
made almost simultaneously in Canada, where the prevailing trend in the
labour market was that of 'an expanding middle' class of service workers
(Myles 1988). Restitution of Bell's gloomy portrayal of popular culture followed shortly afterwards, when David Harvey acknowledged that his
representation in the 1970s was 'probably more accurate than many of the

left attempts to grasp what was happening' (1989: 353). Certainly Bell's glum scenario matches a number of the renditions of the culture of consumption in current post-modern societies.

A generation after its publication, Bell's book continues to have an important message for students of the service economy. First, its integrative challenge offers a precedent for bringing together the isolations of economic, political and socio-cultural processes. Second, the focus upon the knowledge-based, professional and managerial cohorts, the absolutely central role of the bearers of complex information, clearly anticipated a major trend line in economic and social evolution. Third, his dismay at cultural trends has undoubtedly accelerated in the growing lament over consumerism and its fantasy worlds. But Bell's work is important too for its gaps, which become ever more salient. It was written at a time when, in Canada at least, there were no food banks and the central government was still actively involved in social housing construction. The erosion of the welfare state, mean-spirited neo-liberalism, social polarisation, the intensification of globalisation, all lay in the future. But for *our* present they are central elements to understanding the social geography of the service economy. Nowhere is this more self-evident than in the contemporary global city.

From the post-industrial city to the global city

Drawing upon the insights of Bell and Habermas, I developed the themes of post-industrial society toward an interpretation of emerging landscapes in Vancouver in the 1970s (Ley 1980). The paper noted the rise of a white-collar, middle-class work force, the development (and success) of social movements and political parties favouring an expanded sense of the public household, and the socio-cultural ideology of the livable city which all too accurately upheld Bell's call for 'a responsible social ethos ... the demand for more amenities, for greater beauty and a better quality of life in the arrangement of our cities' (1973: 367).

This paper was criticised from a neo-marxist position, steadfast in its consistency, which wanted to bring the whole argument back to the factory and a two-class model of society. But the post-industrial city is a phenomenon in its own right, no matter whether weak or strong linkages can be found connecting it with hinterland production economies. As we shall see, the growth of a middle-class, white-collar work force exercises its own effects. And such growth has most certainly occurred. In Canada, for example, the professional-managerial cohorts, enjoying the highest average incomes, added more than three million jobs from 1961 to 1991, and though under-represented spatially in central cities in 1961 were relatively over-represented by 1991 (Ley 1996). In London (Hamnett 1994, 2003), even in New York (Bailey and Waldinger 1991) and Los Angeles (Clark and McNicholas 1996), the major tendency among the employed was the expansion of this quaternary or advanced service sector. As we shall see, this new middle-class has distinct family arrangements, certain political orientations and well-defined cultural dispositions.

But expansive as it may be, the concept of the post-industrial city does not engage globalisation without some extension; service-based settlements may be as large as New York or as small as college towns in the American Midwest. While a study of small service centres requires no apology, it does not address the theoretical or empirical purposes of this chapter. The concept of the world city or global city, in contrast, carries forward some of the features of the post-industrial city, but places them in the intellectual context of the past 20 years, a globalising economy and a neo-liberal political philosophy that seeks to dismember the public household in an individualistic space of flows. In this respect, John Friedmann's anticipatory short paper on world cities in 1986 moved the argument decisively forward. Although overly directed by political economy, Friedmann did project the economic thrust of the corporate complex of head offices and producer services into a political and social context. It had been known for some time that a dual labour market, of professionals and managers on the one hand, and clerical and other service staff (cleaners, cafeteria staff, janitors, security guards, messengers, etc.) on the other, was part and parcel of the corporate complex. But Friedmann added to that a growing social polarisation emerging not only from the labour market but also from an eviscerated welfare state that reduced or withheld services, and abandoned the Keynesian hope of full employment. Thus it became possible to resolve the debate between upskilling and downskilling by identifying professionalisation in the labour market, but polarisation in society when those outside the active labour force were factored into the calculations.

Another significant addition of Friedmann's framework was the recognition of immigration, fast replacing domestic migration as the primary impulse of demographic growth in world cities. Immigrants, while diverse both culturally and economically, contributed to the growing polarisation, particularly in those cities where refugees, asylum seekers, and undocumented arrivals contributed significantly to new arrivals. These groups were marginalised, often without identification papers, linguistically disadvantaged, and sometimes traumatised by family loss and torture – all attributes that impeded their successful entry to the labour market.

The class formation of global cities

So pronounced were the effects of polarisation both inside and outside the labour market in service-dominated metropolitan areas that Saskia Sassen (1991) identified a reconfiguration of social classes within the global city. Economic restructuring has diminished the significance of manufacturing to minor status in employment (30 per cent of London's work force in the 1950s but less than 10 per cent in the 1990s), while the loss of union jobs has also contributed to income erosion. At the same time the well-known growth of the corporate sector of producer services has created a large number of high-end jobs in financial and business services. Much less discussed is the parallel long-term growth of managerial and professional jobs in the public sector, despite cutbacks in recent years. In Canada, both tertiary and quaternary sectors, the lower and the higher cohorts of the service economy, added more than three million jobs each between 1961 and 1991, and decomposing these results showed three important trends. First, the public service sectors of health, education and welfare (HEW) outpaced the private producer service sectors in job growth by a significant margin of some 450,000 jobs (Ley 1996). Second, while the share of jobs between the genders was approximately equal in the producer service sector (though of course heavily biased by earnings), in HEW women accounted for 75 per cent of jobs, including fast-rising representation in professional and managerial positions. Third, jobs in HEW (and direct government employment) were on the whole good jobs. By the mid-1980s, public administration ranked second and HEW third in an eight-sector comparison of relatively high-paying sectors; in contrast business services ranked seventh. These trends not only helped to produce a new middle-class, but also, fortified by feminism, opened up new career trajectories for women with attendant new models of family formation, favouring small one and two-person households. The co-existence of considerable disposable income and careerism in the downtown labour market in advanced services has contributed significantly to the gentrification of inner city neighbourhoods (Ley 1996).

At the same time the global city, as noted by Friedmann (1986) has been an unkind place for the poor. The price of housing and other services has been driven up by the wealth of the new middle-class, while job loss in traditional sectors, and the creation of what Sassen (1991) has called 'downgraded manufacturing' in new sweatshops, plus the provision of casual, part-time and poorly-paid service jobs has contributed to a growing poverty population among the employed. The limiting of welfare and other state services to this population has formed a spectre of visible poverty in the global city, of food banks, homelessness and growing street crime, prompting the vigorous war on the poor by some big city mayors, led by Mayor Giuliani in New York (Smith 1998).

While the most dramatic examples of first world polarisation are in American cities, growing inequality has spread to other nations where a more complete welfare state had formerly been in place (Hamnett 2003). Toronto, Canada's principal global city, is also the most polarised urban setting in the country, with the top 10 per cent of the population enjoying average incomes of $Cdn261,000 in 2000, 27 times higher than the average income of $9,600 for the bottom 10 per cent (Toronto Campaign 2003; Carey 2003). Among the most at-risk families, the median income for lone-parent households fell by almost 10 per cent between 1995 and 2000, despite a period of economic growth in the regional economy. Both social assistance payments and employment insurance were cut back by over 40 per cent as a result of welfare state policy, while the minimum wage has been fixed since 1995, which after inflation amounts to a real decline in value of 20 per cent. In this manner income polarisation is a result of economic change compounded by regressive state social policies.

Sassen (1991) has observed the functional relationships in place between what she calls the two circuits of the service economy in the global city. In their pursuit of advancement, employees in the advanced services work long hours, and, particularly in the early years of career development, there are advantages to living close to work. With discretionary income, and a premium on time, they contract out significant household tasks to food preparers, cleaners, child minders, security personnel, interior designers and decorators, most of these industries paying wages at, or near, the minimum level. This commodification of everyday life by the new middle-class (that has included a return of domestic servants in the form of the Filipina nanny) has created a web of functional relations between the well-educated and well-paid upper tier of the service economy and the lower tier, commonly comprising immigrants and more likely to be women than men. In Vancouver immigrants in the lower tier of the service economy come primarily from the countries of Asia-Pacific.

Socio-spatial demarcations: the internationalising property market

Social relations shaped by a reconfiguration of economic, political and demographic processes have given rise to changing spatial patterns in the global city. Friedmann (1986) noted that the world city was not simply a basing point for global economic relations, but was also an important site of wealth generation in its own right, not least in its property market. The internationalisation of the property market has been a striking development of globalisation in major cities. London property markets, particularly in the 1970s and 1980s, were significantly impacted by Middle East oil wealth, while the economic miracle in East Asia up to the mid-1990s led to vigorous real estate investment in such cities as New York, Los Angeles, Toronto, Vancouver and Sydney (Olds 2001).

These building projects were typically office towers, hotels or shopping centres, or else large residential redevelopments, usually in or on the margins of downtown. But internationalisation is increasingly being achieved outside the downtown border with the arrival of immigrants to world cities. As many immigrants are now by-passing the old inner city reception areas and moving directly into the suburbs, so commercial development with local and off-shore principals is moving heavily into Canadian suburbs, notably Richmond (Vancouver) and Scarborough-

Markham (Toronto), the location of large immigrant populations from East Asia, many of them landing with appreciable financial resources. David Lai (2003) has identified more than 50 large and small shopping centres in each suburban region that serve a primarily Hong Kong and Taiwanese immigrant market, and the development of such suburban 'Chinatowns' is repeated in Los Angeles and New York.

By 2001 the foreign-born component was as high as 43 per cent of the population of 4.7 million in the Toronto metropolitan area, and 37 per cent of the 2.0 million in metropolitan Vancouver. Under these circumstances we would expect to find that immigration would also affect the residential property market. The impact of immigration upon housing has intensified due to an unforeseen demographic cross-flow in many post-industrial global cities, including New York, Los Angeles, Toronto and Sydney, where there has been a net loss of population through domestic or internal migration and metropolitan numbers have been sustained through elevated immigrant numbers (Frey 1995, 1999; Ley 2003a). This countervailing trend has reached huge proportions in the major global cities in the United States, where the New York and Los Angeles regions both lost over 1,300,000 residents through domestic migration between 1990 and 1996, while gaining over 900,000 new immigrant arrivals (Frey 1999). In contrast, metropolitan areas like Atlanta and Phoenix, with modest immigration, were leading recipients of domestic migrants. Numbers are less dramatic outside the United States, but the trends are similar. Toronto suffered a net loss of a quarter of a million domestic migrants between 1986 and 2001, and population gain was controlled by immigration, which amounted to over 90 per cent of net growth in the 1990s. In Australia, Sydney added 440,000 new residents from immigration between 1976 and 1991 while there was a deficit of 280,000 in net domestic migration (Burnley *et al.* 1997).

There has been considerable speculation about the causes of such population redistribution, with the most common explanations favouring some mix of cultural avoidance (the 'white flight' thesis writ large) and economic competition in the labour market. But empirical analysis is mixed at best on the role of labour market effects; in the Toronto CMA there is a negligible (and negative) correlation of -0.10 between net internal migration and unemployment levels in the quarter century from 1971 to 1996 (Ley 2003a). Indeed out-migration from Toronto is growth averse, *rising* during periods of population growth and inflation.

This leads to a re-focusing of possible causes of countervailing migration flows that stresses the effect of immigration on the housing markets of global cities. Immigrants tend to be located in high cost housing markets; in Toronto the correlation between immigration and house price movements was 0.81 over a 20-year period from 1977 to 1996, and in Vancouver the association reached the surprising heights of 0.96 (Ley and Tutchener 2001). In contrast there were negative correlations in Toronto

(−0.77) and Vancouver (−0.32) between domestic migration and annual house prices over the same period. Oversimplifying the argument, one can see domestic migrants leaving high-priced housing markets. But how can immigrants, many of them poor, afford to enter such markets? In cities like Toronto and Vancouver (especially the latter), some immigrants arrive with sufficient wealth to buy into the market directly. But this is not a solution for most. The evidence is that immigrants are prepared to endure housing affordability burdens two to three times higher than other Canadians (Ley 2003a). So too they tolerate higher rates of crowding. Whereas only 4 per cent of Canadian-born households in Toronto exceeded the crowding threshold of more than one person per room in 1996, this figure rose to 24 per cent among immigrant households landing in the 1980s, and soared to 46 per cent among arrivals in the 1990s (CIC 2000). In this manner, rich and poor immigrants are able to enter the expensive housing markets of the global city, markets sustained in part by these immigrants' own demand structures.

Immigration is part of a much wider internationalisation of the land market in global cities. Analysis of house price movements in Toronto and Vancouver shows that a bundle of indicators of global connectedness shares high correlations with price change. These variables include the levels of foreign direct investment, numbers of overseas visitors, and the exchange rate between the Canadian dollar and other currencies, all of them strong predictors of house prices (Ley and Tutchener 2001). Importantly, a similar bundle of national variables, including the Canadian bank rate, unemployment, domestic migration and even vacancy rates and dwelling starts, show correlations that are low and sometimes even counter-intuitive in their directions. It is hard to avoid a conclusion that the land market in the global city is increasingly 'outer-directed'.

Pointing in the same direction is the correspondence of price trends among global cities in different nation states. Global cities in the United States, Canada and Australia experienced rapid inflation in the 1980s, with a peak in 1989–1990, and a marked correction to the mid-1990s. While this cross-border correspondence is surprising, even more striking is that these profiles were not shared with cities in their own nations. Cross-border affiliation in price behaviour was more pronounced than uniformity within the contained borders of each nation state.

Socio-spatial demarcations: the gentrifying housing market

Rather more attention has been given to the process of gentrification in the service economy of global cities. Here the linkages are more self-evident, for the economic power of advanced service jobs in the labour market is readily transferred to more privileged access to the housing market. Inner city neighbourhoods have a number of advantages to this population; they are close to downtown work sites, have an adult orientation, frequently adjoin the amenities of parks or waterfront settings, and are in neighbourhoods of historical and/or cultural distinction. Not least they border the considerable panoply of downtown and inner city leisure and cultural sites, including the arts, professional sports facilities, clubs and restaurants. The gentrified inner city is well suited to childless adults with discretionary income working downtown or in inner city institutions like hospitals and universities.

Initially the gentrifying cohort was linked to the profile of the young urban professional, and led to speculation that the phenomenon could not expand beyond this sub-market. But more recently there has been growing diversity among the urbane middle-class, as other sub-markets including empty-nesters and even parents with children have been added to their ranks. The strength of this movement has led to sharp social divides in the inner city, for example in Vancouver between the upscale condominiums of the Hong Kong-financed False Creek redevelopment that abut directly against the rooming houses of the Downtown Eastside, Canada's poorest urban neighbourhood. Whether this mixing of income groups will be more than temporary is unlikely; if demand by the middle-class is sustained the operation of the land market will continue to lead to the displacement of poorer households (Smith 2000).

The power of the advanced services in the central city housing market of global cities is measurable. In inner city Toronto there was a net gain of some 60,000 residents with quaternary occupations between 1971 and 1991, and in Vancouver the figure was 33,000 (Ley 1996). A displacement effect was underway, for simultaneously there was a net loss of 75,000 workers in other employment categories in inner city Toronto, though in Vancouver, with extensive construction of inner city social housing, residential losses among those in non-quaternary jobs scarcely occurred. But common to each city was the displacement of a large population of people not in the work force, including the elderly, the handicapped, the long-term unemployed and students. Significant erosion of affordable older apartments and rooming houses occurred, to be replaced by renovated structures with fewer occupants, and new town houses and condominiums. In many global cities talk of a 'crisis' of affordable housing became commonplace from the 1970s onwards, with very low vacancy rates, and the rapid diminution of rental stock, what Hamnett and Randolph (1986) in London called the break-up of the private rental market. Isolated middle-class enclaves close to downtown expanded, and in some cities dominated by the advanced services, including even London, observers see a virtual encirclement of downtown by a cordon of gentrified and gentrifying housing.

It is not surprising, then, to see the growth of the quaternary sector also associated with house price movements. The same analysis that indicated the role of international impulses in inflating prices in Toronto and Vancouver also showed the role of employment restructuring. The growth of the quaternary work force was strongly correlated with rising house prices in Toronto ($r = 0.85$) and Vancouver ($r = 0.89$) during the 1971–1996 period (Ley and Tutchener 2001). Moreover, and indicative of the perverse social polarisation of this recent period, growth of the advanced services was accompanied by a tendency toward rising unemployment in these metropolitan economies overall, with appreciable *positive* correlations linking the two variables in both Toronto ($r = 0.65$) and Vancouver ($r = 0.48$).

The map of metropolitan price changes reveals strong spatial concentration in central cities from the 1970s onwards. As the downtown and inner city places of employment in the quaternary sector have consolidated with income growth, so too has the geography of house prices. The distribution of house price inflation at the intra-metropolitan scale shows the impact of gentrification around the downtown core, wealth deepening (one side of social polarisation) in older elite neighbourhoods, and immigration in a range of districts (Ley, Tutchener and Cunningham 2002). Interestingly, however, there were variations between Canada's global cities both in the ranking of explanatory processes, and even their directions. In Vancouver, immigration was a consistent predictor of rising dwelling values at the census tract level, whereas in Toronto associations were smaller and usually negative. An explanation here has to do with the composition of immigrant cohorts, with Vancouver consistently receiving higher proportions of wealthier business-class immigrants and skilled workers from Asia-Pacific, while Toronto has been a preferred destination for typically poor refugees. In both cities social polarisation correlated positively with changes in dwelling values, while gentrification exerted local effects.

Cultural and political relations in the global city

How have cultural and political realms responded to these labour and housing market transformations in cities dominated by service employment? There is no real research programme that has examined this question to the present but we can piece together the findings of several studies to attain a suggestive profile.

In work that unfortunately has not been replicated with more recent data, Caroline Mills (1989) found some striking variations in family structure and gender relations between 22 metropolitan areas in Canada with industrial and post-industrial employment profiles, evidence that the post-industrial city can be a theoretical as well as an empirical object. The human capital of women was higher in post-industrial cities, with both family partners far more likely to be university-educated in metropolitan areas with an employment concentration in professional-managerial occupations, but much less likely in cities with a manufacturing employment emphasis. Education parity was joined by gender income parity in families, strongly positively correlated with post-industrial employment status but negatively correlated with manufacturing employment. Not surprisingly, with these distinctions came a greater probability of dual income families in cities with a post-industrial employment profile. This career trajectory was then associated with a set of distinctive family characteristics. Cities rich in professional-managerial jobs were characterised by late marriages, higher divorce levels without remarriage, and a higher share of childless families. In contrast a blue-collar industrial profile was associated with opposite trends, including an earlier age for marriage and fewer childless families. In this manner we can see the emergence of a distinctive family profile in the post-industrial city that diverges from the traditional model.

This divergence is associated with other dimensions of identity politics. Working with the same set of 22 Canadian metropolitan areas, an analysis of secularisation showed persistent relations with the presence of members of the new middle-class. A measure of religious disaffiliation drawn from the census showed a remarkably strong correlation of 0.79 with a post-industrial city index, while in the largest cities religious disaffiliation was a consistently reliable predictor of the presence of gentrification (Ley and Martin 1993). The more recent 2001 Census of Canada indicated that inner city neighbourhoods continue to run ahead of the rest of the metropolitan area with in excess of half the adult population lacking any religious affiliation in some Vancouver census tracts. These relations are not true only for Canada, for an ecological analysis in Sydney showed a robust relationship between lack of religious affiliation and the residential location of young urban professionals (Horvath *et al.* 1989).

More liberal lifestyles may be associated with distinctive political responses. In Canadian federal elections centre city constituencies invariably offer a greater orientation toward left-liberal candidates than surrounding suburban and rural regions (Walks 2003), while in Australia inner city professionals are associated with a 'cosmopolitan social agenda' that includes a republican constitution, a re-definition of Australia towards Asia, and liberal minority, multicultural and Aboriginal policies (Betts 1999). While such responses are contingent upon a range of other factors, there can be a similar tendency toward a liberal option in city politics as well. The rise of political reform movements in Toronto, Vancouver and Montreal was closely associated with the onset of gentrification in each city in the early 1970s, with the movement of an articulate cadre of young professionals into advocacy roles (Ley 1996). A similar convergence occurred in the United States, where San Francisco, Boston and Seattle, post-industrial cities with entrenched gentrification, were among the relatively few that could be characterised as having a liberal political regime in the 1970s and 1980s (Stone *et al.* 1991).

Friedmann (1986) observed the vitality of the leisure component of global cities. The presence of artists is strongly correlated with urban concentrations of the new middle-class (Ley 1996, 2003b) and gentrifiers themselves are enthusiastic consumers of the arts (Butler 1997). Indeed the appetite for diversion runs strongly in Canada's major centres, not least in the pursuit of global leisure spectacles, boosting the tourist and hospitality sector of the service economy, and with the hope of more substantial investment and trade to follow. During its period of global aspirations as the (then) largest city in Canada, Montreal secured the 1967 World's Fair, plus the 1976 Olympics. Vancouver successfully lobbied to be the site of the 1986 Fair and, as I write, 18,000 people are crowded into a downtown sports stadium waiting to hear news from Prague via television hook-up as to whether Vancouver-Whistler's bid for the 2010 Winter Olympics will be endorsed. Meanwhile Toronto, since 1976 Canada's major global city, has been rebuffed as an Expo site and twice as a summer Olympics city within the past decade. With Vancouver's award of the 2010 Games, Toronto advocates are suggesting they should delay their next bid until the 2016 Olympics! There is no limit on forward planning it seems for the culture of consumption in global cities.

Discussion and conclusion

This chapter has identified themes in the social geography of the service-dominated metropolis, leading to a potentially richer human geography when economic trends are integrated with socio-cultural and political relationships. It is a strength of both the post-industrial and global cities literatures that such integration has been attempted. We have seen the value of the post-industrial city as a heuristic, though its earlier conceptualisation needs to be updated with trends identified by scholars of the global city.

Moreover, such effort is warranted because both the post-industrial and the global city are more than descriptive terms. The post-industrial metropolis has conceptual significance for it is associated with distinctive forms of family development, with identity politics that favour lifestyle liberalism, and with cultural proclivities that endorse the consumption of an urbane suite of arts and entertainment opportunities, including global sporting and leisure events. To this the global city adds the characteristics of socio-spatial polarisation and a vigorous internationalisation of the population and the property market. Our conclusion is that the large service-dominated metropolis is not simply a new economic form, but more than this a novel geographical region defined by interlocking economic, political and socio-cultural relationships whose reach exceeds the boundaries of the nation state.

But to what extent is this city-region a universal form or culturally contingent upon relationships specific to western nation states? A number of authors have alerted us not to assume a simple transference of western societal trends to East Asia, but rather to be attentive to distinctively Asian forms of modernity, including a Chinese modernity embedded in hybridised Confucian forms that prescribes among other things alternate views of the family and the state (Ong 1999). On the one hand a post-industrial economic profile is rapidly taking shape in some East Asian global cities like Tokyo (Cybriwsky 1998), Hong Kong (Thompson 2000), or even Shanghai (Olds 2001). This tendency is most apparent in Singapore (Perry *et al.* 1997; Baum 1999; Poon 2000), where labour market restructuring is leading to the same professionalisation tendency in the work force as in the major cities of western nations. But the social and cultural accoutrements of western-style post-industrialism are less evident, and have not been adequately studied. As Scott Baum (1999: 1096) has noted, while the economic transition in major urban centres of Southeast Asia has been documented 'there has only been speculation as to the social outcomes which may follow'.

A preliminary assessment of emerging social geographies would see continuities but also significant discontinuities between global cities on opposite sides of the Pacific Ocean. Consumerism, primed both by tourism and internal demand, *is* well-developed in all centres, even reaching the level of the spectacle in the Disneylands existing in Tokyo and anticipated in Hong Kong. But while western-style consumerism is enthusiastically welcomed in Tokyo (Tanaka 1994; Cybriwsky 1998) and to a lesser degree in Hong Kong, its excesses may be officially frowned upon by pronouncements on lifestyle liberalism emanating from Singapore or Kuala Lumpur. Or again, social polarisation may be evident, though its spatial form is dissimilar to the landscapes of Vancouver or Toronto. Deep poverty in Tokyo, for example, is concealed in remote neighbourhoods north of the central business district, and when I presented a lecture on gentrification in Tokyo in 1993 it was an unfamiliar phenomenon to my audience. Since then the massive waterfront residential redevelopment begun by River City 21 has introduced distinctive Asian-style gentrification (Cybriwsky 1998). In Hong Kong high-rise residential intensification in the innermost districts around Central has created a similar landscape, and it is relevant to note the export of that landscape to Vancouver and Toronto through the property development of the wealthy Li family (Olds 2001). But at the same time the widespread public housing projects in Hong Kong and Singapore have mitigated the visible extremes of wealth readily visible in more privatised western global cities.

In contrast, some of the softer quality-of-life issues are beginning to attract policy concern such as air and land pollution in Tokyo, and heritage preservation, with retention and renovation contributing to the invention of tradition in Singapore, including Tanjong Pagar shophouses as well as colonial buildings in the Civic and Cultural District (Yeoh and Kong 1995). More recently, Singapore continues to track closely western innovations in both tourism and an arts policy as part of its calling as a global city (Chang 2000), a rapid response that keeps it ahead of its principal regional competitor, Hong Kong, in style if not always in substance (Jessop and Sum 2000).

But significant elements of the social relations constituting the post-industrial city on the American side of the Pacific are less in evidence in East Asia than these landscape forms. The evolution of family relations that has contributed to greater parity and new social geographies among professional households in Canada is much less developed in the patriarchal societies of the Asia-Pacific. Equally dissimilar are attitudes toward immigration – so massive a presence in the global cities of the west – as part of the imagined community of the nation state in Asia-Pacific where post-colonial status has often led to very uneasy relations around cultural pluralism (Ang 2001). Although 'foreign talents' (and domestic workers) are welcome in the global cities of East Asia their transient and alien identity is institutionalised by this very designation. Both Singapore and Tokyo turn a blind eye (usually) to illegal unskilled migration, but despite labour market needs there is no institutional capacity or will for open immigration leading to citizenship for significant numbers from diverse origins. In these respects at least we can expect only a slow convergence in the social geography of global cities in Asia-Pacific and North America.

References

Ang, I. (2001) *On Not Speaking Chinese: Living Between Asia and the West.* London: Routledge.

Bailey, T. and Waldinger, T. (1991) 'The changing ethnic/racial division of labor', in Mollenkopf, J. and Castells, M. (eds), *Dual City: Restructuring New York.* New York, NY: Russell Sage Foundation.

Baum, S. (1999) 'Social transformations in the global city: Singapore', *Urban Studies*, 36, 1095–117.

Bell, D. (1973) *The Coming of Post-Industrial Society.* New York, NY: Basic Books.

—— (1976) *The Cultural Contradictions of Capitalism.* New York, NY: Basic Books.

Betts, K. (1999) 'The cosmopolitan social agenda and the referendum on the republic', *People and Place*, 7(4), 32–41.

Bravermann, H. (1974) *Labor and Monopoly Capital: The Degradation of Work in the Twentieth Century.* New York, NY: Monthly Review Press.

Burnley, I., Murphy, P. and Fagan, R. (1997) *Immigration and Australian Cities.* Leichhardt, NSW: Federation Press.

Butler, T. (1997) *Gentrification and the Middle Classes.* Aldershot, UK: Ashgate.

Carey, E. (2003) 'Child poverty spreading fast in GTA', *Toronto Star*, 30 June.

Chang, T. C. (2000) 'Renaissance revisited: Singapore as a "Global city for the arts"', *International Journal for Urban and Regional Research*, 24, 818–31.

CIC (Citizenship and Immigration Canada) (2000) *Recent Immigrants in the Toronto Metropolitan Area: A Comparative Profile Based on the 1996 Census.* Ottawa: CIC, Strategic Policy, Planning and Research.

Clark, W. A. V. and McNicholas, M. (1996) 'Re-examining economic and social polarisation in a multi-ethnic metropolitan area: the case of Los Angeles', *Area*, 28, 56–63.

Cybriwsky, R. (1998) *Tokyo: The Shogun's City at the Twenty-First Century.* New York, NY: Wiley.

Frey, W. (1995) 'Immigration and internal migration "flight" from US metropolitan areas: towards a new demographic balkanization', *Urban Studies*, 32, 733–57.

—— (1999) 'Immigration and demographic balkanization: toward one America or two?', in Hughes, J. and Seneca, J. (eds), *America's Demographic Tapestry*. New Brunswick, NJ: Rutgers University Press.

Friedmann, J. (1986) 'The world city hypothesis', *Development and Change*, 17, 69–83.

Habermas, J. (1970) *Toward a Rational Society*. Boston, MA: Beacon Press.

—— (1975) *Legitimation Crisis*. Boston, MA: Beacon Press.

Hamnett, C. (1994) 'Socio-economic change in London: professionalisation not polarisation', *Built Environment*, 20, 182–203.

—— (2003) *Unequal City: London in the Global Arena*. London: Routledge.

Hamnett, C. and Randolph, W. (1986) 'Tenurial transformation and the flat breakup market in London', in Smith, N. and Williams. P. (eds), *Gentrification of the City*. Boston, MA: Allen and Unwin.

Harvey, D. (1989) *The Condition of Postmodernity*. Oxford, UK: Blackwell.

Horvath, R., Harrison, G. and Dowling, R. (1989) *Sydney: A Social Atlas*. Sydney: Sydney University Press.

Jessup, B. and Sum, N.-L. (2000) 'An entrepreneurial city in action: Hong Kong's emerging strategies in and for (inter)urban competition', *Urban Studies*, 37, 2287–313.

Lai, D. (2003) 'From downtown slums to suburban malls: Chinese migration and settlement in Canada', in Ma, L. and Cartier, C. (eds), *The Chinese Diaspora: Place, Space, Mobility and Identity*. Lanham, MD: Rowman and Littlefield.

Ley, D. (1980) 'Liberal ideology and the postindustrial city', *Annals, Association of American Geographers*, 70, 238–58.

—— (1996) *The New Middle Class and the Remaking of the Central City*. Oxford: Oxford University Press.

—— (2003a) *Offsetting Immigration and Domestic Migration in Gateway Cities: Canadian and Australian Reflections on an 'American Dilemma'*, Vancouver Centre of Excellence, Research on Immigration and Integration in the Metropolis, Working Paper No. 03-01 [www.riim.metropolis.net].

—— (2003b) 'Artists, aestheticisation and the field of gentrification', *Urban Studies*, 40, 2525–42.

Ley, D. and Martin, B. (1993) 'Gentrification as secularisation: the status of religious belief in the postindustrial city', *Social Compass: International Review of Sociology of Religion*, 40, 217–32.

Ley, D. and Tutchener, J. (2001) 'Immigration, globalisation and house prices in Canada's gateway cities', *Housing Studies*, 16, 199–223.

Ley, D., Tutchener, J. and Cunningham, G. (2002) 'Immigration, polarization or gentrification? Accounting for changing house prices and dwelling values in gateway cities', *Urban Geography*, 23, 703–27.

Mills, C. (1989) 'Interpreting gentrification: postindustrial, postpatriarchal, postmodern?', unpublished PhD thesis in Geography, University of British Columbia.

Myles, J. (1988) 'The expanding middle: some Canadian evidence on the deskilling debate', *Canadian Review of Sociology and Anthropology*, 25, 335–64.

Olds, K. (2001) *Globalization and Urban Change: Capital, Culture and Pacific Rim Mega-Projects*. Oxford: Oxford University Press.

Ong, A. (1999) *Flexible Citizenship: The Cultural Logics of Transnationality.* Durham, NC: Duke University Press.

Perry, M., Kong, L. and Yeoh, B. (1997) *Singapore: A Developmental City State.* New York, NY: John Wiley.

Poon, J. (2000) 'Reconfiguring regional hierarchy through regional offices in Singapore', in Andersson, A. and Andersson, D. (eds), *Gateways to the Global Economy.* Cheltenham, UK: Edward Elgar.

Sassen, S. (1991) *The Global City: New York, London, Tokyo.* Princeton, NJ: Princeton University Press.

Smith, H. (2000) 'Where worlds collide: social polarisation at the community level in Vancouver's Gastown/Downtown Eastside', unpublished PhD thesis in Geography, University of British Columbia.

Smith, N. (1998) 'Giuliani time: the revanchist 1990s', *Social Text,* 16(4), 1–20.

Stone, C., Orr, M. and Imbroscio, D. (1991) 'The reshaping of urban leadership in US cities: a regime analysis', in Gottdiener, M. and Pickvance, C. (eds), *Urban Life in Transition.* Newbury Park, CA: Sage.

Tanaka, A. (1994) 'Tokyo as a city of consumption: space, media and self-identity in contemporary Japan', unpublished MA thesis in Geography, University of British Columbia.

Thompson, E. (2000) 'Hong Kong as a regional strategic hub for manufacturing multinationals', in Andersson, A. and Andersson, D. (eds), *Gateways to the Global Economy.* Cheltenham, UK: Edward Elgar.

Toronto Campaign 2000 (2003) *Child Poverty Report Card.* Toronto: Toronto Campaign 2000.

Touraine, A. (1971) *The Post-Industrial Society.* New York, NY: Random House.

Walks, R. A. (2003) 'Polling apart: suburbanization and the political polarization of large Canadian cities', unpublished PhD thesis in Geography, University of Toronto.

Wright, E. O. and Martin, B. (1987) 'The transformation of the American class structure, 1960–1980', *American Journal of Sociology,* 87, 1–29.

Yeoh, B. and Kong, L. (eds) (1995) *Portraits of Places: History, Community and Identity in Singapore.* Singapore: Times Editions.

5 Service industries and occupational change

Implications for identity, citizenship and politics

K. C. Ho

The sociology of the service economy

Taylor and Gane commented that 'Pacific-Asia is by far the region of greatest service growth' (2002: 4). This is a conclusion they arrived at after tracking a sample of 100 firms (including 18 in accountancy, 15 in advertising, 23 in banking/finance, 11 in insurance, 16 in law and 17 in management consultancy) and mapping the geography of these firms in terms of staff strength and office function changes, as well as the opening and closing of branches in North America, Western Europe and Pacific Asia between 2000 and 2001. This growth, however, has a particular pattern. As pointed out by Daniels (1998, 2001), while service industries have indeed been growing in Asia, the growing trade in services has been dominated by American and European multinationals. This dominance is reflected in the balance of trade in merchandise and commercial services by regions and in the net growth in services imports by Asia and the Far East in 1985 as well as in 1996 (Daniels, 2001: 218). The continued dominance of European and American service multinationals in the global service industry is an outcome of several factors, as highlighted by Sen and Sen (1997) in a review of the literature. The major service multinationals in accounting and advertising acquired strength and expertise through mergers. There has been a long history of exposure to foreign markets, with the large service companies having followed their major manufacturing clients as these established production bases overseas. In the internationalization process multinational service firms established local brand presence through foreign direct investments rather than by licensing or selling information abroad and, where regulatory barriers were faced, these service firms have resorted to a variety of joint ventures and in the process developed a growing local clientele (Sen and Sen, 1997: 1159, 1163). Besides this dominant presence of multinational service firms lies a layer of smaller, less diversified Asian service firms created by transnational business networks (Yu Zhou, 2000) as well as by the expansion of Japanese, Korean and Taiwanese manufacturing overseas.

Driven by the need to locate close to large client bases and to efficient transport and communications infrastructures in order to coordinate delivery (Dunning and Norman, 1987; Hutton and Ley, 1987; Esparza and Krmenec, 1994; Sassen, 2001), this mix of large multinational American and European service firms, plus their smaller Asian cousins, have a similar tendency to locate in the largest cities in Asia. Drawing on the Dun and Bradstreet database of 2,646 foreign firms in Taiwan, Ko and Tzeng (2002) found that 90 per cent of all producer services firms (474 in total) are located in Taipei City. With the opening up of China, Hong Kong has increasingly lost its manufacturing to the mainland and has been reduced to an intermediary role in managing the trade between China and the rest of the world. Consequently, import and export trade, finance and business services have led the growth of producer services in Hong Kong (Tao and Wong, 2002: 2350–2). Hong Kong's producer services reflect the two-tier trade flows through the system of cities highlighted by Esparza and Krmenec (1994). The first-tier city bias of producer services is also evident in Sirat's (2000) study of producer services firms clustered in Kuala Lumpur, the capital city of Malaysia, where the orientation is clearly intra-metropolitan trade linkages. Faced with rising manufacturing costs and declining competitiveness in manufacturing, Singapore in the late 1980s switched to a services-led strategy of promoting itself as the base for trans-national regional headquarters (Dicken and Kirkpatrick, 1991; Perry, 1992). Yeung *et al.*'s (2001) study of firms with regional headquarters in Singapore found the availability of good quality business services to be a significant factor in the location choices of such establishments. This close relationship between regional headquarter activity and business services suggests a regional service complex centred in Singapore.

In asking the question 'does the increasing concentration of producer services in Pacific Asian cities have any socio-political consequences?', this chapter seeks to extend engagement in three strains of literature. The question of relating social consequences to services development has a long history in the social sciences beginning with theories of information society. For the purposes of this chapter, direct relevance can be found within global cities research, where Sassen's (2001) phrase the 'Sociology of the Service Economy' serves as a good characterization. The intention of this chapter is to extend her analysis to discover if services concentration has effects beyond inequality and polarization.

The literature on consumption and identity has tended to look at issues of raising income or work within the concept of social class as the major explanation behind economic change and social outcomes. But is it possible to create a more precise understanding of consumption behaviour by linking this to specific occupational cultures? I want to build on Lury's (1996) work on consumer cultures to seek a more specific link between the way new forms of work are organized and their relationship to lifestyle. In doing this, the chapter will deal with different types of services by making the link between producer and consumer services, whilst noting that different studies refer to different types of services (see footnote 2 for example).

Finally, I will be highlighting the impacts of services on the urban economy by re-examining the idea of services and its effects within the broader concept of agglomeration. Urbanization economies recognize the beneficial effects of locating in large urban economies for the access to various goods, services and markets. Amin (2000) makes the distinction between fixed assets and non-traded relational assets in theorizing about the economic base of cities, drawing on Michael Storper's work, and refers to untraded interdependencies which stem from the abilities of influential classes and organizations, through their informal associations, to reflexively shape trends, products and fashions. In examining the links between occupational shifts and consumption outcomes, the focus is to read the social into what are essentially economic spatial patterns by revisiting earlier ideas developed by Zukin (1991) on space and consumption and those of Massey (1984) on how place is a unique product of layers of economic activities accumulated from different time periods. The attempt is to establish a connection between producer and consumer services and examine how this interaction ultimately influences the complexion of the city.

Thus, this chapter surveys the socio-political consequences of an expanding service economy in Pacific-Asian cities. While I will illustrate main points with data from Singapore, the patterns suggested will have resonance with major cities in Pacific-Asia. It is important to recognize that the changes described come from a variety of sources and operate at different scales; from economic globalization and the way work is organized across countries as well as the professionalization of services. Consumption and identity, have, of course, multiple causes and consequences, both global as well as local reference points, but the intent in this chapter is to relate these to the growing concentration of services.

Socio-political impacts of the service economy

Evidence of growing services, occupational shifts and income inequality?

Sassen started this line of inquiry with the finding of a new income inequality, resulting from a number of shifts, namely the growing producer services concentration in global cities. Sassen (2001) has demonstrated how the expansion of the global economy in the form of the activities of multinational companies has led to the growing concentration of service professionals in global cities. As transnational corporate networks expand, management, coordination, servicing and financing activities proliferate, increasing specialized skill sets, outsourcing and the increase in demand for service professionals (Sassen, 2000: xx–xxi, 11). Storper (2000: 384) suggests that the expansion of worldwide markets actually increases the demand and remuneration for the top-end professions. These trends create dramatic increases in the pay of these professionals, while at the same time, the number of middle income jobs have shrunk because of the migration of manufacturing to distant shores. These trends become the driving forces for polarization (Sassen, 2000: 244–5, 361–2).

Are these trends necessarily happening in Singapore? Figure 5.1 shows the employment shares by key sectors over the last ten years. The statistics clearly show increasing services growth in financial insurance and real estate services, especially in the mid 1990s, and the growth of personal, community and social services in the later part of the 1990s.

The growing dominance of services in the Singapore economy has also led to occupational shifts. The economic restructuring of Singapore's economy, particularly in terms of technical upgrading, and the emerging role of Singapore as a regional headquarter site, has led to an occupational structure that is increasingly populated by managers and professionals.

However, Table 5.1 shows that financial and business services, compared to manufacturing, are significantly top heavy in the sense that there is a much higher component of professionals and managers. Thus, one could argue that the shift to services has the accompanying trend of skewing the occupational structure to management and professionals. Table 5.1 also shows a trend towards an increasingly higher share of managers, professionals and technical and associated professional workers within the manufacturing sector.

Figure 5.2 shows the average monthly earnings for major sectors. The

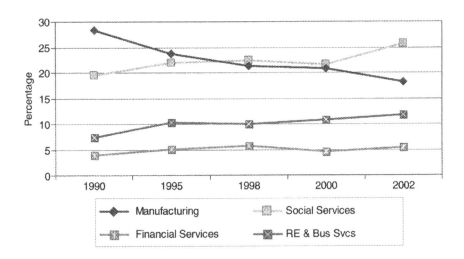

Figure 5.1 Sector employment shares, 1990–2002 (%) (source: Singapore, Department of Statistics, *Yearbook of Statistics Singapore*, various years).

Table 5.1 Manufacturing and financial/business services occupational structure, 1993–2003

Occupation	1993	1995	1997	1999	2001	2003
Manufacturing						
Management and administration	8.7	11.9	11.4	11.9	14.2	14.3
Professionals	4.8	6.5	9.3	11.1	13.0	14.1
Technical and associate professional workers	12.3	15.4	16.8	16.9	17.2	16.9
Clerical workers	11.3	9.7	10.7	11.3	10.8	11.0
Service workers	1.0	1.6	1.1	1.3	1.1	0.9
Production workers, cleaners and labourers	61.9	54.8	50.7	47.5	43.6	42.8
Financial and business services						
Management and administration	12.2	6.6	14.6	13.4	16.5	16.0
Professionals	13.1	16.5	18	18.2	22.8	22.3
Technical and associate professional workers	27.7	27.4	28.5	31.6	27.0	26.9
Clerical workers	29.8	23	27.9	20.8	18.4	18.6
Service workers	5.2	3.9	3.1	5.3	5	5.4
Production workers, cleaners and labourers	12.0	12.5	7.9	10.7	10.3	8.4

Source: Singapore Ministry of Manpower, Report on the Labour Force in Singapore, various years.

remuneration to financial service workers is significantly higher than in the other sectors. Table 5.2 also shows that the average monthly earnings for other service jobs in business and real estate services and in social and personal services are higher than those in manufacturing.

The Ministry of Manpower published a report on wages in Singapore, and the statistics included the remuneration accruing to specific occupational categories. Table 5.2 displays the mean monthly gross wages (including bonuses, retirement contributions, etc.). The purpose of the table is to display the industry salary differentials in management and professional occupations. I therefore included, as far as possible, similar occupational categories represented in the six industry sectors.

The income differentials between the FIRE (financial, insurance and real estate services) and the other sectors is clearest in the top posts of managing director, general manager and company director with salary differentials between 20 to 40 per cent higher than in the other industries. Salaries for sales and marketing managers and for accountants are also highest in FIRE. However, the salary differentials in the other occupations are much lower, and in the case of human resource managers as well as in the case of legal officers, the reported incomes in manufacturing are actually higher.

In reviewing the four sets of data, we saw evidence of how the services

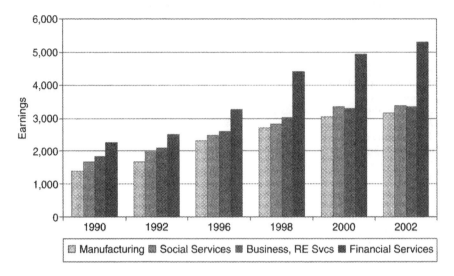

Figure 5.2 Average monthly earnings for major sectors, 1990–2002 (Singapore $)
(source: Singapore, Department of Statistics, *Yearbook of Statistics Singapore*, various years).

economy is clearly growing in Singapore (Figure 5.1), and how services tend to be top heavy in management and professionals (Table 5.1). The aggregate average monthly data indicates (Figure 5.2) salary differentials between services and the other sectors. In Figure 5.2 there are clear wage differences at the very top end of the occupational spectrum, but for other occupational categories, these wage differences are much less pronounced, particularly in the professional categories.

To demonstrate social polarization, one must show not just the appearance of a high-wage service professional class, but also the erosion of a middle income group displaced as manufacturing moves overseas (Sassen, 2001: 361–2). Additionally, there must be an accompanying growth in the lower income group populated by the displaced workers.

Baum (1999) examined this thesis for Singapore using different statistics. Showing occupational shares (but not by industry) for two time periods (1983 and 1996), he was able to demonstrate that there was greater professionalization, a point supported by the greater educational attainment of the local population within the two time periods. He also concluded that income polarization was not occurring in Singapore because the data he used (gross monthly income) showed declining shares of low income groups and increasing shares of middle and high income groups between 1986 and 1996. However, the analysis supporting this last point regarding income shares is flawed on several counts. Baum (1999) based his analysis on fixed income categories instead of

Table 5.2 Mean monthly gross wage of selected occupations in key sectors, June 2002

Occupation	All industries	FIRE	Wholesale retail hotel/ restaurants	Manufacturing	Social and personal services	Transport	Construction
Managers							
Managing Director	14,448	19,076	13,423	14,538	12,318	14,320	8,710
General Manager	11,238	11,799	11,043	11,338	9,625	11,834	6,886
Company Director	10,336	12,273	10,643	11,457	7,757	8,950	5,524
HR Manager	7,104	7,154	7,611	7,382	5,751	6,585	4,293
Sales and Marketing Manager*	6,657	7,684	6,648	6,873	6,830	6,064	3,707
Professionals							
Accountant	3,970	4,239	3,902	3,773	3,436	4,290	3,330
Lawyer/Legal Officer*	6,380	5,705	6,866	6,857	–	5,458	–

Source: Singapore Ministry of Manpower, Report of Wages in Singapore, 2001.

Note
*Where applicable, the higher salary of the two occupational categories was used.

analysing income difference within equal groupings (as Sassen (2001) has done in her update). Because Baum (1999) did not use inflation adjusted income data between two time points, the smaller shares in the lowest income categories could have been predicted merely on cost-of-living difference between the two time periods. Established income inequality measures such as the Gini coefficient may yield more convincing accounts. On this issue, a recent Singapore Department of Statistics paper (2002) indicated that by dividing households into ten equal groups or deciles, the lowest 20 per cent of households suffered declines in average household income on a yearly basis from 1997 to 2000, while the top 10 per cent enjoyed continuous increases in average incomes within the same period. The Gini coefficient for Singapore has increased from 0.436 in 1990 to 0.481 in 2000, reflecting greater income inequality (see Figure 5.3).

While Hamnett (1994: 202) noted that growing income inequality does not necessarily mean social polarization, the situation in Singapore is

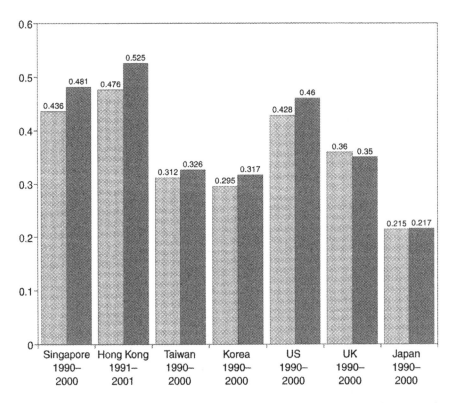

Figure 5.3 Gini-coefficients of selected countries (source: Income Distribution and Inequality Measures in Singapore, Chart 1. Online: http://www.sing-stat.gov.sg/papers/seminar/income2000.pdf (available as of 28 May 2004)).

certainly not as problem-free as Baum claims.[1] While there has been a growing share of managerial and professional workers within both manufacturing and financial and business services, this development has also been coupled with growing income inequality, partly the result of the significantly higher remuneration enjoyed by financial services in general (Figure 5.2), particularly those in top management compared to similar occupations in other industries (Table 5.2). The growing income inequality in Singapore (Figure 5.3) has created widening income gaps between groups, with the lowest income groups experiencing declines in household incomes for the past five years.

If social polarization stems from growing income inequality, then other socio-political effects may follow, as polarization then becomes the catalyst for mobilization for those at the bottom end of the income ladder, what Margit Mayer (1999: 215) terms as *anti-growth movements* which organize against specific developments and in defence of threatened communities and *poor people's movements* organized around welfare rights. The government's response to this new economic climate has also contributed to the polarization, by devoting more attention and resources to economic policies at the expense of social policies (Mayer, 1999: 211–12). Body-Gendrot (2000) uses the examples of the United States and France to show how cities have become the locus of social control of the disorders arising from the growing disparities. Margit Mayer has gone to the extent to suggest that 'social policies have been abandoned in favor of punitive and repressive treatments of a growing segment of marginalized population' (Mayer, 1999: 213).

Are such effects manifested in the cities of the Asia-Pacific? Rodan (1996: 6) calls the state the 'mid-wife of industrialization in Asia'. Where states have been successful, as they have been in several East and Southeast Asian countries, it enjoys a measure of legitimacy and popular support. A successful industrialization phase also created a pro-growth middle-class.

However, the effects of globalization and the increased mobility of the service class are creating two sets of tensions for Asian governments. The poor economic conditions mean that the local populations increasingly view foreign workers, especially those in highly valued jobs, as competitors and growing resentment results.

Place loyalty is another issue that governments have to contend with. Economic globalization requires professional and managerial workers stationed in various global cities to manage production networks, moving as networks reconfigure in response to the shifting of markets and production sites. Daniels (1998) observed that the increased presence of foreign (non-Asian) multinational service firms adds to the floating population of service professionals in Asian cities.

Holston and Appadurai (1999: 12–13) suggests three reasons why this class is likely to have reduced place loyalties:

1   the polarization which Sassen identifies is likely to reduce the possibility of common allegiances across class lines;
2   temporality in serial locations suggests reduced personal and moral commitments to place. In such situations diasporic identities are primary, not place identities;
3   the intense inter-city competition for managers of global capital and the professionals who support them have led the state to relax legal structures as attempts to attract these professions. The new laws tend 'to give special privileges to the managers of global capital, in the sense that it absolves them from local duties and makes them immune to local legal powers'.

In Singapore both problems have surfaced. Singapore's growing base of multinational companies has led to an expanding foreign-born population (see Table 5.3). The large expatriate community in Singapore, as Holston and Appadurai suggests, are sojourning communities which are inward-looking and are not integrated with local society. Local resentments increase as the economy weakens and job prospects dim. At the same time, growing numbers of Singaporean workers have been moved to manage offices in the region as multinational companies increasingly coordinate their operations on a regional scale. Recognizing these twin problems, the government has embarked on a campaign to socialize the population to accept foreigners and to understand that they add to the economic opportunities rather than reduce them. At the same time, initiatives are being created to strengthen place allegiances in the light of increased geographic mobility on the part of Singaporean workers.

Mapping the relationship between services, consumption and identity

In this section, my intention is to look for links between the nature of work among advanced services[2] and how this may possibly translate to issues of consumption and identity, a terrain in which Sassen (2001: 284–8) hints at but never quite develops. To fill this void, we need to turn to a number of other writers.

The idea is that the class of advanced services professionals have three sets of driving forces which create particular consumption patterns.

Table 5.3 Composition of resident population, 1970–2000

	1970	1980	1990	2000
Total population (000's)	2,074.5	2,413.9	3,016.4	4,017.7
Citizens	90.4	90.9	86	74
Permanent residents	6.7	3.6	3.7	7.2
Non citizens, non-PRs	2.9	5.5	10.3	18.8

Professionalization, identity and class formation

Service professionals have assumed greater economic and social importance in society. Using a post-industrial society argument, Friedson (1994:
61–8) identifies the weakening of bureaucracy (the administrative principle) and the rise of the occupational principle; pointing to the increasing need for knowledge as the impetus for this shift of the organizing
principle of work. Producer and social services provide holders of such
jobs with more opportunities for decision-making. Singelman and
Mencken (1992: 841) present data from five countries to show that producer and social services provide employees with more opportunities for
decision-making as well as creating demand for higher status occupations.
Such professionals have greater autonomy within the organization and are
driven by peers' definitions of standards. Lury (1996) has linked the presence of professional autonomy and the increasing significance of the peer
group and shared professional culture to attempts at identity building
against other class fragments, with lifestyle and consumption as the
primary elements for cultivating this identity. More importantly, Lury
(1996: 105) argues that consumption and taste varies according to the
nature of work: bureaucrats have an unremarkable culture, public sector
professionals display an aesthetic lifestyle and high culture, while private
sector professionals have a post-modern lifestyle.

We have also shown earlier how the advanced services class command
much higher salaries. Thrift and Leyshon's (1992) *In the Wake of Money* is
about the professionals in financial services, the money they make and the
money they spend, in the reproduction of life chances. Thus, Thrift and
Leyshon (1992: 295) note that the rich spend 'their money in tasteful
ways, that is, in ways which would accrue social and cultural capital for
themselves and their children'. This appears to be a strategy of the rich,
whether they amass their wealth through financial services, or any other
economic activity.

This observation, while not false, is however not very relevant in the
context of this chapter as money or wealth is the explanatory factor,
rather than the nature of services work.

Service work and consumption

A more relevant line of argument between services work and consumption
evolves around the notion of how consumption supports the nature of services work. Lash and Urry (1994: 164) suggest that the specialized, person-
alized and knowledge intensive nature of advanced services require a high
amount of cultural capital on the part of the service provider. This level of
generality is not very convincing and we need to break this down further
by looking at the actual work of services and its link to cultural capital.
There is some evidence that the services involving design have a greater

tendency to design personal appearance as all part of the pitch. Nixon (1997: 1167–217), for example, provides an account of male executives in advertising that demonstrates the need to style the self at work, as self representation through dress codes. A clear example of the link between the work and consumption comes from a comment made by a creative director from a Hong Kong advertising firm:

> Of course, it is important to dress decently for my job, because what I sell are ideas, my brains ... It is not the same as selling a fax machine or a microwave oven, where the consumer can check out the product, and hold it in his hands and look at it before deciding to buy it. For advertising, if you cannot show the client you have good taste, how can they trust you enough to put their products into your hands?
>
> (field material from Chan, 2000: 121–2)

It is easy to extend this argument to the need to sculpture the body through a fitness-exercise lifestyle regime, all the more necessary because of the sedentary nature of creative work (Florida, 2002: 174, 176, 177).

But can this observation of the link between design services and consumption be extended to other kinds of service professionals? The key point, I think, rests in the quotation above, that the tendency of services to focused client-based project work rather than material products create in service professionals a greater need to develop cultural capital as the foundational asset in the management of diverse clients in national as well as international settings. The importance of developing a personal element in the relations with clients is highlighted in the following quote from Bloomfield and Best's (1992: 544) study of IT consultants:

> I think that the personal element of consultancy is absolutely paramount, its very much a personal thing whereby you develop a relationship with somebody. It may be only over a very short time, but you seem to be talking the same language, he [the client] believes that you understand what his issues are, that he can rely on you to deliver something which adds value to his organization and at the end of the day they trust you, it's very much about that.

It is this need to establish the personal element in client relations which require service professionals to draw on cultural capital.

Drawing on Hutton and Ley's (1987) study of the intensity of client-based relations in the work organization of services, Whimster (1992) has suggested that the 'new professional seeks to maximize her knowledge through networks of contacts, both within and outside the firm'. There are three contextual reasons why the new professional seeks to maximize the building of such networks. The first reason has to do with the tendency that the job mobility of the professional itself is of value to firms since this allows them to tap into comparative insights from previous jobs and projects. Moreover, the client-based nature of producer-services has also meant that professionals often bring their clients with them. The third reason has to do with what Smith and Rubin (1997) term as 'hyper-exploitation'. The hyper-mobility of capital and, by implication, jobs has meant the increase in job insecurity. In such an economic scenario, while Smith and Rubin (1997: 306) show that even educated and highly skilled workers are vulnerable to the same threats faced by their less skilled counterparts, the acquisition of information and upgrading of skills seem to be a rational strategy rather than the commitment to a single employer. For all these reasons, creating individual and portable assets and the enhancement of personal capital both in terms of skills and in terms of social networks have become life-long projects of professionals, part of which includes a link to consumption practices.

What differs are the different styles within different service professions. Advertising, as Nixon (1997) notes, relies on outward appearance, while merchant banking, as McDowell and Court (1994) note, depends on class and educational background. On this note, advanced service professionals are also more likely to have as Warde (1994: 893) suggests, 'a sharing style, have considerable access to relevant knowledge, or a clear sense of rules and source of belonging'.

In stating that 'a reflexive mode of consumption demands a more self-conscious mode of production', Zukin (1991: 259) points to two complementary sets of forces linking producer services to consumer services. The shift to a more reflexive mode of consumption is as much an outgrowth of an advanced service economy that is based on design and ideas as it is an outcome of its work force, that engages in consumption as a quest for meaning (is this authentic?), as an expression of value (going green, helping local producers, etc.), and as a strategy for social differentiation. Zukin's (1991: 202) point is also that these consumption tendencies are being matched by a 'critical cultural infrastructure' staffed by another set of service workers in social and personal services providing an 'aesthetic critique that facilitates upscale consumption'. Thus, in Zukin's work, we find the clearest links between creative segments of producer services and their counterpart in consumption services.[3] Post-fordist systems enable the commodification of taste linking advanced service professionals as producers and consumers (Lury, 1996: 94–5; Lash and Urry, 1994: 164). As a result, as Lash and Urry (1994: 164) point out 'this core of upper professionals possess highly valued intellectual resources, is employed in powerful organizations, and gets involved in the cosmopolitan side of economic and cultural life in the city'.

Cities have therefore been transformed by the concentration of services through the concentration of wealth and its consumption possibilities. Whether through the influence of professional cultures and/or the need to style the self, or enhance life chances, the propensity of service professionals to embrace consumption as a central mechanism to fashion identity make them key players in post-fordist production systems, where the system flexibility allows for such effective demand to be incorporated. Service middle-classes therefore play crucial roles in influencing production processes.

Building the Asian city for the service class

Amin (2001: 1238) points out the need to examine embedded institutional (social) practices because these condition economic action by giving life to, and regulating economic activity. This link between the economic, the social and the spatial is fundamental and represents the economic foundations of the city. By examining the literature on class and consumption and looking at advanced services workers and their consumption habits, we are able to identify a pattern between the growth of producer services and their implications for consumer services. This relationship comes closest to what Parr (2002: 720), in a recent review of agglomeration, terms as urbanization economies, involving the incentives of dissimilar firms co-locating to take advantage of specialized and extensive services and a well-developed infrastructure. There are, however, two important distinctions. There can be clear external economies of scale and scope within the producer services, within consumer services and between producer services which are the result of firm spatial-economic practices (see Figure 5.4). While urbanization economies refer to the benefits to firms of co-location, what is described here is a critical appendage, that of worker consumption practices and the socio-economic circuits created by these behaviours. Thus, while the concept of agglomeration is useful in understanding inter-firm dependencies as a basis for clustering, within a service economy setting, inter-firm dependencies must be broadened to include consumption practices and how these interact with the production process. The other distinction is the need to incorporate a more evolutionary perspective in the development of such embedded linkages. As the service economy grows, the pattern is likely to be an evolution of mixed producer and consumer services, activities which benefit both producers and consumers. Production carries with it a set of social relations of different fluidities, and it is a subset of these relations, which tend to become entrenched over time. This entrenchment over time is what gives places their uniqueness, as Massey points out (1984: 117–18).

The consequences and interdependencies created by the service economy will prompt governments to act in different ways. They must enact strategies of solidarity, holding together the diverse cosmopolitan urban population, working against a growing local sentiment which desires the economic activities brought by multinational service firms, but not the workers that may come with such activities. In building the city for

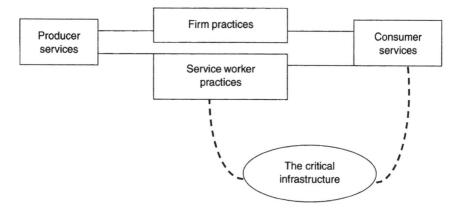

Figure 5.4 Nature of embedded linkages.

the service class, governments operate at two levels: building an infrastructure for hosting services[4] and also providing a set of cultural and lifestyle amenities as a means of attracting and anchoring the international mobile class.

Notes

1 Baum (1999: 1114) concluded that 'the most important finding reported here is that there is no social polarization in the sense portrayed in the work by Sassen and others, but instead strong trends towards professionalization and growing middle- and upper-income strata are evident'.
2 Lash and Urry (1994: 163) include software, personal finance, education and health, business services, the culture industries and parts of hotel, catering and retail services in this category.
3 A common way of making the distinction between producer and consumer services is the nature of the market for these services. Thus, producer services involve the services which are demanded by other firms (accounting, advertising, etc.), while consumer services (social and personal services in medicine, education, tourism, retail and hotel sectors) are used by final users. There are also intermediate categories which involve mixed producer and consumer services such as transport services (Williams, 1997: 2–5).
4 There is a growing literature pertaining to the building of such infrastructures, for example, the development of flagship projects where renowned architects and design teams employed to build trophy buildings or superlative projects (worlds tallest, largest, etc.) as anchors of such urban promotional attempts to attract global capital (Olds, 1995: 1713, 1719; Hubbard and Hall, 1998: 7).

References

Amin, A. (2000) 'The economic base of contemporary cities', in Bridge, G. and Watson, S. (eds), *A Companion to the City*. Oxford: Blackwell.

Baum, S. (1999) 'Social transformations in the global city: Singapore', *Urban Studies*, 36(7), 1095–117.

Bloomfield, B. P. and Best, A. (1992) 'Management consultants: systems development, power and the translation of problems', *Sociological Review*, 40(3), 533–60.

Body-Gendrot, S. (2000) *The Social Control of Cities*. Oxford: Blackwell.

Chan, A. H. N. (2000) 'Middle-class formation and consumption in Hong Kong', in Chua, B. H. (ed.), *Consumption in Asia: Lifestyles and Identities*. London: Routledge.

Daniels, P. W. (1998) 'Economic development and producer services growth: the APEC experience', *Asia Pacific Viewpoint*, 39(2), 145–60.

Daniels, P. W. (2001) 'Globalization, producer services and the city: is Asia a special case?', in Stern, R. (ed.), *Services in the International Economy*. Ann Arbor, Michigan: University of Michigan Press.

Dicken, P. and Kirkpatrick, C. (1991) 'Services-led development in ASEAN: transnational regional headquarters in Singapore', *Pacific Review*, 4(2), 174–84.

Dunning, J. H. and Norman, G. (1987) 'The location choice of offices of international companies', *Environment and Planning A*, 19, 613–31.

Esparza, A. and Krmenec, A. J. (1994) 'Producer service trade in city systems: evidence from Chicago', *Urban Studies*, 31(1), 29–46.

Florida, R. (2002) *The Rise of the Creative Class*. New York: Basic Books.

Friedson, E. (1994) *Professionalism Reborn*. Cambridge: Polity Press.

Hamnett, C. R. (1994) 'Socio-economic change in London: professionalization or polarization', *Built Environment*, 20, 192–204.

Holston, J. and Appadurai, A. (1999) 'Cities and citizenship', in Holston, J. (ed.), *Cities and Citizenship*. Durham: Duke University Press.

Hubbard, P. and Hall, T. (1998) 'The entrepreneurial city and the "new urban politics"', in Hubbard, T. and Hall, T. (eds), *The Entrepreneurial City*. New York: Wiley.

Hutton, T. and Ley, D. (1987) 'Location, linkages and labour: the downtown complex of corporate activities in a medium size city, Vancouver, British Columbia', *Economic Geography*, 63(2), 126–41.

Ko, C. F. and Tzeng, R. (2002) 'Mapping global manufacturing and services in Taiwan', paper presented at the American Association for Chinese Studies, USC, USA October 26–27.

Lash, S. and Urry, J. (1994) *Economies of Signs and Space*. London: Sage.

Lury, C. (1996) *Consumer Culture*. New Brunswick, New Jersey: Rutgers University Press.

Massey, D. (1984) *Spatial Divisions of Labour*. London: Macmillan.

Mayer, M. (1999) 'Urban movements and urban theory in the late 20th century', in Body-Gendrot, S. and Beauregard, R. A. (eds), *The Urban Moment: Cosmopolitan Essays on the Late 20th Century City, Urban Affairs Annual Reviews*. Thousand Oaks, California: Sage Publications.

McDowell, L. and Court, C. (1994) 'Missing subjects: gender, power and sexuality in merchant banking', *Economic Geography*, 70(3), 229–51.

Nixon, S. (1997) 'Advertising executives as modern men: masculinity and the UK advertising industry in the 1980s', in Nava, M., Blake, A., MacRury, I. and Richards, B. (eds), *Buy this Book: Studies in Advertising and Consumption*. London: Routledge.

Olds, K. (1995) 'Globalization and the production of new urban spaces: Pacific

Rim mega projects in the late 20th century', *Environment and Planning A*, 27, 1713–43.

Parr, J. B. (2002) 'Agglomeration economies: ambiguities and confusions', *Environment and Planning A*, 34, 717–31.

Perry, M. (1992) 'Promoting corporate control in Singapore', *Regional Studies*, 26(3), 289–94.

Rodan, G. (1996) 'Theorising political opposition in East and Southeast Asia', in Rodan, G. (ed.), *Political Opposition in Industrialising Asia*. London: Routledge.

Sassen, S. (2001) *The Global City: New York, London, Tokyo*. Princeton, New Jersey: Princeton University Press.

Sen, S. B. and Sen, J. (1997) 'The current state of knowledge in international business in producer services', *Environment and Planning A*, 29, 1153–74.

Singapore Department of Statistics (2002) *Income Distribution and Inequality Measures in Singapore*, online: http://www.singstat.gov.sg/papers/seminar/income2000.pdf.

Singelmann, J. and Mencken, F. C. (1992) 'Job autonomy and the industrial sector in five advanced industrial countries', *Social Science Quarterly*, 73(4), 829–43.

Sirat, M. (2000) 'Globalising Kuala Lumpur and the strategic role of the producer services sector', *Urban Studies*, 37(12), 2217–40.

Smith, B. T. and Rubin, B. A. (1997) 'From displacement to reemployment: job acquisition in the flexible economy', *Social Science Research*, 26, 292–308.

Storper, M. (2000) 'Lived effects of the contemporary economy: globalization, inequality and the consumer society', *Public Culture*, 12(2), 375–409.

Tao, Z. and Wong, R. Y. C. (2002) 'Hong Kong: from an industrialised city to a centre of manufacturing-related services', *Urban Studies*, 39(12), 2345–58.

Taylor, P. G. C. and Gane, N. (2002) 'A geography of global change: services and cities, 2000–01', GaWC Research Bulletin 77, online: http://www.lboro.ac.uk/gawc/rb/rb77.html.

Thrift, N. and Leyshon, A. (1992) 'In the wake of money: the City of London and the accumulation of value', in Budd, L. and Whimster, S. (eds), *Global Finance and Urban Living*. London: Routledge.

Warde, A. (1994) 'Consumption, identity-formation and uncertainty', *Sociology*, 28(4), 877–98.

Whimster, S. (1992) 'Yuppies: a keyword of the 1980s', in Budd, L. and Whimster, S. (eds), *Global Finance and Urban Living*. London: Routledge.

Williams, C. C. (1997) *Consumer Services and Economic Development*. London: Routledge.

Yeung, H. W. C., Poon, J. and Perry, M. (2001) 'Towards a regional strategy: the role of regional headquarters of foreign firms in Singapore', *Urban Studies*, 38(1), 157–83.

Yu Zhou (2000) 'Bridging the continents: the roles of Los Angeles Chinese producer services in the globalization of Chinese business', in Yeung, H. W. C. and Olds, K. (eds), *Globalization of Chinese Business Firms*. Houndmills, Basingstoke: Macmillan.

Zukin, S. (1991) *Landscapes of Power*. Berkeley, California: University of California Press.

6 The bazaar and the normal

Informalization and tertiarization in urban Asia

Michael Leaf

Introduction: the benefits and limits of comparative analysis

The purpose of this chapter is relatively straightforward – to bring questions of informality and informalization to bear upon the discussion of socio-economic change, particularly tertiarization, in the cities of Pacific Asia. Nonetheless, the path I will follow is somewhat circuitous. This is in part because of the contentious and convoluted state of discussions around the concept of informality, although it is complicated as well by the diversity of conditions and developmental experiences of the cities of Pacific Asia[1] and the question of how, in particular, tertiarization may be interpreted in these varying contexts. One stream of thinking, which in some senses propels the main line of inquiry of this volume, is the question of how tertiarization, that is, the continuing growth and differentiation of service industries within the socio-economic make-up of the city, is linked to patterns of progressive modernization. One line of argument from the study of informality and informalization brings a direct critique to bear upon the assumptions of stage-wise modernization, whereby societies progress from traditions of primary production, through the secondary sector of industrial manufacturing to higher value added – tertiary – forms of economic activity as characteristic of 'post-industrial' societies (see Chapter 3). Yet many cities of the world, particularly those of the erstwhile 'third world', have long been characterized by high levels of tertiarization – such as small-scale trading and services – despite the lack of a developed intermediary manufacturing phase, a phenomenon which the study of informality has sought to explain.

In the following sections of this chapter, I will first review the basic relevant aspects of the informality 'debate' as it has been put forward in the literature on development and, from this, discuss the different trajectories of analysis that have emerged out of this debate and how they might be relevant to thinking about urbanization in Asia today, particularly with regard to the nature of local state-society relations. Following this theoretical overview, I will examine some illustrative examples of how informality is articulated under current conditions of urbanization in Asia and conclude with some thoughts on how we might think about the place of informality in the context of the growing systemic linkages within the urban system of the region.

In recent years, theories of informalization have come to figure increasingly in discussions of globalization processes and their implications for capitalist production. Consideration of globalization is also greatly configured by the urban milieu, though how these processes are manifested in any particular setting will be highly dependent upon differentiating social and economic conditions in specific cities. First, however, I think it is pertinent to remind ourselves of how and why we construct our analytical models, here regarding urban socio-economic change. At heart, such models are intended as a basis for comparative analysis, comparative temporally or geographically – to examine changes across time, or between locales, or in some complex combination of the two – in the attempt to deal with intricacies of differences between places and time periods. Every city is unique, and unique to each particular point in history, though if we only emphasize uniqueness we undermine the possibility for learning through comparative study. The purpose of model-building, or theory-building, is therefore to arrive at an interpretive understanding of 'what's going on here and now', potentially in the interest of articulating supportive policy change or intervention. The challenge in this is not to be too easily deceived by superficial similarities, while not throwing up our hands in despair at not being able to identify or understand the most useful commonalities. This caveat may be particularly useful when thinking about the assumptions we make regarding modernization, as ongoing changes in the current phase of globalization pose challenges to these assumptions.

From informal sector to regulatory informality

This concern for the epistemology of theorizing is relevant to the putative 'discovery' of the 'informal sector' in the early 1970s. The original articulation of this concept – first in the work of anthropologist Kenneth Hart in Ghana (Hart 1973), and further regularized by the International Labour Organization in studies in Kenya (ILO 1972) – can be seen in retrospect as an effort to adapt the observed realities of African urbanism to prefigured concepts of the proper ('formal') structures of urban employment (Bromley 1979; Moser 1984). How were people 'employed' or 'unemployed', and with such observed high levels of unemployment as existed in these cities, how were people able to manage their livelihoods? The problem lay in the pre-conceived categories of employment utilized by ILO observers. Hence, the concept of 'under-employment': urban workers could undertake the sorts of proper, 'formal' employment if only such jobs existed. As such work was not broadly available in these contexts (contexts of 'over-urbanization', itself a normatively loaded concept), workers must rely on lower level, or even clandestine, forms of livelihood generation. The informal sector was thus a residual component of labour absorption by the urban economy, with the argument that in the absence of sufficient jobs in the 'formal' (or normatively expected) categories of employment, labour absorption was occurring through some sort of 'other' process, the 'informal sector', the 'shadow economy'. Thus, livelihood generation activities were not defined by what they *were*, but what they *were not* – that is, *not* in adherence to prevailing standards of proper employment.

In the ILO's formulation, the informal sector was defined by seven characteristics, each of which was counterposed to an opposite characteristic which defined the formal sector. Informal sector firms and activities were thus characterized by:

1   ease of entry;
2   reliance on indigenous resources;
3   family ownership of enterprise;
4   small scale of operation;
5   labour-intensive methods of production and adapted technology;
6   skills acquired outside the formal school system;
7   unregulated and competitive markets (ILO 1972; see also Thomas 1992).

In practical terms, this emphasis on symptomatic features for defining the basic parameters of the informal sector was highly problematic, as one can easily find contrary examples – such as the capital-intensive, small-scale family enterprise – which defy inclusion in either 'sector'. Nor, as subsequent research has come to show, is the informal sector particularly informal, in the sense of lacking organization. The backward and forward linkages of even small-scale enterprises may be highly dependent on intricate social networks, raising questions as to the practical meaning of 'ease of entry'. Despite criticism that the 'informal sector' as defined in this manner is neither 'informal' nor a 'sector', the seven characteristics of the ILO have nonetheless shown remarkable persistence as a basis for categorization and analysis of informal activities (see Perlman 1987, as one example).

What has too readily been disregarded here is that this inherently dualistic definition of the informal sector was in essence an attempt to fit a long-standing discourse on the tensions between tradition and modernity into an over-arching theoretical/normative perspective on what constitutes a 'proper' trajectory of socio-economic change or development. In other words, the formal/informal dichotomy essentially updated earlier (colonial and early post-colonial) attempts to understand socio-economic dualism, as seen, for example, in the work of Boeke (1942), Furnivall (1939, 1948) and Geertz (1963). These studies addressed the tensions between tradition and modernity, or the tensions between indigenous socio-economic formations (such as peasant agriculture and traditional notions of statecraft) and imported or imposed forms (urbanization, industrialization and the modern institutional structures of the colonial state), in the interest of furthering the modernization of the newly independent nations. In this respect, one could point to the parallels between the ILO's dualism of informal and formal and Geertz's (1963) distinction between the traditional 'bazaar economy' and the modern 'firm economy' in his study of two Indonesian towns.

Critical writing on the informal sector challenged such dualisms of progressive modernization, portraying the informal sector instead as the persistent expression of Marx's petty commodity mode of production, not separate from dominant capitalist production (the erstwhile formal sector), but fully integrated with and dependent upon it (Moser 1984). Such an alternative conceptualization also challenged the notion of the transitional nature of the informal sector. If the continuing existence of petty commodity production in effect served as a form of subsidy to capital by reducing the costs of reproduction of labour, what interest was there on the part of socio-economic elites to do away with (or to otherwise 'formalize') the informal sector?

In both formulations, dualist and petty commodity production alike, the question of state regulation came to be a central concern. On the one hand, excessive regulation, as proffered by the stereotypically bloated and self-interested bureaucracies of the post-colonial state, was portrayed as the primary obstacle to formalization, with rapacious state agencies thus holding back the entrepreneurial promise of indigenous petty capitalism, a point famously argued by DeSoto (1989). From the Marxian perspective, however, regulation was not an inadvertent problem restraining formalization and thus the advancement to properly regulated forms of production, but rather the mechanism by which the uneasy tensions between formal and informal, that is, between capitalist and petty commodity modes of production, could be held in check. The seemingly obscure question of why a state would create a regulatory framework that it had no intention of enforcing is thus clarified (Gilbert 1990). Such regulatory structures give semblance to the formal operations of capitalist production, while their uneven application to the array of otherwise illegal activities which characterize the Third World city – from unregulated land occupation and real estate development, to petty trading, to the myriad informal forms of production and services – allow for a higher degree of governmental flexibility than would otherwise be the case. The uneven application of law, as seen in the persistence (and reproduction) of informality is thus not arbitrary; it is the very means by which society is governed under such conditions of socio-economic polarization as characterize the rapidly growing cities of post-colonial nations.

The question of regulation also provides the conceptual point of connection between informality as a putative 'Third World' condition and the changing nature of capitalist production in the core regions of the world economy, a connection explored in a series of studies brought together by Portes *et al.* (1989), and furthered through the writings of those working in a world-systems framework (Tabak and Crichlow 2000). The tolerated illegality associated with Third World informality has thus come to be seen as an offshoot of the regulatory restructuring of post-Fordism, in terms of such changes as the expansion of sub-contracting, the growth in temporary and part-time employment and the overall weakening or circumvention of labour protection laws (Sassen 2000). The reconfigured social contract of globalizing capitalism thus prompts the growth of 'informalization', seen as the 'unmaking of once formalized relations' (Tabak 2000: 6) in the core regions, in partial contrast to the assumptions of the continuing reproduction of more 'traditional' conceptualizations of informality in the peripheral regions of the world. In both cases, informality is an outcome of the functions of the state: in core regions of a weakening regulatory state vis-à-vis the interests of footloose capital, and in the periphery, of the persistence and reproduction of weak states vis-à-vis the interests of local elites. These two streams of analysis are, of course, interrelated, as seen both in the observation that the ongoing loss of the rural labour reserves, as is characteristic of the continuing rapid urbanization of the erstwhile Third World, accelerates the breakdown of Fordist relations of production (Tabak 2000: 8), and in the idea that through globalization we may conceive of an overall model of capitalist restructuring consisting of the 'unmaking' of the formal in core regions, the migration of sites of production from core to periphery and the continuing limits to labour absorption in the formal, or state-regulated, urban economies of the periphery.

The obvious problem here, in this attempt to articulate informality within a purportedly global model of capitalist relations, is that there is too much diversity encapsulated within a single concept. Although this has been a critique of the informal sector since the beginning, the essence of this critique has usually been with regard to the diversity of types of production, as the informal sector, broadly defined, contains both the secondary production of small-scale manufacturing as well as a wide range of tertiary services (and arguably as well, certain types of primary production such as semi-clandestine urban agriculture). An alternative formulation would stress not the outputs of production, but the relations of production, and in this respect, one can turn to a categorization by Portes (1983; see also Broad 2000), which distinguishes between three interrelated forms of production, the informality of each determined by the uneven application of state regulation:

1   direct subsistence, which includes various types of production for self-consumption;
2   petty commodity production, whereby self-employed individuals produce or commercialize goods and services for the market;

3   backward capitalist production, including both small enterprises utilizing unprotected wage labour and disguised wage-labourers whose work is subcontracted out from larger firms.

This conceptualization redraws the boundaries around the various forms of livelihood subsumed under informality, and is useful for making the connections between informalization as an outcome of capitalist restructuring in core regions and the persistence of informal relations of production in the face of growing Third World modernization, as seen, for example in the range of efforts to upgrade, modernize or formalize small and medium scale enterprises throughout the cities of the developing world.

A critical point for understanding the broad implications of informality and informalization is what may be labelled as the 'social embeddedness' of informality. The developmental view of the informal sector stresses the value of economic relations, with efforts at formalization undertaken as a means toward legitimizing sub rosa activities in the interest of such socially beneficial goals as the improvement of basic working conditions, improved provision of public services through more equitable financing (achieved through the taxation of small enterprises, among others), clarified property rights resulting in more regularized contractual relations and, for the owners of small enterprises, improved access to credit sources, which thereby encourages growth, job creation and overall economic dynamism. In contrast to such economistic thinking, it can be argued that the 'informal' – that is, the ability to operate in a sub rosa manner in the first place – is first and foremost an expression of social relations, rather than being narrowly premised on an economic rationale.

An early assumption regarding the informal sector in developing countries was that the informal served as an alternative to the formal sector, with actors driven into it by necessity or lack of alternatives (akin to Portes's 'subsistence' production), though this view by no means encapsulated relations or practices of informality in urban economies, especially not the most dynamic of them. Empirical studies (as reviewed by Moser 1984) belied such assumptions and raised the point that if informality was actually a preferred condition for many urban dwellers, it could not be wished away through the continuing expansion of formal employment. This line of argument has been bolstered by findings in more advanced economies that informal practices often tend to build from the formal, in that opportunities for informality arise out of the personal linkages of those in formal employment (Harding and Jenkins 1989: 44). In this sense, formal relations of employment are seen as pre-requisite for successful informal relations, rather than a truly alternative livelihood strategy. Such observations, however, are difficult to generalize as they are highly (and locally) coloured by social relations as articulated through the linkages of the informal economy.

This line of analysis necessarily brings us to what might be termed the 'governance' question, that is, that informality is not so much 'unregulated', broadly speaking, as it is 'unregulated by the state', or in other words by the formal institutional structures of government. This view of the informal as socially regulated (despite being ostensibly unregulated) is also useful in clarifying the boundary between informality and criminality as two forms of illegality, as it is the degree of social acceptance, or social sanction, which distinguishes the illegality of the informal from the illegality of the criminal (Thomas 1992). Clearly there is a cultural component to this, in that it brings into the equation the basic socially accepted understanding of the functions (and thus legitimacy) of the state, an understanding that in most instances is derived over time through the political discourse of the locality, if not the nation. Such an interpretation of informality as, first, a social phenomenon, and second, a manifestation of state-society relations, forces us to think about models of the state, such as that of Migdal (1988), which distinguishes between state penetration and state effectiveness as an outcome of the relative strength of societal forces vis-à-vis the state in the developing countries of the post-colonial world. Here the culturalist argument comes to bear in the form of personalism as a distinguishing factor of state-society relations in contradistinction to liberal ideals regarding the efficacy of the rule of law in determining the fundamental 'rules of the game' for society. For our purposes, this points to the need to see informality as a critical factor in social relations in those societies characterized by persistent personalistic rule, long seen as a traditional characteristic of state-society relations throughout much of Asia.[2]

Current directions

To summarize the argument thus far: although the idea of informality originated out of dualistic roots as a problem posed within an overall framework of modernization, the concept has evolved into a critique of the regulatory function of the state vis-à-vis social and economic relations, thus prompting the need to examine questions of the persistence or growth of informality from the perspective of relations and traditions of governance. The idea of informality thus invokes questions regarding the nature of the state, whether in reference to the shifting relations between state, society and capital in the ongoing post-Fordist restructuring of advanced capitalist systems, or to the potentially more culturalist notions of personalistic governance as they pertain to developing Asian polities.

In general, one may distinguish between three trajectories of analysis and practice that have developed out of the debates regarding the conceptualization of informality. First, one can point to the persistence of modernist thinking in the notion of 'formalizing the informal', a mode of thinking which could be said to dominate development practice despite the many theoretical challenges to dualistic conceptualizations of informality (Sanyal 1996). Here one may discern the developmentalist desire that the continuing expansion of state regulation will eventually bring all socio-economic relations into its regulatory ambit, indicative of an earnest belief in the efficacy of the state and the viability of rule of law to fairly encompass all societal formations. Examples of this thinking may be found in the continuing and growing interest among developmental agencies, if not national governments, in the developmental potential of small and medium enterprises (if not micro-enterprises as well), as a means for economic development that is an alternative to more conventional efforts to encourage formal sector expansion (Turner 2003). This emphasis on formalization as a development strategy tends to set aside the unresolved (and unresolvable?) debates of the informal sector literature around questions of 'evolution vs. involution', that is, with regard to the potential dynamism of small-scale economic activities, and further obscures the distinctions between informality as urban subsistence and unregulated petty capitalism. Here one might extend this point to encompass the current global fascination with micro-credit as a development strategy, as much of the actual practice of the 'micro-credit sector' is comprised of efforts to properly structure and regulate 'traditional' socially-embedded practices of group savings and lending which are perhaps found in every part of the world where the modern, formal banking system has not yet penetrated.[3]

In contrast to the hopeful impulses of persistent modernism, one can find a more critical view of the informal in the post-modern, or at least 'post-developmentalist', literature, couched in the critique of the long-term viability of the modern state in attempting to regulate what should be seen as somewhat 'natural' tendencies of indigenous social formations. In the writings of social theorist Serge Latouche (1993), what we identify as the 'informal' becomes the baseline for understanding how socio-economic relations are structured, while what we now see as the 'formal' is transformed into somewhat of an historic aberration, paralleling the expansion and current hegemonic dominance of the modern nation state. Thus, the academic (and policy) discovery of the informal sector in the 1970s was simply an early acknowledgement of cracks in the facade of the overall 'project of modernity'. The trajectory of development, from this perspective, is not the further strengthening or rationalization of the state in the service of an advancing modernity, but rather its eventual dismantlement. What we interpret as the informal is thus a direct challenge to the state – though a perhaps unselfconscious 'everyday form of resistance' – and, beyond that, to the total enterprise of modernity. To talk of how best to 'formalize the informal' as a way of resolving this dualism thus becomes entirely irrelevant. Almost by definition, any talk of policy becomes an abstraction in the long run, in that policy exists only within the structures of formal governance. Latouche returns to the metaphor of modernity as a luxury liner adrift in the ocean, and policy development, since it originates from within the boat, can at best plug a few leaks. It cannot prevent the ship from foundering and breaking up on the shores of the 'archipelago of the informal'. Thus the expansion, or reassertion, of informality is more properly seen as the retreat of modernity and all that it entails, including the presumption that the inherent logic of the informal is not inconsistent with the profit motives of the marketplace.

A third trajectory of theoretical development is derived from world systems thinking, which sees informality as a characteristic linked to the changing relations between capital and society as mediated through the regulatory functions of the state. It is argued that in the current period the growth and entrenchment of informality is a direct outcome of post-Fordist shifts in the structure of production as the increasingly transnational disarticulation of firms under globalization is changing the basic contractual relationships between capital and labour (Sassen 2000). This line of analysis poses the possibility of a cyclical understanding of informality, with informalization and casualization of labour seen as 'not anomalies, as is often thought, but . . . integral to periodic restructuring in the face of cyclical downturns and systemic impasses that hinder capital accumulation' (Broad 2000: 24). In previous historic periods, the expansion and contraction of informality may be seen in the changing relations between city and countryside, with cyclical downturns prompting periods of increased ruralization of production through processes of 'putting-out' (Tabak 2000), an observation which emphasizes the systemic nature of urban-rural relations in a way which is perhaps instructive for thinking about centre-periphery relations in the current phase of capitalist globalization. Such an interpretation also paints a very different picture of the historic position of informality over the *longue durée*, when compared to Latouche's analysis of 'the informal' as the historic base upon which we examine the coming and going of the 'project of modernity'.

It may also be of passing interest that in thinking about informality we find a concept that, having originated out of the analysis of conditions in the periphery – the 'Third World' – is now finding increasing relevance to our understanding of relations throughout the world economy. In the following sections we look at specific cases of informality in Asian cities in the interest of understanding both the underlying conditions that allow for the persistence (if not growth) of informality, and the value of taking a systemic view of the place of informalization in the regional structures of Asian urbanization.

Informality and tertiarization in Asian cities

Generalizations do not come easily to the wide array of urban conditions across the Asia-Pacific region. One can, of course, point to the commonality of geographic proximity and – arguably – to the growing interconnectedness of urban economies across the region (Smith 2001), which is seen as rooted to an extent in the common historical experiences with Western colonialism well into the last century, and reproduced beyond this through the exigencies of more recent capitalist globalization. To the extent that one may discern a convergence in forms or structures within the urbanization patterns of the region, it is usually couched in terms of the presumed homogenizing influences of trans-local or global connectivities (see Dick and Rimmer 1998 as one example), and it is in this conceptual context where discussions of tertiarization tend to be located. An understanding of the persistence and growth of informality in the cities throughout the region presents a challenge in this regard. First, and undoubtedly the most obvious point, it is in the nature of the informal to be unenumerated, if not clandestine, and thus one may search in vain for a statistical basis for properly evaluating the true extent of informalization within any particular locality, let alone for developing a framework for comparison of cities throughout the region. Second, there is the more fundamental conceptual issue of the great diversity of forms of informal relations, even within one set of socio-economic activities. This diversity may be understood as a direct consequence of the lack of formalized, and thus standardized, institutional structures, or as Tsai describes it, the tendency away from 'institutional isomorphism', derived in practice from the 'micro-logic of institutional choice and market segmentation at the grassroots level' (Tsai 2002: 6). The highly local nature of the informal is thus indicative of its social embeddedness, and raises the question of why, if informality is intrinsically derived from local conditions, it is observed to be a growing global phenomenon.

As described above, the perspective of the world systems theorists presents us with an explanation which lies in the evolving structures of global capitalism, in particular the regulatory restructuring associated with the transition out of Fordist relations between state, capital and society. In the context of Asian polities, one may see the congruence here between these trends and the persistence of traditionalist (or neo-traditionalist, in recognition of the re-articulation of older relations in new socio-economic settings) governance practices. From the perspective of institutional economics, the central concern here is with regard to property rights and how they are secured, particularly under conditions of weak guarantees through the objective rule of law. By way of illustration, one may look at the following two examples of informality in urban contexts in Indonesia and China – examples that relate to such services as real estate and local enterprise finance.

Informal real estate in urban Indonesia

The extent of informality in the residential environments of Indonesian cities has been well studied, with perhaps the most thorough survey of housing development in urban Indonesia indicating that nearly two-thirds of urban housing by the late 1980s was built on lands that were not fully legal, despite the promulgation of a land rights law in 1960 which was intended to unify and codify land rights across the country (Stryuk *et al.* 1990). Explanations for such persistent disparities between formally registered and unregistered lands in Indonesian cities lie in part in problems of insufficient institutional capacity for land registration despite the many efforts for instituting the modern land rights system over the decades (Henssen 1988). Beyond such 'technical' analyses, which focus on the pragmatic challenges of registering urban residential properties throughout the Indonesian national territory, one can also look to the local politics of land rights claims to understand the persistence of land rights informality in Indonesian cities. Understanding how the major proportion of urban households live in one way or another in contravention of formal regulations for property ownership or housing construction requires an assessment of, first, how the dualistic roots of urban property ownership derive from land law of the colonial period, and second, how such dualism has persisted or been re-invented under present conditions.

In the context of the capital city of Jakarta, it may be seen how competition between traditional forms of land ownership and an imported Dutch land rights system had led to an overlayering of land rights by the end of the colonial period (Leaf 1993). In the colonial period, land rights claims through the formal Dutch system, which obtained to Europeans, foreign Asians and the few wealthy Indonesians who had obtained 'same level' (*gelijkstelling*) legal status, took precedence over what were understood to be the 'traditional' (*adat*) claims of the natives, even when (or especially when) such claims were applied to the same parcels of land, a situation which was common across the territory which now constitutes the city of Jakarta. Dismantlement of this dualistic structure was an explicit goal in the formulation of the Basic Agrarian Law (BAL) of 1960, which revoked all previous claims in favour of a unified system of land rights claims. However, the uneven application of this law has, over time, led to the development of a new form of dualism, with some lands fully registered within the formal system, while the majority of lands are still technically 'not yet registered', although a sense of tenure security is still obtainable through what are now archaic land rights categories.

The question therefore arises as to how urban property owners are able to secure their claims in the absence of clear adherence to formal legal structures, a question that can be re-phrased in terms of the underlying source of authority in an extralegal setting (Leaf 1994). Transactions on lands that have not yet been surveyed and registered under the provisions of the BAL are nonetheless still informally recorded at local government offices, though such recording does not enter into the purview of the National Land Agency, the agency charged with overseeing the application of the BAL. This informal 'system', whereby rights are essentially guaranteed by local level administrators despite being in contravention of formal statutes, may be described as a form of 'detached, top-down authority' (Leaf 1994: 14), a situation akin to what has been described in another very different setting (of Caracas, Venezuela) as 'official because those who prevent and resolve conflicts are public servants, (yet) informal because it lies outside of the conventional legal framework' (Perez Perdomo *et al.* 1982: 224). The widespread application of such practices throughout urban Indonesia in turn underpins a thriving industry of land trading and brokerage – an informal real estate industry – that is itself in contravention of the formal legal framework of the state. Although such a situation may be seen as 'official' in that it necessarily involves agents of the state, it is still informal in that through their involvement, local state agents are nonetheless acting outside of the formal stipulations of the law.

An important point here is that security of tenure is obtained through personalistic ties between putative property 'owners' and local officials, in contrast to securing formal claims of property rights through the objective rule of law (as in this case encoded in the BAL). The degree to which such practices may become institutionalized over time is of great consequence, as this underlies the strength (or potential tenuousness) of such extra-legal claims. In the Indonesian context, one can see that a high degree of institutionalization has been possible through the hold-over of practices inherent in the land law dualism of the colonial period, whereby local officials had significant authority in overseeing the 'traditional' claims of local residents. A contrasting example from China, where no such historical depth of practice pertains, may be seen in Zhang's analysis of extralegal tenure claims of rural migrants to village lands on the outskirts of Beijing (Zhang 2001). In this instance, the case of the urbanization of an area informally known as 'Zhejiang Village', there was again a highly personalistic basis to tenure security, as migrant leaders from Zhejiang Province developed ties to local village officials in order to secure property for building housing for tens of thousands of migrant workers. In contrast to the Jakarta example, the authority of the local village officials has been contested from time to time by higher level authorities, resulting in periodic demolitions of the settlement.

Informal finance in urban China

High degrees of informality may also characterize private finance in Asian cities, particularly in instances where access to formal financial systems is constrained, either because of conditions of poverty, whereby significant components of an urban population are not able to secure loans due to lack of secure collateral, or, as in the case of China, because of institutional constraints which shape lending policies. Perhaps the most thorough investigation of the role of informal finance in Asian urbanization is that of Kelee Tsai in her study of China's 'back-alley banking'. Tsai's inquiry was driven by her observation that although over the course of

China's current reform period the most dynamic components of Chinese urban economies lay in the non-state sectors, either in what was officially recognized as the 'collective' sector, which in the early years of reform was often utilized as a means to disguise private enterprise (Tsai 2002: 34), or in the burgeoning private sector itself, despite the fact that lending by the formal finance system was heavily weighted in favour of state-owned enterprises. It is probably no exaggeration to characterize the state-owned sector over this period as essentially a welfare system, used to shore up urban employment during a period of dramatic economic restructuring, while the dynamism of what has been for many localities sustained double digit economic growth was derived from the activities of private enterprise, whether registered as such or operating semi-clandestinely. In this context, the operations of the formal finance system, as a component of the state, have been highly constrained, with lending practices driven more by policy concerns than by economic opportunities.

The informal finance 'system' which developed in this context, and which has, by Tsai's account, underpinned China's impressive economic boom in recent decades, is characterized by a wide array of nongovernmental mechanisms and institutions, ranging from tolerated but unsanctioned activities to those that are perforce illegal. Indeed, the boundaries between legal and illegal forms of lending have shifted over time in response to various central and local state policy pronouncements. Within the Chinese policy context, an important distinction has been between 'private' lending practices, which are implied to be for-profit and thus not acceptable, and 'popular' practices, which are seen to draw upon local traditions of mutual aid and are thus more readily tolerated by officials. In practice, however, the boundaries are often blurred, as for example when interest payments are adapted to traditional forms of borrowing circles. Tsai further emphasizes the 'gray area of quasi-legality' (2002: 35) as a complicating factor, an indication of conflicting interpretations between components of bureaucracy or between central and local governments, whereby certain practices may be tolerated by some officials though deemed unacceptable by others. Despite the questionable legality of much of informal finance in China, Tsai estimates that perhaps as much as three-quarters of all private sector credit over the two decades of the reform period may be attributed to such activities.

This interpretation also places strong emphasis on local differentiation of practices, such as the 'curb market' on which private entrepreneurs must rely for financing their operations and which differs greatly from place to place, reflective of varying degrees of institutionalization for different practices. Institutionalization, in this regard, refers to the degree that specific lending practices, although originally rooted in interpersonal relations, may become accepted in a form that transcends a reliance on personalism. Local differentiation of lending practices in the cases Tsai studied is therefore indicative of varying relations across China between local states and the citizens with which they interact. Rather than finding the absence of a system of property rights to guarantee the economic transactions which finance is dependent upon, local institutionalization of informal, unsanctioned or quasi-legal practices gives further indication of the extent of state fragmentation – or at least local state differentiation – in the diverse setting of reform-era China. Such diversity allows for a wide range of outcomes, from local state corruption to policy flexibility in support of market-enabling developmentalism.

From these two examples; dealing with urban real estate, particularly for housing development, and entrepreneurial finance, it may be apparent that efforts to describe and analyse informality tend to devolve down into somewhat particularistic cases. The point of commonality here, however, is that the persistence of personalistic governance practices in the face of growing pressures for modernization – and the degree to which such practices are institutionalized or otherwise transcend the personalistic basis from which they originate – creates the context for the continual reinvention of informality by providing means for securing property rights which are alternative to the formal rule of law as advanced by ostensibly modernizing states. Here one sees a challenge to the teleological view of modernity; if the prevailing view of state modernization assumes a convergence of practice around the impersonal guarantees of rule of law, does the expansion of informality in the current period indicate the possibility of multiple and divergent modernities across the societies of Pacific Asia? By providing functional alternatives to the monopoly of the state in guaranteeing property rights, personalistic practices are locally reinforced and are able to persist under conditions of state fragmentation, whether such fragmentation is derived from the persistence of weak states vis-à-vis other societal forces or arises from the increasing pressures of globalization under post-Fordist relations of production.

Systemic informality in urban Asia: emphasizing the local

In this chapter I have argued that informalization in the cities of Pacific Asia may be seen to be expanding through two different (although interlinked) processes of socio-economic change. Current discourses regarding change in the global capitalist economy emphasize the re-working of the state's regulatory role in the context of expanding post-Fordist production systems, a set of processes underpinning patterns of increasing socio-economic polarization. This is seen to be particularly apparent in service sector employment, with high levels of remuneration for top-end professionals occurring simultaneously with the decline of the middle and the rapid expansion of low-end, poorly-paid service employment, a characteristic which Ho (this volume) refers to as an integral part of the current 'sociology of the service economy' (see also Sassen 2000). The growth of informalization from this perspective is in contrast to earlier, more conventional views of informality as a persistent (though possibly transitional) condition associated with partial modernization in those developing countries of the region now in the midsts of their transitions to becoming urban societies. A critical point of connectivity between these twin impulses of informalization is the off-loading or international migration of middle income jobs to localities with lower wage rates. The 'hollowing out' of the middle classes in the wealthy nations of the developed world – and the degree of leverage that this allows 'footloose' transnational capital – is thus linked to the persistence of informality in the developing countries of the region, informality which underpins prevailing low-wage structures, thereby offering an indirect subsidy to globalized capitalist production.

Such a broad international trend plays out quite differently in the various urban contexts of the region, although a common concern across all cases is with the potential for the further entrenchment of socioeconomic dualism in the cities of both wealthy and poor nations. For the wealthy nations, such polarization has increasingly come to characterize the growing service sector as the local decline of middle-income manufacturing employment underpins the separation between the high wages of the professional classes and the low incomes of lower order service workers. For the poorer countries, polarization is again apparent in the segmentation between the growing professional classes, low-wage manufacturing employment, and the legions of 'informal' service workers whose unregulated labour is necessary for keeping formal wage rates competitively low.

Such observations prompt us to think about how we might categorize or typologize the cities of the region, as the assumptions inherent in conventional notions that dichotomize cities of 'developed' from 'developing' countries are now inadequate for dealing with an increasingly interconnected transnational urban system. As Smith (2001) has summarized, current patterns of urbanization in the region emphasize not only the increasing concentration of national development in primary urban settings as the region continues to urbanize, but the increasing differentiation, as well, of transnationally linked hierarchies of cities. The identification of such 'growth triangles' and 'growth corridors' helps us to understand regional, transborder hierarchies of cities, linked through flows of investment, goods, ideas and people, and shaped by regional patterns of investment decision-making. One may distinguish between the poorer and still rapidly urbanizing cities of the region and those wealthier centres that have completed their urban transitions, though the interlinkages through transborder flows tend to downplay such distinctions with regard to overall patterns of spatial development. When looked at in terms of an overall urban system, both processes of informalization – through pressures for the regulatory flexibility favoured under post-Fordist production, and through the urban absorption of erstwhile rural labour – are clearly relevant.

Typological classifications of the cities that comprise the increasingly interlinked urban system of the Asia-Pacific region have conventionally stressed the differentiation of urban functions related to the transborder connectivities of globalization, with first-, second-, third- and fourth-order global cities distinguished by such factors as the concentration of corporate centres, the development of higher order services, and other factors related to control functions and operations of the globalizing capitalist system (Hutton 2001). Critical to such interpretations of urban systems have been the nodal functions of specific cities in the overall flows of globalization, with regard to capital (market interlinkages), goods (port facilities), people (airports and airline routing) and, increasingly, cultural and information flows (advanced communications facilities, the growth of place marketing through cultural events, and so forth).

At the lower end of such interpretive hierarchical structures, one would locate those cities that tend to be characterized as late industrializers, whose contribution to the global system is seen primarily in terms of their low-wage productive capacity, with concomitantly lower levels of control functions in the overall urban system. It is in such locales that one may find the greatest concern for the persistent effects of informality, as these are the cities of societies still in the midsts of their urban transitions, with continuing high levels of informal employment and development, and relatively weaker regimes of state regulation. Understanding informalization as driven by twin impulses – linked to the continuing urban transition of peripheral regions, coupled with the 'unmaking of the formal' in core regions through shifts to post-Fordist production – raises the question of possible convergence, here seen as the increasing informalization of lower-order services throughout the regional urban hierarchy. The implication from this is that further analysis of the region's urban hierarchies, whether separated out into discrete transnational urban corridors or with regard to the urban system overall, needs to take careful consideration of what may be highly localized regimes of urban governance, and how these interact across borders.

Within such regional urban hierarchies, informalization is most readily apparent at the level of the local state – as illustrated by the examples given above – for as one could argue, it is the disparity between application of policy between higher and lower levels of government that allows for the persistence or re-creation of informality. Formal structures may be put in place through national-level initiatives, as seen in policy formulation, legal reform efforts, and institutional design in the name of 'good governance', yet it is when such structures are articulated at local levels that one finds the regulatory flexibility which underlies informalization, whether driven by the necessity for administrative tolerance in rapidly urbanizing societies or by the pressures of globalizing capital in the context of expanding neo-liberalism.

In this article, I have focused on cases more peripheral to the urban hierarchies of the region by drawing upon the examples of informal real estate in Indonesia and informal finance in China. One may argue that in pointing to examples from societies in the midsts of their urban transitions a number of other factors are brought to bear upon thinking about the underpinnings of informality, factors related to the pressures of rapid urbanization and the lack of institutional and administrative capacity to respond, as for example with regard to urban employment creation, the provision of urban services, environmental monitoring, and so forth. I would argue, however, that the numbers alone do not account for everything; the high levels of informalization associated with the region's urban transitions cannot fully be explained in terms of insufficient governmental capacity. The nature of the local state and how it intersects with local socio-economic forces is key to understanding informalization. As seen in the cases above, the persistence of personalistic governance practices – in contrast to the presumed impersonal guarantees of objective rule of law – underpin the flexibility and informality of local level governance. And the degree to which such personalism may become institutionalized, either through cultures of governance rooted in traditional practices (as seen in the Indonesian case) or more recently created as an off-shoot of higher level policies (as with the case of Chinese finance), will ensure the persistence of such relationships beyond the urban transitions of these and other similar societies in the region.

Notes

1 Here I will use the term 'Pacific Asia' to refer specifically to East and Southeast Asia as a territorial subset of 'the Asia-Pacific', which comprises the overall scope of this volume. As a generalization, Pacific Asia contains the majority of the cities in the erstwhile 'developing countries' of the overall region, the analysis of which is my principal concern in this chapter.
2 Scott (1977), for example, argues that personalistic rule through webs of patron-client relations is deeply rooted in Southeast Asian contexts and has become a core characteristic in the post-colonial period; the persistence and re-creation of personalistic relations of governance in state socialist and reform era China is also a recurring theme in a number of diverse works, as for example, by Walder (1986), Ding (1994), Zhang (2001) and Potter (2001).
3 For an early analysis of the potential for basing local credit strategies on traditional forms of rotating savings and credit, see Geertz (1962).

References

Boeke, J. H. (1942) *The Structure of Netherlands Indian Economy.* New York: International Secretariat, Institute of Pacific Relations.
Broad, D. (2000) 'The periodic casualization of work', in Tabak, F. and Crichlow, M. A. (eds), *Informalization: Process and Structure.* Baltimore: Johns Hopkins University Press, 23–46.
Bromley, R. (1979) 'Introduction – the urban informal sector: why is it worth

discussing?', in Bromley, R. (ed.), *The Urban Informal Sector: Critical Perspectives on Employment and Housing Policies*. Oxford and Toronto: Pergamon Press.

DeSoto, H. (1989) *The Other Path: The Invisible Revolution in the Third World*. New York: Harper & Row.

Dick, H. W. and Rimmer, P. J. (1998) 'Beyond the Third World city: the new urban geography of Southeast Asia', *Urban Studies*, 35(12), 2303–21.

Ding, X. L. (1994) 'Institutional amphibiousness and the transition from Communism: the case of China', *British Journal of Political Science*, 24, 293–318.

Furnivall, J. S. (1939) *Netherlands India: A Study of Plural Economy*. Cambridge: Cambridge University Press.

Furnivall, J. S. (1948) *Colonial Policy and Practice: A Comparative Study of Burma and Netherlands India*. Cambridge: Cambridge University Press.

Geertz, C. (1962) 'The rotating credit association: a "middle rung" in development', *Economic Development and Cultural Change*, 10, 241–63.

Geertz, C. (1963) *Peddlers and Princes: Social Change and Economic Modernization in Two Indonesian Towns*. Chicago: University of Chicago Press.

Gilbert, A. (1990) 'The costs and benefits of illegality and irregularity in the supply of land', in Baross, P. and van der Linden, J. (eds), *The Transformation of Land Supply Systems in Third World Cities*. Aldershot: Gower Publishing, 17–36.

Harding, P. and Jenkins, R. (1989) *The Myth of the Hidden Economy: Towards a New Understanding of Informal Economic Activity*. Milton Keynes and Philadelphia: Open University Press.

Hart, K. (1973) 'Informal income opportunities and urban employment in Ghana', *Journal of Modern African Studies*, 11, 61–89.

Henssen, J. (1988) *Indonesia: Issues of Land Policy for Housing and Urban Settlements*. Ministry of Housing, Physical Planning and the Environment in the Netherlands.

Hutton, T. A. (2001) 'Service industries and the transformation of Asia-Pacific city-regions', CAS Research Paper Series, Singapore: National University of Singapore, Centre for Advanced Studies.

ILO (1972) *Employment, Incomes and Equality: A Strategy for Increasing Productive Employment in Kenya*. Geneva: International Labour Organisation.

Latouche, S. (1993) *In the Wake of the Affluent Society*. London: Zed Books.

Leaf, M. (1993) 'Land rights for residential development in Jakarta, Indonesia: the colonial roots of contemporary urban dualism', *International Journal of Urban and Regional Research*, 17(4), 477–91.

Leaf, M. (1994) 'Legal authority in an extralegal setting: the case of land rights in Jakarta, Indonesia', *Journal of Planning Education and Research*, 14, 12–18.

Migdal, J. (1988) *Strong Societies and Weak States: State-Society Relations and State Capabilities in the Third World*. Princeton: Princeton University Press.

Moser, C. O. N. (1984) 'The informal sector reworked: viability and vulnerability in urban development', *Regional Development Dialogue*, 5(3).

Perez Perdomo, R., Nikken, P., Fassano, E. and Vilera, M. (1982) 'The law and home ownership in the barrios of Caracas', in Gilbert, A. (ed.), *Urbanization in Contemporary Latin America*. Chichester, UK: John Wiley and Sons.

Perlman, J. (1987) 'Misconceptions about the urban poor and the dynamics of housing policy evolution', *Journal of Planing Education and Research*, 6(3),187–96.

Portes, A. (1983) 'The informal sector: definition, controversy and relation to national development', *Review* 7 (Summer 1983).

Portes, A., Castells, M. and Benton, L. A. (eds) (1989) *The Informal Economy: Studies in Advanced and Less Developed Countries*. Baltimore: Johns Hopkins University Press.

Potter, P. B. (2001) *The Chinese Legal System: Globalization and Local Legal Culture*, New York: Routledge.

Sanyal, B. (1996) 'Intention and outcome: formalization and its consequences', *Regional Development Dialogue*, 17(1), 161–78.

Sassen, S. (2000) 'The demise of Pax Americana and the emergence of informalization as a systemic trend', in Tabak, F. and Crichlow, M. A. (eds), *Informalization: Process and Structure*. Baltimore: Johns Hopkins University Press, 91–115.

Scott, J. C. (1977) 'Political clientelism: a bibliographical essay', in Schmidt, S. W., Lande, K., Guasti, L. and Scott, J. C. (eds), *Friends, Followers and Factions*. Berkeley: University of California Press, 483–505.

Smith, D. W. (2001) 'Cities in Pacific Asia', in Paddison, R. (ed.), *Handbook of Urban Studies*. London: Sage Publications, 419–50.

Struyk, R. J., Hoffman, M. L. and Katsura, H. M. (1990) *The Market for Shelter in Indonesian Cities*. Jakarta, Indonesia: Hasfarm Dian Konsultan; Washington, D.C.: Urban Institute Press; Lanham: Distributed by University Press of America.

Tabak, F. (2000) 'Introduction: informalization and the long term', in Tabak, F. and Crichlow, M. A. (eds), *Informalization: Process and Structure*. Baltimore: Johns Hopkins University Press, 1–19.

Tabak, F. and Crichlow, M. A. (eds) (2000) *Informalization: Process and Structure*. Baltimore: Johns Hopkins University Press.

Thomas, J. J. (1992) *Informal Economic Activity*. New York: Harvester Wheatsheaf.

Tsai, K. S. (2002) *Back-alley Banking: Private Entrepreneurs in China*. Ithaca: Cornell University Press.

Turner, S. (2003) *Indonesia's Small Entrepreneurs: Trading on the Margins*. London and New York: RoutledgeCurzon.

Walder, A. (1986) *Communist Neo-traditionalism: Work and Authority in Chinese Society*. Berkeley: University of California Press.

Zhang, L. (2001) *Strangers in the City: Reconfigurations of Space, Power, and Social Networks within China's Floating Population*. Stanford: Stanford University Press.

Part II

Services and urban development in the Asia-Pacific

Sectoral perspectives

7 Localizing international business services investment

The advertising industry in Southeast Asia

Larissa Muller

A small number of transnational corporations (TNCs) dominate the world's advanced business services, making it one of the most internationalized industries in the world. These corporations have penetrated local country markets across the globe through an extensive network of branch and affiliate offices. These offices tend to be located in key cities in each region. As a result of this spatial clustering effect, TNC activity has created important global business service centres in every region of the world. Asia-Pacific has the highest concentration of these centres among the developing country regions (Beaverstock *et al.* 1999). Within Southeast Asia, the cities of Singapore, Bangkok and Manila have emerged as important second and third tier hubs in this global business services network. Despite their prominence, however, there has been relatively little study on the business service industry in these Southeast Asian capitals.

This oversight reflects an ongoing focus in the internationalized services literature on global cities where the advanced business service TNCs are headquartered. Representative of this work is Sassen's *The Global City* (1991) and Nachum's work on service multinationals (1997, 1999, 2000), which explain roles of global cities in delivery of business and producer services based on a command and control model that emphasizes the role of headquarter offices located in advanced economies, assumes a hierarchical division of labour, and home country advantage. Because no developing country has produced transnational firms in the advanced business service industry, second and third tier business hubs in these countries are typically ignored as passive recipients of foreign investment. I argue that this description and explanation of internationalized advanced business service dynamics is wrong, at least in the case of Southeast Asia.

The appropriate construct for understanding advanced business services on the periphery is one of distributed intelligence. Many business and professional services, such as advertising, are too complex, have too much cultural content, and require too much extensive and rapid nuanced adaptation to local conditions to be produced under a command and control hierarchical structure. Transnational business service corporations need the periphery for functions far more important than standardized production. Accordingly, this role gives the periphery more power and importance than in the stereotypical command and control construct.

Using the case of the advertising industry in Southeast Asia, this study explores factors that enable the local development of unique, innovative and internationally competitive products and services in an industry dominated by transnational corporations. The findings challenge the passive periphery assumption in the advanced business services literature.

Advertising growth in the Asia-Pacific region

The advertising sector is expected to undergo tremendous growth over the next 20 years in the Asia-Pacific region. Conservative forecasts estimate that the region's advertising spending will rival European levels by 2020, and be one and a half times larger than the rest of the developing world combined (see Figure 7.1). As of 2000, advertising spending in Asia, excluding Japan, had already reached US$ 34 billion. The Philippines, Thailand and Singapore have similar levels of advertising expenditure. In 2000, the Philippines and Thailand spent well over US$ 1 billion on advertising, with Singapore close behind at US$ 0.83 billion – a significant component of their respective urban economies. As the Southeast Asian region's macro economic structure moves increasingly toward consumer-oriented growth (as opposed to the export-driven growth of the past), advertising in these countries is gaining importance.[1]

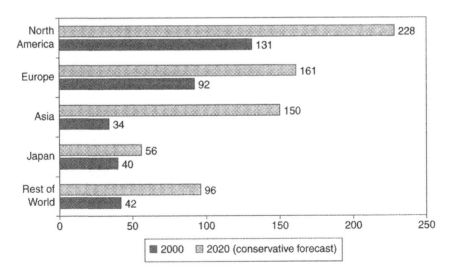

Figure 7.1 Global advertising expenditure (US$ billions) (source: Batey (2002)).

The Bangkok puzzle: unexpected performance of a peripheral node

Approximately 24, predominantly American, transnational corporate networks dominate the global advertising scene (Daniels 1995). These transnational advertising agencies (TNAAs) overwhelmingly dominate the advertising industry in the three case study cities of Bangkok, Manila and Singapore. The TNAA branches/affiliates in Bangkok control upwards of 75 per cent of total advertising spending in Thailand. But equating foreign dominance with a passive, foreign-controlled local advertising industry fails to explain much of the differential performance among peripheral hubs.

Consider Bangkok's advertising industry. Thai-led TNAAs are outperforming other country offices in the region and in their global corporate network. Lowe Lintas Thailand was the only Asian agency ranked in the world's top ten by the global industry magazine *Ad Age International* in 1999. Moreover, the chairperson and CEO of this agency, Charuvarn Vanasin, is Thai. Under the deft steerage of another Thai, Bhanu Inkawat, Leo Burnett's Bangkok office was awarded the corporation's 1998 'Agency of the Year', and won their 'Brightest Star in Asia' award for several years running. Ogilvy & Mather's Bangkok agency became one of the top five global profit centres in its corporate network. On the strength of the office's performance, the agency's Thai Chair and CEO, Sunandha Tulayadha, became the first Asian representative to be appointed to the Ogilvy & Mather Worldwide Board in 1996 (*The Nation [Thailand]* 2001).[2]

Bangkok has also distinguished itself creatively. Bangkok's advertising industry has gained a reputation for locally produced, international award-winning advertisements (Gearing 2001; Mertens 2001). A number of Thai creative directors have won the prestigious Cannes, Clio and London Advertising Awards, which contributes to the international image and reputation of the corporate networks they represent, and attests to the ability of Bangkok's TNAA subsidiaries to add value locally. In fact, winning an international award has become the new measure of success among Bangkok's creative advertising people.[3]

More importantly, these winning ads reflect Bangkok's own distinct style of advertising, characterized by strong visuals, minimal copy and a quirky sense of humour. This successful style often involves fusing Western ideas and technology with local creative talent. 'Thai creators focus on goods and customers, to devise campaigns that make products part of people's lives in ways not distorted by foreign artifice', wrote one industry observer (Gearing 2001: 43). Such successful innovations and adaptations of advertising campaigns is the hallmark of a creative milieu, not a passive periphery.

As another sign of dynamism in Bangkok's advertising industry, Thailand hosts the Annual Asian Advertising Festival (AdFest). The brainchild of veteran Thai ad-maker, Vinit Suraphongchai, AdFest represents one of the more tangible examples of the Asian advertising industry developing its own independent style and identity relative to the West. The festival is gaining international recognition. The 6th annual festival, held in 2003, attracted more than 2,400 entries from 33 cities. That the annual competition is always held in Thailand underscores Bangkok's central role in helping to define and sustain a distinct Asian advertising identity and dynamic.

Bangkok's unexpectedly dynamic performance calls into question the passive periphery thesis of advanced business services, which emphasizes an international division of labour among knowledge workers in so-called global cities (the location of the TNAA headquarters) and the peripheral centres. Locals in Bangkok (and other cities in Asia) have demonstrated their ability to develop innovative and internationally competitive advertising products and services in an industry dominated by foreign transnational corporations. This would suggest that certain conditions and processes exist that allow for interaction between the foreign investment in the advertising industry and local factors of production and endowments, in a mutually reinforcing way.

Understanding how this occurs is of critical importance to transnational corporations and host cities seeking to create dynamic business service industries. I have attempted to address the 'how' question by examining the advertising centre in Bangkok, and drawing comparisons with Manila and Singapore, in terms of the extent of localization, which is associated with the ability to tap local sources of knowledge, and second, in terms of the extent of global connectivity, which refers to the ability to tap into international know-how and currents.

Characteristics of Bangkok's advertising industry

This section looks at the key characteristics of Bangkok's advertising industry which contribute to its strong performance. The analysis focuses on two scales: within the TNAA subsidiaries, and within the wider industry cluster in which the TNAA is embedded. Not surprisingly, the Bangkok industry displays a significant degree of localization. But its success has as much to do with its ability to link these local factors of production to global market and industry networks.

Staffing dynamics

In professional business services, people are the main assets, and talent, the measure of labour skill. The capability of the work force determines the firm's ability to create and innovate, and thereby set itself apart from competitors (*The Economist* 2001). The passive periphery thesis would suggest that innovation in local products and firm management would be carried out by foreigners. But this is clearly not the case in Bangkok, where foreign advertising agencies are being run by Thais.

In the case of Bangkok's advertising industry, the human assets – agency talent – are increasingly locals. Expatriates account for just 1.6 per cent of TNAA staffing (see Table 7.1), and less than 1 per cent of the total advertising work force in Thailand.[4] The Thais have rapidly ascended into every position in the TNAA subsidiaries, including the upper echelons of management. As early as the late 1970s, less than ten years after the widespread appearance of TNAAs on the local advertising scene, the expatriate hierarchy in the internal structure of the firm had disappeared. The first generation of Thai advertising professionals had been intensively trained by the TNAAs, and quickly rose through the ranks of the early foreign advertising agencies. By the early 1990s, the most successful TNAAs were being led by experienced Thai executives and, as of 2000, seven out of every nine top management positions in the major TNAAs were held by Thai nationals. Thus, to judge the Thai advertising industry solely on whether a Thai-owned firm has joined the ranks of the international advertising agencies is to overlook the significant changes that have taken place within the walls of the TNAA.

Most TNAAs have an explicit policy to train and promote local staff to key positions, but success varies across countries. Recognizing the importance of local knowledge in order to get into the hearts and minds of local consumers, expatriate managers make a concerted push to hire locals. This has certainly been the case in Thailand. But, whereas in many other developing economies expats continue to run the day-to-day operation of foreign advertising agencies, in Thailand locals have mostly taken over the management of the agencies. In China, for example, the Chinese deputy manager plays a nominal role (Wang 2000), although this is likely to change with time, and in Singapore, which suffers labour shortages, the rising stars of an agency are often lured away to client firms before they reach top management positions. Thai advertising professionals have clearly seized the opportunity and performed.

A special characteristic of these successful Thai ad-makers is that the majority are returnees from the US and Britain. These 'internationals' or 'bridgeheads' are people who understand and operate in both the western and local worlds and serve to bridge the two.[5] Because the level of formal training in advertising was limited in the early years of the industry, the first generation of Thai 'internationals' benefited from personal mentorship by foreign managers. Such intensive interaction greatly enhanced the exchange of cultural, management and technical knowledge, exposing Thais not only to a global view but also to a global network of people. Especially influential in shaping this and subsequent generations of local advertising professional are localized expatriates, who adopted Thailand as home. With both an insider's knowledge, and an outsider's perspective, these individuals were uniquely positioned to see the strengths within the local culture, and worked with the locals to create a unique Thai style.

Table 7.1 Profile of Thailand's top 21 transnational advertising agencies, 2000

Transnational advertising agency	Year founded	Total no. of staff	Total no. executives/ directors	Total no. of expats	Income (1999) (mil. Baht)	Ownership
Lowe Lintas	1970	270	10	2	451.8	Wholly owned by Interpublic Group of Companies (NY)
Ogilvy & Mather	1972	287	11	7	397.1	Wholly owned subsidiary of WPP Group
JWT	1983	208	10	1	303.0	Wholly owned subsidiary of WPP Group
Dentsu (Thailand)	1977	195	18	2	277.6	Jointly owned affiliate: Dentsu has minority share
Euro-RSCG	1980	37	6	1	261.6	Wholly owned
Leo Burnett	1965	225	13	2	255.1	Wholly owned by Leo Burnett Co. (Chicago)
Dai-Ichi Kikaku	1974	115	12	3	233.9	Jointly owned: Dai-Ichi Kikaku has minority share
Dentsu, Young & Rubicam	1985	165	9	3	225.9	A WPP company joint venture with Dentsu (Japan)
Far East DDB	1964	163	12	0	217.4	Jointly owned: DDB (Omni-com) has minority share
Thai Hakuhodo	1973	65	9	2	213.9	Jointly owned: Hakuhodo has minority share
Grey Worldwide (Thailand)	1992	80	6	2	203.3	Jointly owned: Grey Global Inc. has majority share (acquired 100% in 2003)
Prakit/FCB [FCB Worldwide (Thailand)]	1978	164	6	1	199.5	Jointly Owned: FCB Worldwide has minority share
Prakit Publicis	1998	100	8	2	189.6	Jointly Owned: Publicis has minority share
McCann-Erickson	1965	115	10	3	188.2	Wholly owned by Interpublic Group of Companies (NY)
Chuo Senko (Thailand)	1963	105	14	3	149.3	Jointly owned affiliate: Chuo Senko has minority share
TBWA Next & Triplet	1996	60	11	0	139.9	Jointly owned: TBWA has majority share
Bates Thailand	1989	75	7	1	130.6	Jointly owned: Bates Worldwide has minority share
BBDO Bangkok	1987	80	6	0	108.9	Wholly owned
Saatchi & Saatchi	1988	70	6	4	75.4	Wholly owned
Hakuhodo Bangkok	1988	64	6	3	55.8	Jointly owned: Hakuhodo hold minority share
Siamese DMB&B	1998	70	5	1	45.5	Jointly owned until recently: DMB&B acquired 100%
Total		2,713	195	43	4,323	

Source: AP Publications (2001), Adarco (2001).

This working community of localized-foreigners and globalized-locals (foreign-educated Thais) enabled the local and the international to come together, bond and spawn a unique, and highly successful, hybrid. These 'internationals' have a unique perspective which enables them to fuse the western techniques with the distinctiveness of the local culture to create powerful hybrid systems and products, both on the creative and administrative side. They are also instrumental in forging and linking both local and global networks of people which become a key medium of exchange of ideas and learning.

As localization deepens and expatriate executives have been replaced with locals, several TNAA Thai CEOs have begun recruiting foreign expatriates to one of the senior position as a means of 'injecting new blood' into the agency. An infusion of expatriate executives into the staffing mix is seen by local managers as necessary to keep local innovation in Thailand's advertising industry on the cutting edge. The mixed workforce becomes an effective way to maintain an 'interpretive community'[6] through which the international and the local can continue to directly inform and interact with each other, sharing know-how.

Relationships within the corporate network: distributed intelligence

Possibly the single most important factor contributing to the localization of the TNAA is decentralized decision-making within the corporate network. When transnational corporations retain too much control over their networks, technology and techniques, it prevents the development of indigenous production (Landau 2001: 143). For the most part, transnational advertising networks appear to adopt a considerably decentralized structure. Executives from almost all of the top twenty TNAAs interviewed stated that they operated with a high degree of autonomy from headquarters.[7] For the vast majority, headquarter oversight is limited to quarterly financial reporting, hiring approvals for top executive positions only, and new client approvals as a formality to check against global client conflicts.[8] Agencies that meet financial targets, or make a profit are typically given freer reign.

Corporate advertising networks recognize that insight into local markets and the culture requires local knowledge and understanding, necessitating deep involvement of local people, firms and institutions in the creation of advertising products and services. Thus, there is generally no oversight on locally produced marketing and advertising campaigns. The branch firm has creative license to develop its own pitches, local contracts and local campaigns, and to adapt global or regional campaigns to local markets.

In light of considerable local autonomy, training is often the most influential method of achieving corporate standardization within the network. All of the big TNAAs invest heavily in training in order to groom their own high quality professionals. In Bangkok, it is a practice that has earned the top TNAAs the moniker 'advertising university'. Many of the alumni of the top TNAAs have gone on to become major players in Bangkok's advertising industry – often as executives or entrepreneurs of a competitor firm. Thus, TNAA in-house training improves the quality of the labour pool for the whole industry, and labour movement within the advertising industry ensures substantial knowledge spillovers beyond the firm.

Industry competition

Development of the labour pool is not the only way in which TNAAs have contributed to strengthening Bangkok's advertising industry. They have also played an important role in increasing competition and rivalry among agencies in Bangkok, and directly benefited from the vibrant entrepreneurial environment that they helped to spawn.

Entrepreneurship has always been strong in Bangkok's advertising industry. Beginning in the late 1970s, and accelerating through the 1988–1997 period, the industry grew very rapidly, as a number of Thai start-ups entered the market to capture a share of the steadily rising advertising expenditures. The primary impulse driving entrepreneurship in the industry were Thai advertising professionals, usually working in one of the TNAAs, who either were unable to move up inside the transnational agency or were simply ready to strike out on their own. Early success stories include Prakit & Associates, established in 1978, and presently involved in two joint ventures, Prakit Publicis and Prakit-FCB, and CVT & Bercia (now CVT Advertising) which was established in 1984 and remains one of the top locally-owned advertising agencies. Other alumni followed, a number of them successfully competing head-to-head with TNAAs. As recently as 2000, two of the top ten agencies by income in Thailand were Thai-owned (AP Publications 2002).

A great number of transnationals also entered the market during the 1988–1997 period, usually by partnering with or buying out one of the larger and more successful Thai-owned agencies. This entry strategy enables TNAA's to acquire proven talent critical to their success. Typically, for a long time after the name and ownership change takes place, the staff and management of a newly acquired Thai agency remain essentially unchanged. In these instances, the entrance and apparent dominance of TNAAs in the Thai advertising market can, and should, be interpreted as local agencies 'making it big'.

Joining global networks tends to be part of an explicit growth strategy by successful local agencies. Local agencies have much to gain from partnering with a TNAA. As one founder and managing director of a local agency explained:

> It's impossible to grow an advertising agency without being part of an international network, for two reasons

1 you need to retain good quality people, and to do that you need;
2 good, stable income. Having international client accounts secures certain revenues.[9]

In fact, financial security was the reason most often cited by founders of local advertising agencies for partnering with a TNAA. It enables the firm to plan and invest in order to differentiate itself in an increasingly competitive market. Or as another founder put it, 'With financial security you know you're not going to die, so the question becomes one of how much more will you gain [grow]. Psychologically, you are starting at a different place'.[10] Another cited advantage is access to resources, particularly creative resources within the global corporate network.

The symbiosis between the vigorous start-up environment and TNAAs occurs on other levels as well. While a number of local agencies provide direct competition (in full service and niche sectors) and partner and acquisition possibilities for TNAAs, the vast majority of local start-ups remain small players that target a complementary market. These boutique firms provide a broad range of advertising services to clients with limited advertising budgets. As a result, they tend to operate in a completely different sphere from the TNAAs. Nonetheless, the presence of TNAAs is having a positive impact on the quality of services available to this second tier of clients. One reason is that more often than not, the owner/director of the boutique firm previously worked in a TNAA. Thus, the experience and training garnered in the TNAA becomes available to local clientele via small ad shops. Combined with a substantial level of competition at this level, this dynamic has led to increasing sophistication among boutique agencies. A trickle-up effect, from boutique firm to TNAA, is also in play. New graduates, usually from business or art schools, typically gain their first hands-on experience at local boutique shops before being hired at Bangkok's TNAAs.

Production networks: building up the node

A host of new firms and organizations in complementary industries have emerged and developed alongside the foreign advertising industry, deepening the TNAA local supplier linkages. The key supporting industries are:

1 graphic or print production;
2 television commercial (TVC) or film commercial production and post-production (editing);
3 market research and consultancy;
4 media;
5 new media advertising support, such as web design and software.

Expansion in the co-production and supplier industry has outpaced growth of advertising agencies, filling an earlier void. The bulk of this expansion occurred during the 1990–1997 period of rapid growth in the advertising industry, when the number of co-producer and supplier firms more than doubled for most categories (AP Publications 1990 and 2001). The fastest growing area was in TVC post-production studios, which experienced a five-fold increase since 1990.

Growth in these complementary industries has significant localization impacts because of the high rates of local ownership, particularly among the graphic and TVC production firms. The absence of extensive international networks in these advertising support industries is probably due to the fact that it is at this level of the production process that many of the cultural nuances are incorporated into the advertisement product, whether it be a locally produced advertisement or a local adaptation of an international ad. Thus locals have the advantage in knowing what music to append, which backdrops to use, and how to employ cultural references, in order to evoke certain emotions and moods or conjure desired associations.[11] The need for local involvement on the production side has only increased as the advertising in Thailand has become more sophisticated, shifting toward a more subtle, emotional and mood oriented style to communicate their message.

The growing sophistication of advertising in Thailand corresponds to the strengthening of the advertising production complex, notably its television commercial (TVC) production and post-production industry. This industry has experienced a 180 degree turnaround. As late as the 1980s, poor quality local production led Bangkok's TNAAs to farm out their high-end TVC production to Hong Kong. But in the last ten years, the quality of local production has improved dramatically, fuelled by a wave of creative advertising people who left the transnational ad agencies to enter the production side of the industry. Bangkok now has several first-rate, international-style production houses, which rank among the best in the region. Their creative input and high quality production have been credited for Thailand's successes at international advertising competitions. Tellingly, the Thai production houses have begun exporting their services to Hong Kong, as well as Germany, Japan, Australia, Indonesia and Vietnam. Through their export work, the TVC producers are constantly exposed to new international techniques and styles. Their learning then feeds back into the local ad industry through collaboration with creative advertising directors. Thus, a sophisticated market for Thai production is growing as the advertising industry becomes more locally integrated, and with knowledge transfers from multinational and overseas customers, local production firms are likely to develop further, benefiting the overall quality of Thai advertising.

Clusters are made more cohesive by industry associations and other formal and informal networks. Thailand's advertising production complex has arguably benefited most from the Advertising Association of Thailand (AAT), which has been a part of the Thai advertising establishment since

1966 (Chirapravati 1996). Because both local and TNAA agencies are members of the association, it provides an important forum for joint action and knowledge exchange. In response to recent events, the AAT has taken on a much more active role in the industry. The association was instrumental in developing guidelines for a new fee and billing structure in the wake of the economic downturn in 1998 which led to a major industry restructuring. It has also become a much stronger lobbying force. In 2001, it successfully convinced the government to rescind a policy prohibiting government offices from using foreign-owned advertising firms, and it recently persuaded one of the government regulatory agencies, the Food and Drug Administration, to strictly limit its scope for regulation and censorship.

The industry also benefits greatly from Bangkok's three award institutions created to promote creative excellence. These are:

1   the Bangkok Art Directors (BAD) which in addition to award recognition, offers a ten-week training course, the graduates of which are in high demand by TNAAs seeking new creative talent;
2   the Top Advertising Commercial of Thailand (TACT) award established by the academic community;
3   the previously mentioned Asia/Pacific Advertising Festival (AdFest), an Asia-wide competition based in Thailand, whose mandate is to inspire, reward, and raise the standard of creativity in the region while encouraging the rediscovery and use of Asian cultural heritage and values in creative work.

In sum, contrary to what the passive periphery paradigm would predict, the Bangkok advertising industry is becoming increasingly endogenized, i.e., more embedded in the local economy. A key indicator is the presence of innovation dynamics and the decentralization of local knowledge production to the branch agency of major advertising corporations in this peripheral hub, which are under the management and creative direction of indigenous staff. A second key indicator is the integration of the internationalized advertising industry locally, through mutually reinforcing interaction, often on multiple levels, with local competing firms, suppliers and supporting institutions, stimulating a successively stronger and deeper advertising industry. But at the same time as local linkages have deepened, the industry has developed and maintained global networks, which allow for a continuous infusion of new ideas and influences that keep the industry on the cutting edge. The findings suggest that the most successful advertising firms operating on the periphery are those that are most hybridized. This means fostering, and tapping into, local knowledge creation capabilities through both local and global networks. Where the industry as a whole is highly hybridized, that urban region will have a highly vibrant and successful advertising cluster.

Ingredients for success: Bangkok in contrast to Manila and Singapore

Contributing to the strong performance of Bangkok's advertising industry are at least four key factors of hybridization: a labour force dominated by internationalized locals; a distinct creative advertising style that fuses western ideas with local cultural influences; an integrated and highly networked local industry; and strong global industry linkages. Bangkok's most successful firms, and the most innovative sectors of the cluster, are those that share these characteristics.

To better understand the importance of hybridization dynamics in the Bangkok case, these factors are contrasted to the advertising industry dynamics in Manila and Singapore. In all three Southeast Asian advertising industries passivity vis-à-vis international headquarters of TNAAs is far from the norm. However, comparative analysis indicates interesting variations in cluster dynamics and performance, due to different historical, institutional and socio-cultural contexts, leading to differences in levels of endogenization and global connectivity of the advertising industry in each city.

Hybridized work force

First, the Bangkok case indicates that bridgeheads (internationalized locals and localized expatriates) are key to the success of the cluster, driving the hybridization process which enabled the Thai advertising industry to stand out in a global arena. These individuals often work within the TNAAs but may act outside them as consultants, start-up entrepreneurs and so forth. They interpret global learning and practices to local actors, and vice versa, creating an effective bridge for global-local interaction and knowledge sharing. As the number of internationalized locals among new hires declines, mixed work forces of expatriates and locals have become an effective means of maintaining this interpretative community, fresh perspectives, and global-local exchange.

The advertising industries in Manila and Singapore have not achieved a similar mixed work force, for contrasting reasons. Written into the constitution of the Philippines is a clause preventing expatriates from holding key executive management positions in advertising agencies. In minimizing expatriate numbers in the work force, these hiring restrictions unwittingly reduce opportunities for the latest thinking and technical skills to be transferred through the highly effective method of mentoring and learning-by-doing. 'This wastes talent', according to one Filipino ad-maker, 'because we [the Philippines] have a lot of good raw potential talent that needs training'.[12]

In Singapore, by contrast, not a single TNAA is expatriate free. It is relatively easy for foreign advertising professionals to obtain work permits in Singapore, and to navigate the local political and cultural scene, leading to a much higher portion of expatriates in management positions in Singapore's TNAAs as compared to Bangkok's. Singapore's regional headquarter status in Southeast Asia's advertising industry, and intense competition for a limited local labour pool exacerbate the situation. As the gathering point of regional information, training and administration for one of the most dynamic developing regions in the world, advertising corporations place trusted executives there. Talented locals, meanwhile, are in high demand and often lured to better paying positions in the marketing departments of other MNCs before reaching the upper stratum of management in a TNAA.

Hybridized product

Second, Bangkok's successful advertising style is based on high cultural awareness which blends local and western influences to create new and unique styles. Two conditions fuel this trend. First, having never been colonized, Thai consumers exhibit less of a western influence than their Southeast Asian counterparts, necessitating adaptation of western-style advertising (Davies 2002). Second, unlike Malaysia, Singapore and the Philippines, Thailand has a sophisticated indigenous popular culture, providing a rich resource for creators to tap into. The development of a uniquely Thai style is the outcome of close collaboration between creative directors in the TNAAs and in the locally-owned and staffed production houses. The strength of the local production house industry has been a critical element in Bangkok's creative successes.

By comparison, in Manila, greater levels of localization within the TNAA have failed to give rise to significant innovations, particularly in the creative milieu. Its advertising is characterized by imitation rather than innovation. While cautious clients and creators frustrate the process, the country's cultural situation is a serious obstacle. The same colonial history that gave rise to the nationalization of the advertising industry also produced a population that values, and culturally identifies with, Americana. Imitating Uncle Sam is thus easier and less risky than digging deep into the culture to identify that which is uniquely Filipino. As a result, Manila has fallen behind its neighbours in creative advertising.

Singapore is a creative powerhouse, but primarily because it imports talent to create regional advertising campaigns, not because it has developed an identifiable style that it can call its own. Singapore's total openness – as the world's most globalized country – works against the localization of the advertising product. Consumer purchasing patterns are similar to that of the west, and English is an official language, making it easy to adopt Madison Avenue style ads with minimal adaptation. The country's small market size adds an economic rationale to do so.

Extensive local networks

A third factor of Bangkok's success is the movement of labour throughout the industry. Local boutique firms are training grounds for advertising professionals that move on to TNAAs. TNAA-trained professionals head off to create new start-ups in advertising and supplier industries. And TNAAs recruit from each other. This movement of people creates strong industry-wide networks and knowledge spillovers, promoting exchange of new developments and ideas throughout the whole industry. The informal networks form the basis for more formalized interaction, including strong industry associations and award institutions, which reinforce this dynamic. The result is a highly locally-integrated industry.

One repercussion of Singapore's high levels of expatriate staffing is more exclusively expatriate social networks within the industry, weakening localized linkages. Short term assignments lower the commitment of expatriate agency executives to invest in the development of the local industry. At the same time, career opportunities for talented local TNAA staff dramatically lower the incidence of entrepreneurship among them, creating a sharper schism between boutique and large-size agencies, in terms of clients, knowledge flows, etc.

Despite having the highest local staffing levels, Manila's industry is not as well networked as Bangkok's, primarily because labour movement between the two tiers is lower. Ironically, this reflects Manila's stronger local educational system. TNAA and larger local firms tend to hire new graduates straight out of school. As a result, boutique firms experience more difficulty attracting raw talent, weakening this second tier. And with a greatly diminished role in developing a skilled labour pool, the smaller agencies and suppliers have fewer personal networks with TNAAs to facilitate intra-industry interaction and knowledge flows.

Global connectivity

In Bangkok, the creation of strong local advertising services is the result of a mutually reinforcing interaction between local factors of production and transnational firms and global markets. Bangkok's openness and exposure to outside ideas has fostered a blending of western and local influences to create a unique, competitive style. Effective mechanisms of global interaction contributing to flows of ideas and knowledge have primarily taken the form of skill transfers via foreign owned agencies, international labour flows, development of export markets and hosting and participating in international competitions.

Manila's advertising industry, by contrast, is much more parochial, despite the presence of many of the same TNAAs. Constitutional restrictions on foreign staffing and ownership in advertising firms, and a weak and underfinanced supporting industry, result in too few opportunities for global interaction, stunting the industry's development.

The Singaporean industry is characterized by high levels of global connectivity, through its TNAA regional headquarter status and large foreign work force. But these connections are largely restricted to the transnational agencies and their suppliers. In the absence of a more localized industry, there are limited opportunities for international ideas and technologies to stimulate, and fuse with, local creative energies.

In sum, Bangkok has the least passive advertising cluster of the three case studies in terms of distinctiveness and innovation based on a high level of hybridization, which, in turn, is built on a balance of localization and globalization within the industry. High localization in Bangkok is associated with embedding knowledge generation and creative innovation capabilities within the local cluster, through mutual reinforcing relationships between the TNAAs and local professionals, firms and institutions (associations, government) – networks and relationships deeply embedded in the local culture. High globalization is associated with a continuous infusion of new ideas and influences, through networks and relationships between the local cluster and international work force, clientele (export of services) and TNAA corporate networks.

Manila has a high level of localization, but lacks the hybridization, and therefore has become parochial. At the opposite extreme, Singapore has a sophisticated advertising cluster with high levels of international interaction, but it is expatriate driven, thus localization, and by extension, local knowledge generation and creativity, is comparatively weak.

Two key factors accounting for variation across these three Southeast Asian advertising centres are socio-economic distance and cultural distance. As a mediator between the clients' product and the market, the advertising agency requires local knowledge and understanding of the local market characteristics and trends. All cities differ from one another in terms of characteristics and conventions that influence local production and consumption, and therefore all subsidiaries of TNAAs will undergo some degree of localization (Storper 1997: 3). But the more significant the socio-economic and cultural differences between the TNAA's headquarters and the host country cities the greater the importance of locally generated knowledge and the higher degree of adaptation of local advertising products and content. This places pressure on TNAAs to become more locally embedded than might otherwise be the case.

Two additional factors are the global hub status and government policy. Places with regional headquarter status and more liberal business investment policies are more strongly connected into global advertising networks and outward-oriented.

Conclusion

Viewing the periphery as passive, in the case of business services such as advertising, is the wrong starting assumption. Peripheral nodes in the global advertising network can be highly dynamic, thrive and innovate, even if the local industry is dominated by foreign-owned TNAAs. Researchers need to get past the ownership structure of TNAAs, and look within the firm, within the industry, and beyond, for evidence of localization dynamics within, and global interaction between, industry nodes.

The evidence presented indicates that the most dynamic peripheries do not follow a development trajectory from expatriate dominance in the cluster to 100 per cent localization, but rather are consistent with movement toward ever more sophisticated hybridization. Hybridization requires serious and deep involvement of local people, firms and institutions, embedded in the local culture, in the creative process. But it also thrives on continual infusion and interaction with outside influences. One of the most effective mechanisms of global interaction is via 'internationals' – people who operate both within and outside the host country, although other mechanisms are also important. Internationals can be returnees from studying and working abroad, or localized foreigners who have adopted the host country as home.

The process of developing a sophisticated, hybridized industry is a cumulative and mutually reinforcing one. Strong networks within the cluster, and between the cluster and the international industry, are a sign of dynamism, but also reinforce it. The development process is not linear. It can take up to 20 years for the cycle to gain momentum, and for the local cluster to 'take off'. This is the time it takes to train a cadre of local skilled workers, for this generation to move up into positions of influence in the major agencies and to diffuse to other sectors of the cluster, and finally to take on leadership roles in the broader industry. But hybridization is not necessarily a steady state. The networks are based on people, and each generation of local skilled workers needs to educate and mentor new ones in order to sustain the high levels of interaction that produce unique combinations of ideas and innovations. Moreover, as local capabilities replace international interaction (e.g. local advertising education, local management, local production), new mechanisms of local-global interactions need to be actively sought out, in order to sustain cross-fertilization between a wide diversity of ideas and innovations.

The importance of transnational interactions with local economies in inducing regional production districts in 'peripheries', and in particular, the role of transnational knowledge workers and entrepreneurs as a key medium for this interaction, is already well understood by scholars of high-tech/knowledge regions (c.f. Saxenian 2000). As this case study illustrates, this mechanism of interaction is also critical in developing strong business service clusters in second and third tier centres.

Differentiation in performance and roles of different cities can be explained to a large degree by the extent to which localization and global connectivity of the cluster is deepened. This requires balancing what may at first glance appear to be opposing forces. Some factors affecting localization and global connectivity are influenced by government policy, or lack thereof. Equally important are other factors, often outside the realm of public policy, that are difficult to change in the short run such as:

1   local economic and social status (demographics, income distribution, preferences) reflected in consumption patterns;
2   the history and distinctiveness of the local culture (cultural distance) and cultural openness;
3   the role of the TNAA subsidiary in corporate networks, often corresponding to the status of the advertising centre, which is a given in the short run but can change in the longer run;
4   the prevalence of cross-cultural mediators ('internationals'), which requires a steady flow of people studying and working overseas and returning home.

Southeast Asian governments have been slow to promote business services in their economic development policies. Where the state has intervened to any significant degree in the industrial policy governing advertising so as to have some impact on localization dynamics, the net effects have been mostly negative. Such outcomes signal the need for a better understanding of internationalized service dynamics among government policy makers. The increasing importance of business services to Asian urban economies makes it imperative. The growing dominance of China in the global manufacturing system threatens Southeast Asia's current economic base and is forcing 'tiger cubs' to shift towards more knowledge-based, often service oriented economies. Toward this end, professional business services are a primary source of quality job creation, and represent an alternative or complement to high-technology industries in developing a knowledge-based economy. In addition, the availability of sophisticated, world-class business services can deepen and strengthen the economy of urban regions by supporting innovation and productivity in other sectors, including manufacturing, which contributes to, and positions Southeast Asian urban economies for strong economic growth.

Notes

1   As another indicator of advertising's growing economic importance, advertising firms are the only local business service sector firms that are listed on the Stock Exchange of Thailand (SET).
2   Tulayadha retired in 2002.
3   Interview with retired creative director of a TNAA in Bangkok, Thailand, conducted by the author, August 2001.
4   Estimate of 2001 President of the Advertising Association of Thailand.
5   Anderson (1984) uses the term 'bridgehead', but he portrays the role of these local elite in a negative light, as a mechanism of foreign penetration based on a dependency theory framework.

6 Wang (2000) considers TNAAs to be cultural brokers. Standing between the global manufacturer and the local consumer, the TNAA produces advertising of global products for local consumption (p. 81). He believes that a mixed workforce of expatriates and locals creates 'an effective juncture' between the foreign product and the local consumer. Expats, Wang contends, have a better understanding of the needs of international advertisers but their knowledge of local cultural conditions is limited. Local staff is needed to achieve authenticity in advertising messages.

7 The interviews were conducted by the author between October 2001 and April 2002 in Bangkok.

8 Marketing strategies of corporations are highly secretive, so as a courtesy to their clients, advertising agencies will not take on the advertising account for a competing product.

9 Interview with a Managing Director of a TNAA in Bangkok, Thailand, February 2002.

10 Interview with the Chair of a local advertising agency, in Bangkok, Thailand, February 2002.

11 Interview with an executive of a TNAA in Manila, Philippines, conducted by the author on February 2002.

12 Interview conducted by the author, February 2002, Manila, Philippines.

References

Adarco (2001) *Advertising Agency Directory, Thailand.* Vol 1–2. Bangkok: Adarco.

Anderson, M. H. (1984) *Madison Avenue in Asia: Politics and Transnational Advertising.* Canbury, NJ: Associated University Press.

AP Publications (1984–2003 – various years) *The Advertising Book: Thailand Advertising, Marketing and Media Guide.* Vol 1–15. Nonthaburi, Thailand: AP Publications.

Batey, I. (2002) *Asian Branding: A Great Way to Fly.* Singapore: Prentice Hall.

Beaverstock, J. V., Taylor, P. J. and Smith, R. G. (1999) 'A roster of world cities', *Cities,* 16(6), 445–58.

Chirapravati, V. (1996) 'Advertising in Thailand', in Toland Frith, K. (ed.), *Advertising in Asia: Communication, Culture and Consumption.* Ames, IA: Iowa State University Press.

Daniels, P. W. (1995) 'The internationalisation of advertising services in a changing regulatory environment', *The Service Industries Journal,* 15(3), 276–94.

Davies, R. (2002) Thailand Capsule. Orient Pacific Century. June 16, 2002 [cited March 30, 2004] Available from http://www.asiamarketresearch.com/thailand/.

Economist, The (2001) 'A matter of choice', December 22, 2001, 72–4.

Gearing, J. (2001) 'That Thai touch', *Asiaweek* (February 9), 42–3.

Landau, A. (2001) *Redrawing the Global Economy.* New York, NY: Palgrave.

Mertens, B. (2001) 'A woman with appeal', *Asia Business* (February), 12–14.

Nachum, L. (1997) 'Home countries' effects on the competitive position of advertising agencies: contrasting success and failure', in Aharoni, Y. (ed.), *Changing Roles of State Intervention in Services in an Era of Open International Markets.* Albany, NY: State University of New York Press.

Nachum, L. (1999) *The Origins of the International Competitiveness of Firms: The Impact of Location and Ownership in Professional Services.* Cheltenham: Edward Elgar.

Nachum, L. (2000) 'FDI, the location advantages of countries and the competitiveness of TNCs: US FDI in professional service industries', in Aharoni, Y. and Nachum, L. (eds), *Globalization of Services: Some Implications for Theory and Practice.* New York and London: Routledge.

Nation, The [Thailand] (2001) 'O&M's Sunandha stepping down', Bangkok, November 30, 2001, 1B.

Sassen, S. (1991) *The Global City: New York, London, Tokyo.* Princeton, NJ: Princeton University Press.

Saxenian, A. (2000) 'Transnational entrepreneurs and regional industrialization: the Silicon Valley-Hsinchu connection', in Tzeng, R. and Uzzi, B. (eds), *Embeddedness and Corporate Change in a Global Economy.* New York, NY: Peter Lang Publishing, Inc.

Storper, M. (1997) 'The city: centre of economic reflexivity', *The Service Industries Journal,* 17(1), 1–27.

Wang, J. (2000) *Foreign Advertising in China: Becoming Global, Becoming Local.* Ames, IA: Iowa State University Press.

8 Producer services and industrial linkages in the Hong Kong–Pearl River Delta region

Anthony G. O. Yeh

Introduction

In 1978, China began to experiment with economic reforms and the 'open door' policy. A number of new development areas, such as the Shenzhen Special Economic Zone, were set up to encourage foreign investments (Wong and Chu, 1985; Phillips and Yeh, 1989). This has substantially affected the development of Hong Kong and the Pearl River Delta (PRD) region. Many Hong Kong industries began to set up branch plants or to undertake subcontracting processes in China. A gradual 'regional division of labour' between Hong Kong and China's new development areas, especially the PRD region, began to take shape (Sit and Yang, 1997). This was quite common among labour-intensive industries such as garment, toys and electronics that were seeking to sub-divide their production processes by shifting labour-intensive production processes which can utilize unskilled labour to the neighbouring PRD in China. Outsourcing arrangements were quite often used to ensure operational efficiency and management control in Hong Kong. Many of these subcontracting arrangements were based on personal contacts because many people in Hong Kong have originated from the PRD (Leung, 1993).

Although economic reform and economic development in China, especially the PRD, have contributed to a decline in Hong Kong's industries, it has simultaneously led to the growth of the service and office sectors. Hong Kong industries that are outsourcing some production functions to the PRD are still using Hong Kong as a base for research, marketing and distribution. Hong Kong is also providing a hub function to China; notably its function as an entrepôt for the rapidly growing areas in the PRD. Re-exports to and from China have increased sharply since 1990. Apart from providing transhipment services by the world's second largest and highly efficient container port in Kwai Chung, Hong Kong also has a sophisticated banking system that can handle letters of credit for China. Many international firms are also using Hong Kong as a stepping stone for doing business with China, making use of local expertise and professional services.

The effect of economic restructuring in Hong Kong is reflected in changes in the employment structure between 1961 and 2002. The manufacturing sector increased steadily from 43 per cent in 1961 to 47 per cent in 1971 followed by a marked decline from 41.2 per cent in 1981 to 28.2 per cent in 1991 and to 6 per cent in 2002. There was a steady increase in the service sector (wholesale and retail trade, restaurants and hotels) from 14.4 per cent in 1961 to 31.8 per cent in 2002 and a more modest increase in the social service sector (community, social and personal services) from 18.3 per cent to 26.2 per cent during the same period. The marked increase in the finance, business and professional services sector mainly occurred after 1981. It increased from 4.8 per cent in 1981 to 10.6 per cent in 1991 and to 15.1 per cent in 2002.

The decline in the manufacturing sector has not only occurred in relation to its share of total employment but there has also been a decline in its absolute share. Employment in the manufacturing sector dropped from 0.9 million at its peak in 1981 to about 0.6 million in 1991 and to less than 0.2 million in 2002. Its contribution to GDP also dropped from 26.9 per cent to 15.9 per cent and then to 4.5 per cent over the same time span. It is anticipated that manufacturing employment will further decrease in the future under the prevailing economic and production environment. The industrial structure of Hong Kong has changed from labour-intensive to capital-intensive and high value-added modes of production, accompanied by large increases in subcontracting to the neighbouring PRD region in China. Apart from its effect on the economy, such dramatic sectoral restructuring has also affected land use in Hong Kong. There is an ongoing spatial shift of industries from the main urban area to the new towns which are closer to the border with mainland China and also offer lower cost industrial land. The demand for all types of industrial buildings in the main urban areas has declined (Yeh, 1997).

While the contribution of the manufacturing sector has been declining, the producer services are becoming an important pillar of Hong Kong's economy, providing crucial services not only to the local community but also to industries in the PRD hinterland. This chapter examines some of the reasons for these changes, and provides a case study of the linkages between Hong Kong's producer services and the industries in the PRD, and considers likely trends in the future development of such linkages.

Industrialization and the development of producer services

Increasingly, producer services are serving other services industries rather than manufacturing industries. For example, according to Coe's study (1998), computer services in the UK are primarily dependent upon other service activities (some 72.4 per cent of their clients) rather than manufacturing sector (19.7 per cent of their clients). Despite this, development of producer services still has close relationship with industrialization. This is especially true in the PRD where, for historical reasons, the service sector is relatively under-developed. During the pre-reform period, most parts of the Delta region were farming communities, mainly engaging in agricultural activities. Guangzhou was the only city to embark on socialist industrialization using a very limited capital allocation from the central government. However, under state socialism, the service sector in Guangzhou was not well developed because of the past policy bias towards heavy industry. Local development, therefore, focused more on production rather than consumption. Each factory/work unit provided basic services to its employees and was responsible for its own services if there was an identified need. The demand for outsourced producer services (e.g. banking, insurance and marketing) was rare because the central government served as the only commander in deciding production orders and was the sole agent in resource allocation. Therefore, producer services were significantly underdeveloped. During the reform period, even though the quantity and quality of service industries have improved greatly in the PRD, the service sector, and in particular producer services, is still largely under-developed by comparison with the astonishing speed of industrialization. There is still much room for expansion of producer services in the Delta.

Based on the experience of other countries, a note of some of the characteristics of producer services is useful when we study the linkages between producer services in Hong Kong and industries in the PRD. First, as noted earlier, producer services are increasingly having closer relationships with other services. However, the manufacturing industry is still an important sector to producer services. This is especially true in the PRD when the service industries are not well developed. Second, producer services centres worldwide are usually in close proximity to a major seaport and/or international airport, where there are also concentrations of professional and business services as well as banking and financial services with convenient land, road and railway transport. Many of them are world cities, such as London, New York, Tokyo, Hong Kong, Moscow and Mexico City. Recently, the Globalization and World Cities (GaWC) Research Group and Network put forward the idea that a hierarchy of world cities can be derived using advanced producer services as an index (including accounting, advertisement, financing, banking and legal services), exemplifying the importance of producer services in the world city system and economy (GaWC, 2003). Moreover, a healthy international producer services centre should be based on its economic hinterland and then provide services to the world as a whole. Third, business cities can provide producer services as their main function; whereas manufacturing centres can also develop into producer services centres. Fourth, there are two kinds of producer services: those produced internally and those available externally. The former refers to the producer services provided by the enterprises themselves; the latter is the services from the freestanding professional services firms. Generally, because of their specialization and expertise, the externalized producer services are more economical and efficient, and more likely to provide state-of-the-art knowledge and expertise than those provided internally.

Competitive advantages have been identified as major factors affecting the growth of cities and nations (Porter, 1990). Certainly, competitiveness is a key factor for the future success of Hong Kong and the PRD, especially in the era of the global economy and the entry of China into the World Trade Organization (WTO). Despite still possessing some advantages over its competitors, such as an open and fair market system, an efficient and well connected sea and air cargo handling infrastructure, a good standing in fund management and as an international financial centre, and a role as an international communication hub (Enright *et al.*, 1997), Hong Kong has still lost its manufacturing industries to the PRD after it became much more open to foreign investment in 1978. Although a large number of manufacturing firms have moved to the PRD, Hong Kong still provides producer services to these industries, providing them with banking, marketing, trading and shipping services. A 'Front Shops, Back Factories' regional division of labour has been formed in the PRD region with Hong Kong providing the front-end marketing and office services whereas factory production is done at the back of Hong Kong – the PRD (Sit and Yang, 1997). Hong Kong businessmen, who used to be the receivers of multinational firms and orders in the 1960s and 1970s, have now become the major source of foreign investment, utilizing the cheap labour and land in the PRD as its production base and hinterland. In the late 1970s, Hong Kong manufacturers were already facing competition from the newly industrializing countries in Asia which had much lower production costs. Economic reform and the open door policy in China in 1978 had again given businessmen in Hong Kong the competitive advantage in the global economy. Instead of 'Made in Hong Kong', it is now 'Made in China' with Hong Kong's capital and management. As the headquarters of Hong Kong's manufacturing industries in the PRD are still in Hong Kong, there has been a major growth of producer services. Because of the 'shop' function, it has grown from 4.1 per cent of the total employment in 1981 to 10.9 per cent in 1991 and to 20 per cent in 2001. However, with very high labour and operating costs in Hong Kong and socio-economic changes in the PRD and other parts of China, especially Shanghai (Yeung and Sung, 1996; Yeh, 1996), there is doubt as to whether the 'shop' function, on which the economy is increasingly heavily dependent, can be sustained.

Because of globalization and China's accession to WTO, China has become the world's most important manufacturing nation. This brings unprecedented challenges and opportunities for the Greater PRD Region, which includes Hong Kong, Macao and the PRD in Guangdong (Figure 8.1).[1] The Greater PRD Region as a whole has a strong economic base, an enormous market, and a relatively well-developed infrastructure and service system. With a population of 47 million, the Greater PRD is one of the world's largest metropolitan regions; the portion in Guangdong Province alone is the world's largest manufacturing workshop, and also one of China's most important economic centres. The GDP for the PRD portion in Guangdong reached US$90 billion in 2002 with GDP per capita as high as US$4,000; this makes the PRD one of the richest regions in mainland China. If the PRD, Hong Kong and Macao are considered together as one economic region, their economic output and wealth is much higher than that for the Yangtze River Delta (YRD) centred on Shanghai. Rapid economic development of the PRD has greatly benefited from the large inflow of foreign investment, particularly investment from Hong Kong and Macao. Eighty per cent of the overseas capital from Hong Kong and Macao is directed to the PRD, accounting for nearly half of the total amount of foreign capital in the country. The huge inflow of foreign capital has enabled the Delta to develop its export-oriented economy and become the most important exporter in mainland China. Concomitant with the rapid industrialization in the PRD, the relocation of industrial capital from Hong Kong and Macao has triggered a process of economic tertiarization in these two former colonies, especially in Hong Kong. The synergy so created has formed the basis of the regional division of labour described by the 'Front Shops, Back Factories' (FSBF) model. Hong Kong and the PRD have thus become two highly dependent and mutually complementary entities.

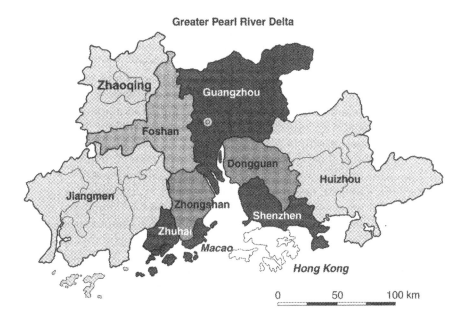

Figure 8.1 The Greater Pearl River Delta Region.

The collaborative relationship described by the FSBF model has brought benefits to both Hong Kong and the PRD. But can this synergy last forever? This question is particularly important to Hong Kong, as many people argued that Hong Kong is now in a process of decline as a result of the rapid development in its PRD hinterland. Today, the latter is not simply a 'world factory' at the back, but also a major market that has developed very quickly and offered immense business opportunities. With the remarkable improvements in service industries, the PRD has made gains in sectors such as logistics, finance, high-tech industry and manufacturing as well as tourism. It is catching up quickly, if not yet competing fiercely, with Hong Kong and Macao in the supply of services. The traditional roles of Hong Kong, Macao and cities in the PRD have to be reassessed in this new regional context. For example, many cities in the PRD have constructed large infrastructure facilities, such as Yantian Port in Shenzhen and New Baiyun International Airport in Guangzhou. These link the region more directly with global players and effectively bypass the equivalent facilities available in Hong Kong or Macao (Xu and Yeh, 2003). Manufacturing industries in the PRD no longer have to use the expensive services available from Hong Kong service providers.

In addition to the increasing inter-city competition within the region, the Greater PRD is also facing external competition. The most notable is the YRD, centred on Shanghai, which has developed rapidly to become a new economic powerhouse in China, attracting a large amount of overseas capital. In 2002, the State Economic and Trade Commission, the Shanghai Committee of Chinese People's Political Consultative Conference, and the Shanghai Academy of Social Science co-organized a conference entitled 'Regional Interactive Development in the Yangtze River Delta'. During the conference, Shanghai, Jiangsu Province and Zhejiang Province reached a consensus on the promotion of regional interactive development through embarking on six major agreements to improve the regional environment, to reduce development costs and to promote the global competitiveness of the region as a whole. In terms of regional cooperation, the YRD has moved ahead of the Greater PRD. Meanwhile, service industries in Shanghai are also developing very rapidly (Ning, 2000). In view of these challenges, the Greater PRD region, including Hong Kong and Macao, urgently needs to speed up the pace of regional cooperation, and move on from the FSBF model to promote regional competitiveness and to explore domestic and international markets. Only in this way can the Greater PRD region enlarge the scope of the tripartite partnership and develop into another economic powerhouse in China together with the YRD. To achieve this objective, various parties within the Greater PRD should utilize their respective advantages to promote complementary development. It is through the combination of the PRD's advantage in manufacturing industry and that of Hong Kong in business services that can promote economic competitiveness of the Greater PRD region. Only in this way can the manufacturing industry in the Delta become more competitive and the role of Hong Kong as a regional service provider be enhanced.

Linkages between the PRD's manufacturing industries and Hong Kong's producer services

To examine the linkages of Hong Kong producer services with the Pearl River Delta, a questionnaire survey of factories was conducted in 2002 in Dongguan and Zhongshan, two major industrial cities in the PRD with a lot of foreign-invested factories from Hong Kong, Taiwan, Japan and other countries. Dongguan is at the eastern side of the PRD and Zhongshan is at the western side, giving a good geographical coverage of the PRD. As Hong Kong's industries in the PRD are mainly those producing toys, electronics and clothing, the survey targeted on factories producing these products in both cities. Six towns in Dongguan and one town in Zhongshan with a high concentration of toys, electronics and clothing industries were chosen for the questionnaire survey. An aerial random sampling was carried out in these selected towns. 784 successful interviews have been obtained.

Producer services in this survey are divided into four categories (Table 8.1), which include financial services, business and professional services, information and communication services, and trade-related services. Most of the factories were using in-house producer services. Only 19.8 per cent used external producer services outside their factories (Table 8.2). The use of external producer services was the lowest among domestic factories,[2] indicating that they were still doing most of their business in-house. This may be because most of them were producing products for the domestic China market and not export-oriented and did not need advanced professional services provided by external service suppliers. The use of external producer services increased from 20 per cent of Hong Kong's factories to 47.6 per cent of factories from other countries, a reflection of their export-oriented nature. The use of external producer services was also related to the size of factories. Small factories used less external producer services than very large factories (Table 8.3), indicating the relationship between the scale of operation and the need for external professional producer services.

Factories in the PRD used more domestic producer services offered by service providers within the Delta and from other parts of mainland China than overseas producer services provided by service firms from Hong Kong or other countries (Figure 8.2). 38 per cent used producer services from domestic providers, whereas only 13 per cent used overseas suppliers (8 per cent for Hong Kong services). Among all kinds of producer services, the most frequently used producer services in Hong Kong was trade-related services (13.4 per cent) which include import/export trade,

Table 8.1 Classification of producer services

Financial services	Banking
	Non-banking
	Insurance
Business and professional services	Legal
	Accounting
	Management consulting
	Building and construction
	Design
	Engineering
Information and communication services	IT services
	Communication
Trade-related services	Import/export trade
	Publishing
	Advertising and market research
	Sea transport
	Air transport
	Services related transport
	Convention and exhibition
	Testing
	Arbitration and mediation

Table 8.2 Use of external producer services outside the factory by sources of investment

Source of investment	% of factories using external producer services outside the factory
Domestic	10.2
Hong Kong	20.1
Taiwan	32.1
Japan	25.0
Other countries	47.6
Total	19.8

Table 8.3 Use of external producer services outside the factory by factory sizes

Factory size	% of factories using external producer services outside the factory
Small (0–49)	10.2
Medium (50–199)	18.3
Large (200–499)	27.9
Very large (500 or more)	29.9

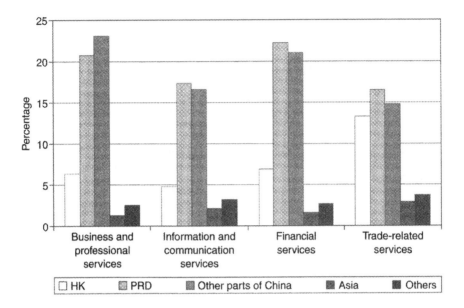

Figure 8.2 Location of producer services by type.

advertising and marketing, and logistics. This is followed by financial (6.9 per cent) and business and professional services (6.4 per cent).

The result of the survey also shows that the main reasons why manufacturing firms in the PRD use Hong Kong producer services are factors that have little to do with cost. These non-cost considerations, such as the need to stay abreast of changing technologies, the need to stay abreast of rapidly changing economic conditions, lack of technical expertise, and increasing complexity of doing business in an international context, were the main reasons. Cost factors, including downsizing by clients or a desire to be 'lean and mean', the perception that outside firms can perform a task more cheaply, were a second consideration after non-cost factors. This result is in line with the findings of another survey into Hong Kong's producer services firms by the author (Yeh, 2003). According to that survey, Hong Kong's producer services firms also considered that clients from the PRD used their services mainly because of non-cost considerations. Similarly, factors such as the increasing complexity of doing business in an international context, a lack of technical expertise, the need to stay abreast of changing technologies or of rapidly changing economic conditions, were generally considered to be the crucial considerations. The credibility, efficiency and international perspective provided by Hong Kong's professional services were regarded as the main attraction for the PRD's manufacturing firms. Under the 'One Country, Two Systems' principle, Hong Kong's legal and financial systems, and comparatively clean and efficient government, give Hong Kong, overseas and local businessmen in the PRD a great deal of confidence in using Hong Kong's producer services.

According to our survey, in the PRD, the utilization of Hong Kong's producer services is not as high as what the 'Front Shops, Back Factories' would have suggested. Twenty-nine per cent of Hong Kong manufacturing firms and 15–20 per cent of firms from other parts of Asia used services from Hong Kong, whereas the figure for PRD's domestic firms using Hong Kong's services is only 10 per cent (Figure 8.3). They are using a lot of local services in the Pearl River Delta and other parts of China, especially in Shanghai and Beijing. The use of producer services by Hong Kong firms is not as high as what the 'Front Shops, Back Factories' model has suggested. The highest use of Hong Kong's producer services is trade-related services which is 38.7 per cent (Table 8.4). This suggests that the 'Front Shops, Back Factories' model is under challenge by local producer services. This is partly because these firms are operating in the mainland and they have to use Chinese producer services because of proximity, familiarity with the Chinese system and language. Furthermore, with rapid advancement in social and economic development in the PRD since 1978, although the services in the PRD are still cheap compared to Hong Kong's prices, the service level has improved. If the survey was done ten years ago, the percentage of the use of Hong Kong's producer services would be much higher because of the relative underdevelopment of producer services in the PRD. But, increasingly, this is not the case. The quality of services is catching up very fast in the PRD.

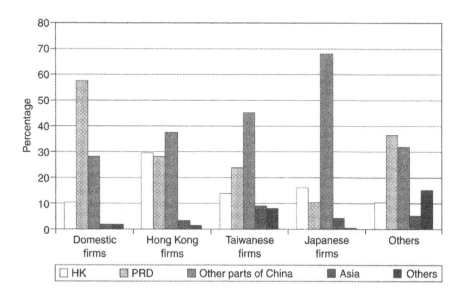

Figure 8.3 Location of producer services by source of investment of manufacturing firms in the PRD.

Table 8.4 Location of producer services used by Hong Kong-invested factories

Producer services	Location (in %)					
	HK	PRD	Other parts of China	Asia	Others	Total
Business and professional services	22.6	28.2	45.8	2.2	1.3	100
Information and communication services	24.7	29.5	39.5	4.6	1.7	100
Financial services	26.8	28.6	39.7	3.6	1.2	100
Trade-related services	38.7	27.0	28.6	3.5	2.2	100
All producer services	29.3	28.1	37.7	3.3	1.6	100

In 2002, there were 16,049 firms in the PRD in total, among which 7,443 (46.4 per cent) were domestic firms, 7,149 (44.5 per cent) were firms invested by Hong Kong, Macao and Taiwan, and 1,457 (9.1 per cent) were foreign invested firms (China INFOBANK, 2004). There are a large number of domestic firms in the PRD, which together with Hong Kong's firms form a potential market for Hong Kong's producer services. However, domestic manufacturing firms in the PRD have barely used Hong Kong's producer services. The use of Hong Kong's producer services was only 10 per cent. Trade-related services, such as trade and logistics services, are the most frequently used of Hong Kong's services (15 per cent) (Table 8.5). Hong Kong has not yet fully utilized its competitive advantages in services in general and producer services in particular to provide their highly efficient services to the Delta hinterland, as exemplified by its low market share in the region. As such, Hong Kong is also not very successful in benefiting from the robust development of domestic manufacturing industries in the Delta to further develop its own economy through producer services. There is still much room for mutual coopera-

Table 8.5 Location of producer services used by domestic factories

Producer services	Location (in %)					
	HK	PRD	Other parts of China	Asia	Others	Total
Business and professional services	8.0	59.9	30.8	0.7	0.6	100
Information and communication services	9.1	58.3	29.6	1.9	1.2	100
Financial services	9.3	63.8	24.9	1.4	0.6	100
Trade-related services	15.0	50.6	26.9	3.6	3.9	100
All producer services	10.6	57.8	28.0	2.0	1.7	100

tion between Hong Kong and the PRD in this regard, especially when the domestic manufacturing industries in the PRD are to be further developed in firm size and to be more export oriented.

Combining advantages of the PRD's manufacturing industry and Hong Kong's producer services to develop world-class manufacturing and business service base in the Greater PRD region

According to Porter (1990), regional competitive advantage is comprised of four main determinants: factor conditions; demand conditions; related and supporting industries; and firm strategy, structure and rivalry. After many years of development, the PRD has achieved several competitive advantages in manufacturing industry. First, the Delta is now endowed with basic and advanced production factors, which include a cheap but highly efficient labour force, relatively cheap land resource, and a well-developed infrastructure and educational system. Second, the PRD has become a major base for light manufacturing industry in China and even in the world. It is also developing into a world production centre of high-tech products (Enright *et al.*, 2003). In many aspects, the Delta's dominant manufacturing role is significant. For example, the outputs of electric fans, sound boxes and telephone sets in the PRD account for 88 per cent, 80 per cent, and 79 per cent respectively in the total production of China. The Delta is developing a broad range of extremely competitive clusters of light-manufacturing and electronics goods that produce the inputs and components in the Delta to support its world-leading producers of toy, footwear and electrical and electronics goods (Enright *et al.*, 2003). Third, the PRD is a huge consumer market. With a population of 40 million, the Delta has experienced a transformation in its social structure with the middle income/class households rising up quickly as the main consumption elites. In addition, numerous factories located in the Delta have made this region an immense market for raw materials and industrial products. Fourth, local governments in the Delta are very active in city development and policy-making to promote economic restructuring to attract footloose capital and talents. As a result, the service sector in the Delta has been greatly improved with more and more export channels, sales networks and information service being set up very quickly (Yan, 1998). In the near future, it is possible for the Delta to develop into a secondary regional business centre. More and more foreign investors have chosen the PRD rather than Hong Kong as a platform to invest in mainland China. Their market strategy has also changed to focus more on domestic rather than international markets. Guangzhou, which is a traditional business and trade nucleus in southern China, is more advantageous than Hong Kong in terms of domestic market networks and proximity to the industrial base in the PRD without the need to go through customs. It thus has the potential to become another platform after Shanghai for foreign investors to penetrate the Chinese market. It is even predicted by some foreign firms that within five years the intermediate and low-end service sector in the PRD will reach the current quality level of Hong Kong. At that time, many firms would like to use local services which are cheap and easily accessible instead of Hong Kong services which are expensive and not easily available in the local area. If this was true, could Hong Kong continue to serve as a regional business centre? Some even argued that services in Hong Kong would follow the propensity of manufacturing industry to hollow out the city. As such, the role of Hong Kong as a regional business and trade centre would then be under severe challenge.

Nevertheless, in the short and medium term at least, Hong Kong still has many advantages over the PRD. Its major competitive advantage lies in its high-end producer services. It not only has the world-class infrastructure facilities, but also has an excellent business environment, including a strong institutional environment and a highly efficient government. At present, Hong Kong has the busiest and the most efficient container port in the world. Hong Kong International Airport is also the world's busiest freight airport and the third busiest passenger airport. All these facilities have made it the major transportation hub in the Asia-Pacific region. Many foreign companies have committed investment and set up regional headquarters in Hong Kong. Currently, the city is the fourth largest source of foreign direct investment, the eighth largest trade economy, and the ninth largest service exporter in the world. In terms of the financial sector, Hong Kong is the world's fourth largest banking centre and fifth largest currency market, having the highest density of fund managers and the largest number of authorized insurance firms in Asia. It is also the biggest centre for the allocation of credits for enterprises in Asia and the sixth largest security market and twelfth largest stock market in the world. In terms of the information sector, Hong Kong is also the biggest teleport in Asia and the first city in the world to have a fully operational digital telephone network. The city also has the busiest international telephone lines in Asia and the second highest usage rates of fax in the world. It ranks first in Asia in terms of density of fibre optic cable and ranks first in the world in terms of speed in the growth of Internet usage.

At the same time, Hong Kong's institutional environment also confers many advantages on its high-end producer services. The principle of these is the 'One Country, Two Systems' which other cities in China do not have. For example, under the 'One Country, Two Systems', the Hong Kong SAR government has the right to negotiate directly with foreign countries for air freedom. No government in the PRD enjoys such a privilege. After the return of Hong Kong to China in 1997, it still retains its efficient and transparent government, well-developed legal system, and wide international perspective that have, for example, given it an advantage in the recent development of high-end time critical logistics services.

Although Hong Kong is doing very well in most economic and competitiveness indicators, one of the main problems is its high land price and operating costs (Yeh, 2001 and 2002). It is eroding the competitiveness of Hong Kong in the development of manufacturing and service industries. Although the salary in the PRD is on the rise, Hong Kong's salary on average is still 7–8 times that of the salary in the PRD, making Hong Kong's producer services less attractive and competitive than those of the PRD. Some companies have already moved their operations away from Hong Kong. For example, the data processing work of Cathay Pacific has moved to Perth in Australia and that of the Hong Kong and Shanghai Bank to Guangzhou in China.

After China's accession to the WTO and the recent signing of CEPA (Closer Economic Partnership Arrangement)[3] between Hong Kong and the Mainland, the de facto tariff and non-tariff barriers for cargo trade between the PRD and Hong Kong would be decreased or eliminated to gradually realize a free trade zone under WTO rules. This would lead to eradication of all discriminative policies and increasing the potential of Hong Kong to become the centre of logistics and business services in southern China, and also creating unprecedented opportunities for Hong Kong and the PRD to embark on a complementary development strategy.

With the accession of China to WTO, regional development in the PRD may have a brighter future, which will lead to a larger demand for producer services. Our survey also showed that there is a great potential market for Hong Kong's producer services in the PRD. More mainland firms need to extend their relationship with the international business community. This has brought many business opportunities for the service sector in Hong Kong. As indicated in the China's Tenth Five-Year Plan, the service sector will become the motor for promoting further economic growth in China, which implies a dramatic increase in demand for producer services. Manufacturing industry in the PRD is well developed, whereas Hong Kong's service sector can provide high quality and very efficient services with a wider international perspective. Using these combined advantages, the Hong Kong-PRD (GPRD) region can be developed into a world class manufacturing and business centre. Currently, Hong Kong still has prominent advantage in producer services in the Greater PRD region. Although Hong Kong's population is only 30 per cent of that of the PRD, the employment in producer services is 1.05 times higher than that in the Delta (Table 8.6). Guangzhou, the second largest business centre following Hong Kong in the Greater PRD region, has only one tenth, one third, and one eighth of Hong Kong's GDP, employment and value-added in producer services respectively (Table 8.7). It is promising to use Hong Kong's producer services to promote PRD's competitiveness in manufacturing industry.

Table 8.6 PRD's employment in producer services in 2002 (in 000's)

City	Producer services	Service industries	% of producer services in service industries	% of producer services in total employment
Guangzhou	57.97	221.05	26.23	11.28
Shenzhen	24.11	154.96	15.56	6.71
Zhuhai	7.43	43.06	17.24	8.41
Huizhou	9.49	38.11	24.9	9.12
Dongguan	8.77	42.75	20.53	7.06
Zhongshan	9.07	66.12	13.72	4.32
Jiangmen	15.39	72.26	21.29	7.48
Foushan	8.45	63.75	13.25	4.13
Zhaoqing	1.65	8.26	19.98	6.14
PRD (1)	142.32	710.31	20.04	7.75
Guangdong (2)	227.54	1,358.53	16.75	5.50
Percentage of PRD in Guangdong Province (1)/(2)	62.55	52.29	–	–
Hong Kong (3)	91.36	293.26	31.1	26.2
Percentage of HK in PRD	64.19	41.29	–	–
Percentage of HK in Guangdong Province	40.15	21.59	–	–

Source: Guangdong Provincial Bureau of Statistics (2003), Hong Kong General Chamber of Commerce (2003).

Table 8.7 Comparison of producer services in Hong Kong and Guangzhou

Item	Year	Item	Unit	Guangzhou	Hong Kong
GDP	2001	Total services	RMB billion	146.34	1,164.45
		Producer services	RMB billion	69.63	672.44
		% of producer services in total services	%	47.6	52.9
Employment	2002	Total services	1,000 people	221.04	2,932.6
		Producer services	1,000 people	41.38	913.6
		% of producer services in total services	%	18.7	31.2

Source: Guangzhou Statistics Bureau (2002), HKGCC (2003).

Combined advantages in Hong Kong and the PRD: challenges and opportunities

The signing of CEPA and China's accession to the WTO has presented Hong Kong with both challenges and opportunities. Many studies have tackled these issues in details (Enright *et al.*, 2003; BPF, 2002; HKGCC, 2000). This chapter will not repeat their arguments, but for the Greater PRD region, the biggest challenge is how to coordinate internal competition and face those external to the region in order to promote regional competitiveness through combining the advantages of different places. As far as Hong Kong is concerned, it is imperative to eliminate cross-border traffic barriers and to promote restructuring of the service sector to offer better services to the PRD.

Eliminate the cross-border barriers as soon as possible

'One Country, Two Systems' makes Hong Kong different from mainland cities. It is a unique advantage of Hong Kong. To sustain this advantage, the border between Hong Kong and mainland China is necessary. On the other hand, the border has also become the barrier hindering cross-border integration (Yeh, 1995). Government policies and other administrative measures, such as planning and building cross-border infrastructure facilities, streamlining and simplifying cross-border procedure, are a possible solution to minimize the negative effects caused by the border.

Cross-border transport facilities are always being planned or are under construction in Hong Kong. Since 1989 the new international airport and container port have become the busiest transport terminals in the world with world-class facilities and services. Hong Kong-Shenzhen Western Corridor connecting Yuen Long and Shekou of Shenzhen will be completed in 2005 to relieve the current pressure on freight transport. Hong Kong has also been active in proposing and building a bridge to link the city with Macao and Zhuhai in the western part of the PRD. Even though new cross-border infrastructure facilities are constantly being proposed and planned it will take at least seven to eight years to plan and complete the projects. This long lead time may hinder the process of Hong Kong-PRD integration. The pressing issue is therefore how to fully utilize existing cross-border transport facilities to augment the commercial and trade linkage between Hong Kong and the Delta.

The Hong Kong government has always paid great attention to improving cross-border freight traffic. Progress has been made in the implementation of co-located immigration and custom facilities to shorten and simplify the procedure in customs clearance for freight traffic. Hong Kong has reached consensus with the mainland side to implement this co-location arrangement at Huanggang and the new control point on

Shenzhen Western Corridor to be completed by mid-2005. The current 24-hour container truck and passenger clearance operation has also been another milestone achievement in further enhancing cross-boundary facilities. The 'SuperLink China Direct' provided by 'SuperTerminal 1' of the Hong Kong International Airport offers tailor-made services to enhance connectivity with China. It is an innovative intermodal air-land-sea transportation network designed to significantly streamline the traditional supply chain. Through simplified customs clearance arrangements, pre-registered bonded trucks can pass directly through the Hong Kong and Chinese border control points to be cleared by customs at a cargo's destination in China. This arrangement speeds up the logistics process. Traditionally, it often takes between two and seven days to move cargoes from Hong Kong International Airport to destinations in China. But this time span has been reduced to less than 13.5 hours under the new arrangement. These examples show that there is still much room for Hong Kong to improve the efficiency of the existing cross-border transport facilities. Without further improving border-crossing mechanisms, relieving cross-border traffic pressure on its own by building up more infrastructure will not be sufficient.

In terms of cross-border transport, the Hong Kong government has paid more attention to freight traffic than passenger traffic. This is understandable because the revenue streams from the logistics industry has contributed greatly to the local economy because of the presence of the world-class sea navigation and air cargo services provided by the container port and international airport. Despite advanced communication technology, face-to-face contact is nevertheless still important for establishing business network (Daniels and Moulaert, 1991; Illeris, 1996). Cross-border business passenger flow is very important to both Hong Kong and the Delta, but there exist many cross-boundary traffic problems that not only tie the hands of the government to provide significant support for business travellers, but also hinder further economic integration. The most pressing task is to streamline the procedure of custom clearance to ease the discomfort for business travellers; for example, it is still a time-consuming process for business people in the PRD to apply for entry permits to Hong Kong. Applications have to be sent to the relevant departments in the mainland, and it normally takes 15 working days before entry permits are issued to eligible applicants. The time span involved is even longer than that required when applying for visas to enter countries in Europe and North America. This situation has inevitably impaired the role of Hong Kong as a regional producer services centre. It is thus imperative to set up a mechanism to provide prompt entry permit application for business travellers from the mainland. To do this, Hong Kong and the Delta should first set up a Greater PRD Chamber of Commerce and encourage qualified commercial organizations to join the Chamber. This can extend commercial linkage between Hong Kong and the Delta, and at the same time provide just-in-time visa application services for the member companies, similar to the APEC arrangements. Hong Kong and the Delta should be more cooperative in formulating policy for cross-boundary passenger movements than is presently the case.

In addition to speeding up entry permit application process, a lot of effort should also be made to relieve the physical problems caused by a very congested border crossing. Currently, the Luohu checkpoint is seriously overloaded; during high season it handles as many as 230,000 cross-border passengers each day during weekends. This over-crowded border already hinders normal cross-border business activities. If Hong Kong wishes to become the regional business centre in the PRD, it should consider incorporating a separate checkpoint facility for business travellers, thus helping to further facilitate commercial integration between Hong Kong and the PRD.

Furthermore, it is also necessary to improve the current cross-boundary transportation network. First, the service frequency of Guangzhou-Kowloon Through Trains should be extended, especially the operation of late evening services in order to facilitate same-day business trips and dinners. Second, water transport can be better utilized in the Greater PRD region. Currently, Hong Kong and Macao are connected mainly by a highly efficient jet boat service operating at 15-minute intervals. If such services with high efficiency and frequency could also be provided to connect Hong Kong with major cities in the PRD, water transport would become an important alternative to the Through Trains for business passengers and would also help to ease the pressure on overcrowded passenger facilities at the land borders.

Transform the Hong Kong service industries to meet the demand from the PRD

Any prosperous world city must have efficient interactions with its economic hinterland, and then extend its services to the world. Large cities, such as London, Paris, New York, Sydney and Tokyo, have strong linkages with their hinterlands. They are capable of attracting a large number of visitors from these hinterlands, which in turn consolidates the function of these cities as regional service centres (Table 8.8). Hong Kong does not stand up to scrutiny in this respect. For a long time, service firms in Hong Kong have had a very limited regional market, mainly providing services to overseas companies located in Hong Kong. Neither local industries nor Hong Kong industries in the Delta are major clients. It is widely acknowledged that service functions in Hong Kong, such as finance, legal and professional services, have a wide international perspective, which is an advantage unmatched by their mainland counterparts. Thus, Hong Kong's service firms have the ability to offer essential assistance for industries in the mainland to explore global markets.

Table 8.8 Comparison of foreign and domestic visitors in major cities, 1995–1999 (million visitors/year)

	Hong Kong	New York	London	Paris	Sydney	Tokyo	Shanghai
International	7.0	6.0	13.5	10.9	2.6	2.5	1.5
Visitors	(73%)	(18%)	(54%)	(55%)	(8%)	(2%)	(2%)
Domestic	2.6	27.3	11.6	9.0	31.6	150	71
Visitors	(27%)	(82%)	(46%)	(45%)	(92%)	(98%)	(98%)
Total	9.6	33.3	25.1	19.9	34.2	152.5	72.5

Source: Ng and Hills (2000).

In order to attract more mainland businesses to consume and purchase services in Hong Kong, it is necessary to transform the service sector to cope with the mainland market. There is a need to find out more about the demand and needs of producer services in the PRD to see how these can best be captured. This requires many additional efforts. First, the human resources in Hong Kong need to improve their multi-language ability in written and spoken Chinese (putonghua) and languages other than English, such as Japanese, German and French. Second, there is a need to readjust manpower training in Hong Kong to meet the needs of the Chinese market, especially culture, professional practice and government laws in the mainland. At the same time, it is also essential to become familiar with international business practice. In the past, most of Hong Kong's professionals in business services were only familiar with local laws and lacked knowledge of Chinese or international law. For example, increasing globalization has greatly increased the need of firms in China for international arbitration and practice. To establish Hong Kong as an international as well as regional producer services centre, some of the human resources need to be trained professionally in this field, so that they can be competent in international arbitration and practice in the future. Furthermore, Hong Kong should actively develop an effective marketing strategy to introduce Hong Kong's advantages in finance, international trade, quality control and market promotion, to manufacturing firms in the mainland. Simultaneously, Hong Kong should play an important role in promoting industrial firms of the mainland to the international business community. At present, there are many domestic factories in the PRD that have the potential to enter the international market. Hong Kong has yet to play an active role in promoting these factories and their products to the international business community. In addition, Hong Kong's government should consider setting up a business service information centre to provide a one-stop information service to firms from mainland China and overseas countries. At some major checkpoints (such as Luohu and Huanggang) and ferry and train terminals it is still rare to see tourist and trade information available for arriving passengers.

In addition to all these issues, there are some other barriers to cross-border cooperation. These include, for example, problems resulting from foreign currency control in the mainland. It is believed, however, that these difficulties can eventually be resolved as part of the ongoing process of further integration between Hong Kong and the Delta.

Conclusion

For years, the economic integration between Hong Kong and the PRD has been a market-led spontaneous process. Since the return of Hong Kong to China in 1997, this has forged ahead within the context of a more favourable institutional environment. Better communication channels with mainland China have been established gradually. The recently signed CEPA agreements provide easier access for Hong Kong manufacturers and service providers in China's market. Manufacturers benefit from a zero tariff on 90 per cent of Hong Kong domestic exports to the mainland. The 18 service sectors gain faster and easier market access in China too. A series of convenient measures will also be taken to simplify procedures in bilateral trade arrangements. It is anticipated that the Greater PRD region can develop into a world-class manufacturing and business centre if it can combine its advantages in manufacturing with those of Hong Kong's producer services. However, according to our survey, Hong Kong's producer services do not have much linkage with the domestic manufacturing firms in the PRD which is growing in size and surpassing Hong Kong manufacturing firms in numbers. The past development model of 'Front Shops, Back Factories' is under challenge. The use of producer services by factories in the PRD is not as high as the model would have suggested. To serve the region better and to further develop its producer services, Hong Kong has to transform from a city that just performs the role of a middleman to a truly regional business centre. To realize this transformation, the provision of world-class 'hardware' or infrastructure facilities is not in itself sufficient. More importantly, within the context of 'One Country, Two Systems' Hong Kong has to eradicate all the barriers impeding its function as a regional service centre by improving cross-border traffic and transforming its producer service industries to better meet the needs of the firms in the PRD. If it fails to accomplish these measures, Hong Kong may lose out to the rapidly growing producer services in other cities in the PRD. At present, like its industries before the open door policy of China in 1978 which had a predominant advantage over the industries in the PRD, its producer services still have predominant advantages over those of the PRD, although they are relatively expensive. However, because of its high costs, like its industry, it is quite possible that Hong Kong's producer services may lose out to the PRD which is rapidly developing its service industries, especially if Hong Kong cannot make good use of its competitive advantages of 'One Country, Two Systems' which other cities in China do not enjoy. This is a big challenge to the producer services in Hong Kong and the Hong Kong SAR Government.

Acknowledgements

I would like to thank the Hong Kong Research Grants Council for funding the study and the assistance of Professors Yan Xiaopei and Zhou Chunshan and the students of the Centre of Urban and Regional Studies of Zhongshan University Guangzhou in carrying out the survey in the Pearl River Delta and Misses Sandy Lam, Florence Lee, Christina Lo, Hong Yi and Dr Jiang Xu of the Centre in processing and analysing the data.

Notes

1 The PRD refers to the Pearl River Delta Economic Region officially demarcated by the Guangdong Provincial Government in 1994. It includes nine municipalities which are Guangzhou, Shenzhen, Zhuhai, Foshan, Jiangmen, Dongguan, Zhongshan, part of Huizhou (city proper, Guidong County, Boluo County) and Zhaoqing (city proper, Gaoyao City and Sihui City).
2 Domestic factories are local factories in the PRD, including state enterprises, collective enterprises, shareholding enterprises and private enterprises.
3 CEPA stands for 'Closer Economic Partnership Arrangement between Hong Kong and the Chinese mainland'. CEPA was signed by the Central and the SAR Government on 29 June 2003, and is effective from 1 January 2004. It is basically a bilateral free trade agreement under WTO rules with the purpose of enhancing economic cooperation and integration between Hong Kong and the Chinese mainland. CEPA intends to give preferential access to mainland market for Hong Kong companies, which exceeds China's WTO entry commitments. Under CEPA, Hong Kong manufacturers are exempt from tariffs when exporting their 'Made in Hong Kong' products to the Chinese mainland to enjoy greater flexibility to access mainland market, while 18 service sectors also get faster and easier market access by reducing or removing geographical, financial and ownership restraints.

References

BPF (Business and Professionals Federation of Hong Kong) (2002) *Impact of Cross Border Economic Activities on Hong Kong*. Hong Kong: Business and Professionals Federation of Hong Kong.

China INFOBANK (2004) 'Statistics of the Pearl River Delta Economic Region 2002', available: http://www.chinainfobank.com (accessed on 8 January 2004) *(in Chinese)*.

Coe, N. M. (1998) 'Exploring uneven development in producer service sectors: detailed evidence from the computer service industry in Britain', *Environment and Planning A*, 30, 2041–68.

Daniels, P. W. and Moulaert, F. (eds) (1991) *The Changing Geography of Advanced Producer Services*. London and New York: Belhaven Press.

Enright, M. J., Chang, K. M., Scott, E. and Zhu, W. H. (2003) *Hong Kong and the Pearl River Delta: the Economic Interaction*. Hong Kong: 2022 Foundation.

Enright, M. J. and Dodwell, D. *et al.* (1997) *The Hong Kong Advantage.* Hong Kong: Oxford University Press.

GaWC (The Globalization and World Cities (GaWC) Research Group and Network) (2003) available: http://www.lboro.ac.uk/gawc/ (accessed 3 July 2003).

Guangdong Provincial Bureau of Statistics (ed.) (2003) *Guangdong Statistical Yearbook 2003.* Beijing: China Statistics Press *(in Chinese).*

Guangzhou Statistic Bureau (ed.) (2002) *Guangzhou Statistical Yearbook 2003.* Beijing: China Statistics Press *(in Chinese).*

Hartshorn, T. A. (1992) *Interpreting the City: An Urban Geography,* 2nd edn. New York: John Wiley & Sons, Inc.

HKCSI (Hong Kong Coalition of Service Industries) (2003) Statistical Data of Services Sector in Hong Kong, available: http://www.hkcsi.org.hk/publications/statistics/stat2003.pdf (accessed 3 July 2003).

HKGCC (Hong Kong General Chamber of Commerce) (2000) *China's Entry into the WTO and the Impact on Hong Kong Business: A Business Perspective.* Hong Kong: HKGCC.

Ho, Y. P. (1992) *Trade, Industrial Restructuring and Development in Hong Kong.* London: Macmillan Press.

Illeris, S. (1996) *The Service Economy: A Geographical Approach.* Chichester, England: John Wiley.

Kwok, R. Y. W. and So, A. Y. (eds) (1995) *The Hong Kong-Guangdong Link – Partnership in Flux.* Armonk: Sharpe.

Leung, C. K. (1993) 'Personal contacts, subcontracting linkages, and development in the Hong Kong-Zhujian Delta Region', *Annals of the Association of American Geographers,* 83, 272–302.

Ng, M. K. and Hills, P. (2000) *Hong Kong: World City or Great City of the World (Provisional Report).* Hong Kong: Centre of Urban Planning and Environmental Management, The University of Hong Kong.

Ning, Y. M. (2000) 'The study on the location of industrial service and office buildings in Shanghai', *City Planning Review,* 24, 9–12 *(in Chinese).*

Phillips, D. R. and Yeh, A. G. O. (1989), 'Special economic zones', in Goodman, D. (ed.), *China's Regional Development.* London: Routledge (with the Royal Institute of International Affairs), 112–34.

Porter, M. (1990) *The Competitive Advantage of Nations.* New York: Free Press.

Sit, V. F. S. (2001) 'Economic integration of Guangdong Province and Hong Kong: implications for China's opening and its accession to the WTO', *Regional Development Studies,* 7, 129–42.

Sit, V. F. S. and Yang, C. (1997) 'Foreign-investment-induced exo-urbanization in the Pearl River Delta, China', *Urban Studies,* 34, 647–77.

Wong, K. Y. and Chu, D. K. Y. (eds) (1985) *Modernization in China: The Case of the Shenzhen Special Economic Zone.* Hong Kong: Oxford University Press.

Xu, J. and Yeh, A. G. O. (2003) 'Guangzhou', *Cities,* 20, 361–74.

Yan, X. P. (1998) 'Analysis of regional differences of the development of information industries in Guangzhou', *Economic Geography,* 18, 5–10 *(in Chinese).*

Yeh, A. G. O. (1996) 'Pudong – remaking Shanghai as a world city', in Yeung, Y. M. and Sung, Y. W. (eds), *Shanghai: Transformation and Modernization under China's Open Policy.* Hong Kong: The Chinese University Press.

Yeh, A. G. O. (1997) 'Economic restructuring and land use planning in Hong Kong', *Land Use Policy,* 14, 25–39.

Yeh, A. G. O. (2001) 'Hong Kong and the Pearl River Delta: Competition or cooperation?', *Built Environment*, 27, 129–45.

Yeh, A. G. O. (2002) 'Further cooperation between Hong Kong and the Pearl River Delta in creating a more competitive region', in Yeh, A. G. O., Lee, Y. S., Lee, D. and Lee, T. (eds), *Building a Competitive Pearl River Delta Region: Cooperation, Coordination and Planning*. Hong Kong: Centre of Urban Planning and Environmental Management.

Yeh, A. G. O. (2002) 'Hong Kong's producer services linkages with the Pearl River Delta', *The Hong Kong Servicing Economy Newsletter*. Hong Kong Coalition of Service Industries, December 2003 Issue, 2–3.

Yeung, Y. M. and Sun, Y. W. (eds) (1996) *Shanghai: Transformation and Modernization under China's Open Policy*. Hong Kong: The Chinese University Press.

9 Hong Kong's changing role as a global city

The perspective of Japanese electronic MNCs

David W. Edgington and Roger Hayter

Introduction

Hong Kong has shifted to a service-based economy at a more rapid rate than many other places in Asia-Pacific. This in large part was due to a favourable location, one which is distinctive in both a global and regional context. On top of this, Hong Kong has long been known for its strength in entrepot trade, financial connections and other services. Today, it is one of the largest financial centres in the world and the largest container port in Asia. Indeed, by the 1980s it had attained the status of a 'global city' – a strategic site for the management of the global economy, together with the production of advanced services and financial operations (Sassen, 2000, 2001).

Recently, the global city phenomenon, as described by Friedman (1986), has received critical appraisal (Scott, 2001; Sassen, 2002), a contemplation extended to the emerging world cities of Asia-Pacific (Lo and Yeung, 1996). In this debate, the meaning of global cities and the global city hierarchy has come under scrutiny. Yet, however defined, global cities are typically portrayed as controlling global processes, reaching out and shaping developments elsewhere. That global cities may themselves be vulnerable to globalization has received little attention. The reasons for this neglect are not clear. After all, among the 'big three' global cities, London and New York were comprehensively deindustrialized in the 1970s and 1980s and land values in Tokyo have declined considerably from the property boom of the late-1980s. Even the most powerful cities are apparently not all-powerful. This chapter addresses the vulnerability of global cities to globalization, with specific reference to Hong Kong.

Hong Kong is widely recognized as a 'second tier' global city, especially important as a gateway city of the dynamic Asia-Pacific region, most notably south China. Indeed, this regional context is critical to understanding the evolution of Hong Kong from its foundation in the 1840s, to its emergence as a global city in the 1970s and 1980s and its present state of uncertainty (So, 1999). In this evolution, trading, distribution and service activities have been paramount, as Hong Kong has sought to inter-connect South China and the rest of the world. During the post-war boom beginning in the early 1950s, the manufacturing sector, centred around textiles, metal fabricating and electronics, certainly played a vital role in Hong Kong's economy, fuelled by immigrant entrepreneur's escaping from communist China and subsequently by direct foreign investment (DFI). However, following China's commitment to a more open economy and, in the years leading up to Hong Kong's repatriation to China in 1997, Hong Kong firms and foreign firms increasingly locating manufacturing operations in South China. Such location dynamics have reinforced and enriched Hong Kong's traditional trading, distribution and service functions. Simultaneously, there are signs of disquiet as integration with China has threatened, as well as stimulated, Hong Kong's status as service centre. China's entry into the World Trade Organization (WTO) in 2001 has underlined this debate (Panitchpakdi and Clifford, 2002).

This chapter analyses the debate over Hong Kong's role from the perspective of Japanese electronic multinational corporations (MNCs). The mandate and scope of MNCs is often used to assess global city hypotheses, and the role of Japanese service MNCs in shaping global city functions within the Asia-Pacific region has already been documented (Edgington and Haga, 1998). In this chapter, recent changes in the structures and strategies of Japanese electronics firms in China and Hong Kong are offered as nuanced indicators on the evolving status of Hong Kong as a global city. In particular, we adopt a disaggregated view of the 'control functions' of these Japanese MNCs and examine the specific changes in their operations in Hong Kong, and to related movements within the so-called 'Greater China Circle' of China, Taiwan and Hong Kong (Mukoyama, 2002). We are aware that such an analysis only represents a partial perspective on Hong Kong's role as a global city. Nevertheless, changes in control functions are typically 'long run' and reflect new directions in corporate strategies. Japan has also been a principle source of DFI for the Greater China Circle and electronics has been a leading sector for Japanese DFI across the Asia-Pacific region.

Information for this analysis was derived from corporate reports, and interviews with head office and regional managers obtained in Hong Kong and Japan during 2001 and 2002 respectively. The companies contacted are listed in Table 9.1, and comprise a similar array of 15 major electronic firms that the authors researched in a previous project examining direct foreign investment by Japanese electronics in East Asia (see Edgington and Hayter, 2000; Hayter and Edgington, 2004). Information from all 15 firms was utilized to assess the evolution of Hong Kong as a service hub and, in the light of competition from Shanghai and Beijing, their perceptions of the contemporary merits and demerits of Hong Kong. The study then highlights the subsidiaries of three electronics firms to examine their contemporary functions in Hong Kong, including sales

Table 9.1 Fifteen major Japanese electronics firms in Hong Kong, surveyed in 2001

Category	No. of subsidiaries, joint ventures and representative offices in Hong Kong	Major functions[1]
General electronics firms		
Hitachi	1	Sales and marketing to Hong Kong, PRC and Korea
Toshiba	2	Sales into Hong Kong and PRC; finance and logistics; IPO
Mitsubishi Electric	2	Sales and marketing into Hong Kong and PRC; finance and logistics
Consumer electronics firms		
Matsushita Electric Industrial	4	Sales and marketing into Hong Kong and PRC; control over South China factories; finance and logistics
Sony	2	Sales and marketing into Hong Kong and PRC; regional HQ functions; warehousing and logistics; IPO and OEM business
Sanyo	5	Sales and marketing into Hong Kong and PRC; sales support; IPO and OEM business
Sharp	3	Sales and marketing into Hong Kong and PRC; sales support, IPO and OEM business
Pioneer	2	Sales and marketing into Hong Kong; IPO
Casio	1	Sales, marketing into Hong Kong and PRC, IPO and OEM business
JVC (Japan Victor Co.)	1	Sales and marketing into Hong Kong and PRC
High-tech firms		
NEC	3	Sales and marketing into Hong Kong and PRC; sales support; R & D, design and IPO
Fujitsu	1	Sales and marketing into Hong Kong and PRC
Fuji Electric	1	Sales and marketing to Hong Kong and PRC
Oki	1	Sales and marketing to Hong Kong and PRC
Omron	1	Sales and marketing into Hong Kong and PRC; logistics, IPO and OEM business

Sources: Authors' field surveys in Hong Kong (2001) and Japan (2002).

Notes
1 For commentary on the major functions carried out in Hong Kong see Table 9.2.
IPO = International Purchasing Office; OEM = Original Equipment Manufacture.

into China, procurement of local parts and final goods from southern China on an OEM (original equipment manufacture) basis, as well as local finance and logistics functions. It also presents comments from regional managers as to their companies' future intentions regarding Hong Kong. The final section attempts an assessment of these trends for Hong Kong's future in the form of two scenarios, one optimistic and one pessimistic. By way of context, the next section of the chapter briefly connects Hong Kong to the global city hypothesis with particular respect to its role in the Greater China Circle.

Hong Kong and the Greater China Circle

Conceptually, our analysis follows So's (1999) arguments that a contemporary appreciation of Hong Kong as a global city requires an analysis of its *regional* dynamics and context. With its development as a colonial city-state within the British empire to become one of the world's largest ports and an Asian Tiger that exported secondary manufactured goods and a magnet for FDI, Hong Kong has truly global connections. Yet, Hong Kong's fortunes are fundamentally driven by its regional situation in East Asia, most particularly by its connections to China, especially south China, and Taiwan. Hong Kong's current status and hopes for the future have to be placed within this Greater China Circle. This increasingly integrated region has become a driving force behind the dramatic growth of investment, production and trade in East Asia, especially since the slowing of the Japanese economy (see Taylor, 1996; Naughton, 1997a; *Business Week*, 2002).

We recognize the importance of the legacy of Hong Kong as a colonial city-state. In 1842, the British created Hong Kong out of a virtually 'empty land' and:

> instituted the rule of law. They established an efficient bureaucracy free of corruption, set up a modern infrastructure, instituted a low tax system, and, most important, adopted a *laissez-faire* economic policy that attracted foreign investment and generated a spirit of capitalism in Hong Kong.
>
> (So, 1999: 3)

An educational system that included the teaching of English further helped Hong Kong become a capitalist 'paradise'. Moreover, even as a colonial city Hong Kong enjoyed considerable autonomy with powers that approached those of a nation state. In this regard, it might also be noted that in comparison to its rival city-state, Singapore, Hong Kong has enjoyed a much bigger land base. We also recognize the vibrancy and distinctiveness of Chinese entrepreneurialism, rooted in neo-Confucian familism, strong kinship ties, use of family labour and speedy decision making (Yeung and Olds, 2000). In Hong Kong, the rules and regulations of the 'colonial' city were a good 'institutional match' for vigorous Chinese entrepreneuerialism.

Yet, as So (1999) emphasizes, this legacy of, and match between, colonialism and neo-Confucian entrepreneurialism, cannot by themselves explain the structural changes in Hong Kong's economy over the last four or five decades. As he notes, in this period, Hong Kong became an industrial city between the late 1950s and early 1970s, a service centre in the early 1980s and, with more uncertainty, Hong Kong now seeks status as a high-tech city. For So (1999), these changes have to be interpreted in the context of regional dynamics, notably the Cold War between the US and China, China's openness and national reunification attempts, the Asian financial crisis and increased competition for the role of global metropolis within Asia. This chapter compliments So's (1999) analysis by focusing on the electronics sector and the role of Japanese FDI. The electronics sector provides a significant lens on Hong Kong's emergence in the post-1950 period, first as an industrial city, second as a service centre, and third with respect to Hong Kong's high tech-based metropolitan hopes. Moreover, the behaviour of Japanese FDI in the electronics sector has been strongly influenced by the regional dynamics mentioned above, to which should be added China's entry into the WTO in 2001. This decision has dramatically raised the stakes for China's Asian neighbours and rivals, including rival 'world cities' (Hong Kong Trade Development Council, 2000; Panitchpakdi and Clifford, 2002).

The uncertainties generated by China's continued exploding growth are underlined by Ohmae's prediction that China's rise and 'super-competitiveness' will cause a 'second Asian crisis', more severe and drawn-out than that of 1997–1998 (*The Economist*, 2001a). On the other hand, Naughton (1997b: 5) is more optimistic. He notes that the Greater China Circle comprises a series of concentric economic circles centred on Hong Kong and that 'Hong Kong's multiple economic functions give it unambiguous centrality no matter what definition of the China Circle is used'.

Of course, Japanese DFI is not the only basis for assessing Hong Kong's global status but it provides important insights from a business perspectives (see also Sung, 1997). In particular, the evolution of local offices of Japanese MNCs are an interesting proxy for the development of Hong Kong's higher order functions associated with the regional offices and regional headquarters of international firms in general. Thus, Hong Kong's prosperity has traditionally been associated with its role as a trading post and top exporter in a wide range of consumer products, then as a major locus of mobile international capital from the 1970s on, which led it to emerge as a major international finance centre. The opening up of mainland China's mainland economy 25 years ago subsequently resulted in Hong Kong evolving into an all-round service centre, managing and coordinating a large share of world trading and manufacturing activities (Lo, 1992). Thereafter, direct foreign investment (including Japanese electronics investment) spread beyond southern China (Pearl River Delta) to other regions, including the Yangtze River Region (Shanghai and its surrounding provinces of Jiangsu, Zhejiang and Shandong).[1] Consequently, Hong Kong's port was no longer the only gateway to the mainland. Indeed, Shanghai, with its advantageous location in the central coastal region, is increasingly seen as Hong Kong's competitor due to its superior position as a distribution and centre of commerce (Hong Kong Development Council, 2001).[2]

Japanese electronic firms in Hong Kong, 1951–2001

Japanese electronics companies have had a varied history in Hong Kong. In order to understand well the way in which they contemplate Hong Kong's role, especially after China's entry into the WTO, it is worth looking back on the last 50 years of their activities in this location. Initially, a small but dynamic export industry emerged during the 1950s along with growth in the labour force through mass immigration. Hong Kong's first industries were plastic flowers, wigs and 'one-dollar blouses' (Lo, 1992). Japanese electronics firms sold into the expanding consumer market (and also that of neighbouring Macao) using a mixture of Japanese trading houses (*sogo shosha*) and local agents and representatives. The Sony Corporation showed its 'maverick' status within the Japanese array of firms doing business in Hong Kong by starting a transistor radio assembly factory in 1959 through a local firm. This was Sony's first production facility abroad.

As the market for consumer electronics and industrial machinery expanded, so Japanese companies entered the market directly. In the 1960s and 1970s they began to set up their own local sales liaison offices (representative offices) in Hong Kong to provide after-sales service and back up for local agents and distributors. Apart from sales of consumer items (audio products, colour televisions and sophisticated household products, including refrigerators and air conditioning units) firms such as Oki, NEC and Toshiba sold integrated circuits (silicon chips) to Hong Kong's assembly factories of radios, digital watches and other electronic goods. Japanese firms also used local Hong Kong firms for the supply of low-end OEM (original equipment manufacturing) products (e.g. audio-visual products and household equipment) for Japan as well as worldwide markets.

The opening up of mainland China in 1978 fundamentally changed Hong Kong's regional dynamics. The People's Republic of China (PRC) now allowed the growth of consumer imports, while Hong Kong's manufacturing firms themselves found advantages in producing across the border in Shenzhen, and later on in cities such as Dongguan in the Pearl River Delta area of Guangzhou province (Vogel, 1989; Lin, 1997). At that time, Hong Kong played the role of an gateway for China; an entry point for technology, capital, management, skills and ideas (Overholt, 1993; Hayter and Han, 1998). For Japanese electronics firms, the explosion in sales of 'big box' items to the PRC (e.g. refrigerators, colour televisions, washing machines and air conditioners) led to the role of their Hong Kong offices changing in two important ways. First, there was a geographic expansion in the territory assigned by their head office in Japan; sometimes not merely encompassing the entire Chinese mainland, but also including nearby Taiwan and also the Philippines (the nearest Southeast Asian nation to Hong Kong). Second, because of the increased volumes of imports, both directly from Japan and Japanese factories located in Southeast Asia (e.g. Thailand, Malaysia and Singapore), Hong Kong representative offices were significantly upgraded to full sales office status. Prior to the early 1980s all sales into Hong Kong and mainland China were small enough to be handled by head offices in Japan (Tokyo or Osaka), who also controlled financial settlements, as well as delivery and logistics functions. An increase in sales to China brought marketing and other operations responsible for regional business activities (e.g. local warehousing and accounting functions), with a commensurate increase in staffing levels and skills.

From 1980 to 1990, Hong Kong continued to play its special role as a unique gateway to mainland China (Skeldon, 1997; Yeung, 1997; Meyer, 2000, 2002). Many Japanese managers commented that at the time 'the biggest problem was in terms of ordering sufficient supplies, because anything you shipped to the mainland could be sold'. For Japanese electronics companies these were often seen as the 'golden years' where competition in the market with both western and other East Asian countries was fairly minimal. The establishment of a full marketing team and their Hong Kong offices by the end of the 1980s allowed more sophisticated product lines to be introduced during the early 1990s (e.g. integrated circuits sold to Hong Kong and Chinese assembly factories). To achieve further sales, branch showrooms of Hong Kong subsidiaries were added in the centres of major Chinese consumer and industrial markets (Beijing, Shenyang, Suzhou, Chengdu and Guangzhou). To these activities were often added International Procurement Offices (IPOs); parts procurement operations located in Hong Kong and organized by the various divisions of Japanese headquarters to purchase cheap materials and equipment from the south China industrial region for Japanese factories in various parts of the world.

The 1990s saw further changes in operations conducted from Hong Kong, mainly due to shifts in the Japanese firms' strategies towards the mainland. Following the Tiananmen Square incident in 1989, Japanese enthusiasm for investing directly in China cooled; but from 1993 onwards the increasing cost of producing electronics goods in both Japan and Southeast Asia led to renewed interest. By that time Chinese production and labour costs were the very cheapest both in Asia and the rest of the world. Competitive pressures, therefore, forced Japanese electronics firms to open up factories in a number of Chinese coastal locations, mainly for export, but also aimed at domestic markets. Up to the end of the decade every Japanese electronic firm established their own factory and sales operation within China. Usually this involved joint ventures with local PRC firms, and for the major companies a Chinese regional head office (RHQ) was set up, either in Shanghai (because of its central location along the eastern coast) or in Beijing (due to proximity to government clients). Figure 9.1 shows that over the 1990s Guangdong, adjoining Hong Kong, was the favoured location overall, with significant other concentrations in Laoning, Shenyang, Tianjin and Beijing in the north, and Shanghai and Jiangsu in the central coastal region. Yet another trend in the 1990s was the growing ability of local PRC firms themselves to copy Japanese and other investors' production, and manufacture low cost medium-quality consumer household electronics and audio-visual goods, such as colour televisions.

Sales activity by the Hong Kong offices of Japanese electronics firms peaked in 1997 for a number of reasons. First, imports of consumer electronics into China fell off because of competition from cheaper production originating from local PRC factories, as well as a new round of factory investments inside China by Japanese firms themselves to supply the local PRC market. For instance, Sharp Corporation opened a major consumer goods factory in Shanghai, and Sanyo Corporation opened several in Dalian, Shenyang and Shanghai. This meant that, for certain products, the sales territory of the Hong Kong office was downsized back to Hong Kong and Macao, as before the mid-1980s. Second, the Asian financial crisis of 1997–1998 impacted upon the Hong Kong market itself, and after this time Japanese firms lost competitiveness in Hong Kong against Taiwanese and Korean importers, except for sophisticated electronics products, such as digital camcorders and 'robot dogs' (Sony Corporation's 'Aibo' series). By the year 2001 the only products being imported into mainland China through Hong Kong were sophisticated industrial products, such as integrated circuits for the assembly factories of PRC, East Asian, US and European assembly factories that had expanded rapidly in China during the previous 15 years or so. Most recently, Japanese firms have begun to manage links with those Chinese/Hong Kong/Taiwanese and Korean-based factories in China, through sub-contracting arrangements set up on an OEM basis, or even retreated from the consumer products sector completely due to increased competition from Chinese and other East Asian competitors.

Increasing DFI in China, and the growing accumulation of industry and commerce in China's coastal regions, has changed not only the industrial map of Asia, but also traditional production chains and links to regional clusters of producer services located elsewhere. The levels of employment

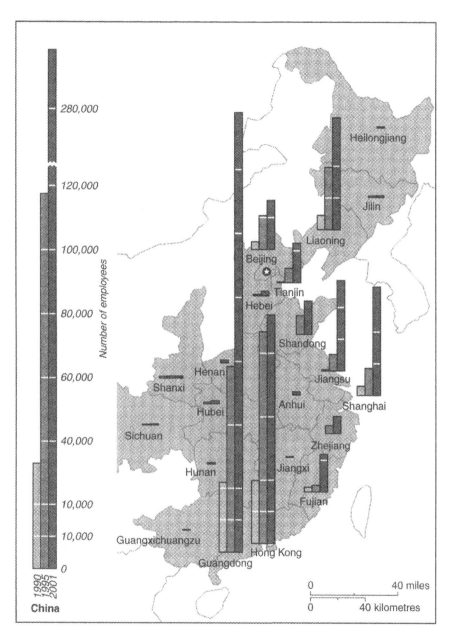

Figure 9.1 Japanese electronics firms: employment levels by Chinese province, 1990–2001.

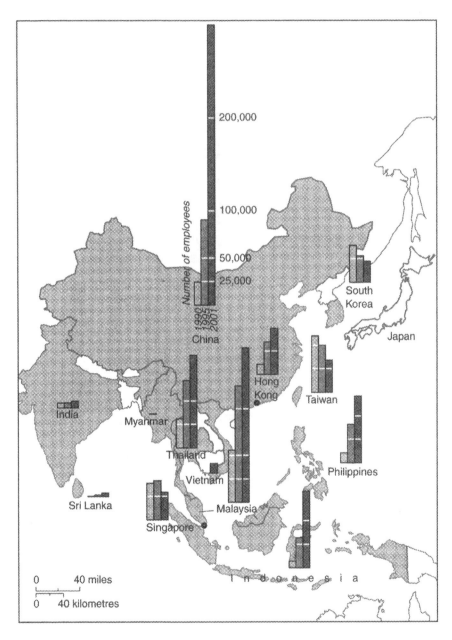

Figure 9.2 Levels of employment in Japanese electronics companies' Asian factory subsidiaries, by country and year.

in Japanese electronics factories in the countries of mainland Asia is shown in Figure 9.2, indicating the eclipse of Malaysia as the first ranked country by mainland China between 1990 and year 2001. The Association of Southeast Asian Nations (ASEAN) countries, being relatively close geographically to China and at a similar state of development, have felt the impact of increased competition with China. Indeed, a sense of crisis is growing along with China's economy, and this is having severe implications for ASEAN and also Singapore's role as a regional headquarters (RHQ) for MNCs (*The Economist*, 2001b). This sense of unease has also affected Hong Kong, and if China's entry into the WTO in 2001 is consistent with Hong Kong's laissez-faire thinking it also adds new vulnerabilities (Hong Kong Trade Development Council, 2000; Panitchpakdi and Clifford, 2002).

Contemporary functions of Japanese electronic MNCs in Hong Kong

There are several major functions that the headquarters of Japanese MNCs have given their Hong Kong offices in the spatial division of labour emerging between Japan and the 'Great China Circle' (Table 9.2). These are a combination of 'sales' activities into various China/Hong Kong markets (Taiwan typically has its own Japanese sales operations because of obvious political issues) and 'production-side' activities involving IPO/OEM operations, as well as supply chain management and delivery logistics from China to Japanese and global destinations. In many cases individual registered subsidiaries of Japanese electronics companies were organized in Hong Kong for the different functions listed in Table 9.2. These are shown for each of the 15 firms in the authors' sample in Table 9.1.

Pertaining to the sales and import dimension of Japanese MNCs based in Hong Kong, permission by the Chinese government to sell consumer products into mainland China from Japan is often contingent upon investment in production facilities. Consequently, this function has lost the most activity in Hong Kong during the past ten years or so. Even though Hong Kong is still competitive as a location for sales of industrial components and machinery into the PRC. Integrated circuits, for instance, are supplied to assembly factories (making mobile phones, computers and colour televisions) located either across the border in Shenzhen, or further north in Shanghai and Beijing. Customers comprise either American or European firms such as Motorola and Nokia, IBM and Dell, other Japanese firms such as NEC's telecommunications factories, or Taiwanese, South Korean or mainland Chinese firms. At present, while the Japanese headquarters provides all R & D and design functions for these products, the Hong Kong office plays an important role in providing 'back up' services, and so requires a strong application team who can work with

Table 9.2 Roles of 15 Japanese electronics firms operations in Hong Kong, 2001

1 *Sales office back-up and marketing operations for consumer products in Hong Kong,*
   Macao (and often the Philippines)
   The mission of these offices is to increase sales, into what is considered a 'thin',
   mature and relatively slow growing market
2 *Sales office back-up and marketing operations for industrial products to factories in*
   mainland China
   This function is usually confined to sophisticated industrial products (e.g.
   integrated circuits) that cannot be produced in Chinese factories.
   Representative offices with engineers and sales staff are maintained in Chinese
   production centres, such as Beijing, Shanghai and Shenzhen/Pearl River Delta
3 *Supplying the parent company in Japan with products and components made in China*
   This is a relatively new business for Japanese electronics firms, and is important
   to maintaining their competitiveness against Taiwanese and South Korean
   rivals. The Hong Kong office works with Chinese, Hong Kong or Taiwanese
   partners in south China (Shenzhen/Pearl River Delta) to purchase Chinese-
   made components (International Purchasing Office functions) and consumer
   electronic products (Original Equipment Manufacture business), and raise
   both production quality and scheduling to Japanese standards
4 *Distribution and logistics operations of Chinese products through the port of Hong Kong*
   Chinese-made products are distributed to Japanese sales and production
   operations in other Asian and North American markets. Even 25 years after the
   Chinese mainland 'opening-up', Hong Kong's port/and airport continues to
   capture about 40 per cent of China's trade to and from the rest of the world
5 *Financial settlements on sales and purchases made in China*
   Hong Kong has traditionally offered its local companies ease of settling
   international transactions. The Hong Kong dollar has been 'pegged' to the US
   dollar and there are no restrictions on repatriating funds or injecting local
   operations with new capital funds. This is often perceived as Hong Kong's
   greatest competitive edge over business conducted in mainland China

Source: Authors' surveys, 2001, 2002.

Chinese-based customers and adapt semiconductors to particular products and applications. Japanese managers reported that following the
WTO these sales and service functions could well take place at, or closer
to, other industrial clusters within China, rather than in Hong Kong.[3] In
fact, they also indicated that the need to achieve financial settlements in
Hong Kong was perhaps the only persisting reason why this function was
still held by their local Hong Kong operations. If the WTO were to allow
confidence in the full convertibility of the local PRC currency (the *renminbi,* or RMB), then this function, currently concentrated in Hong Kong,
would come increasingly under threat.

Concerning the many 'production-side' activities carried out by electronics firms in Hong Kong it is interesting to note that the OEM business
is due to the need for more flexibility, especially in light of market volatility following the 1997–1998 Asian financial crisis. Thus, in contrast to continuous growth from the late-1970s onwards, contemporary markets have
become extremely volatile and cannot guarantee efficient operations in

large-scale assembly plants. The worst possible case scenario reported by local and headquarters managers was where a factory might have to be closed down, should demand for a particular product decline sharply. In this case it would be easier to do this with a Chinese OEM partner than with a 100 per cent owned and operated Japanese factory.

International Purchasing Office (IPO) functions have also been developed in Hong Kong as most parts production takes place in the PRC within the southern production cluster. However, managers reported that IPO functions were even more advanced in Taiwan because of the greater availability of high-technology products from component firms there. In Hong Kong or the PRC, Japanese firms could only purchase mid-to-low range electronic products, such as metal casings and transistors. As Japanese production standards were rigorously high, many local IPO managers complained that it was often difficult to secure local products and PRC producers who were acceptable to headquarters engineering specifications. Nonetheless, the cost of purchasing even standard components from Taiwan and South Korea was also very high, so it was hoped that PRC suppliers could be 'trained' to meet the exacting Japanese standards.[4]

To facilitate OEM and IPO functions, engineers usually came from the various 'mother' or 'focal' factories located in Japan.[5] Local Japanese managers doubted whether manpower in Hong Kong itself could ever replace Japanese engineering. They also doubted whether Hong Kong could become a technology control centre for Chinese-based production, either arising from Japanese direct foreign investment or in relation to Taiwanese or locally owned PRC factories. Because of a dearth of local engineering talent and technology 'back up' in Hong Kong, the 'mother factory' in Japan would invariably maintain the strongest links with any Japanese branch factory or OEM company in China. Consequently, Hong Kong would continually be by-passed in that regard. Still, unlike technology transfer, supply chain management of distribution lines from southern China's factories out through Hong Kong into Japanese, North American and European markets, had to be localized in Hong Kong itself. In the case where overseas factories or sales offices placed an order, the Hong Kong office was well located to address the logistics of supply through the local port and airport system, and at present was also the best place to organize associated financial transactions, such as foreign exchange.

Table 9.3 presents a synthesis of comments from the 15 Japanese electronics firms regarding their appreciation of the contemporary strengths and weaknesses of Hong Kong. The results suggest that Hong Kong is certainly appreciated because of its convenience vis-à-vis other PRC port locations and Beijing. Yet a major drawback is the cost of operating in Hong Kong. Beyond these array of 'static' strengths and weaknesses, managers in Japan were asked about coming plans, and all mentioned that their head office was re-evaluating the future of their Hong Kong operations, especially since their companies were planning how to enter the Chinese market in a more substantial way. This strategy was necessary in order to 'catch up' with previous rounds of investment and sales operations from competing European and North American electronics firms (see also Amano, 2002; Mukoyama, 2003). A more extensive and localized Chinese network (i.e. beyond Hong Kong throughout the eastern and northeastern coastal region) would increase the credibility and technical support to all their PRC customers. Common amongst responses was the acknowledgement that 'China is a very big country: we cannot hope to serve all parts of China from Hong Kong alone'. This points to the realization that after China's entry into the WTO is settled, part of Hong Kong's current role is slowly but surely likely to transfer away from being the only entry point into the PRC. Hong Kong's future role is likely to be one that pivots around being the sales and service for the southern China region. Other sales and service operations will likely be set up in the middle and northern parts of China, as operations there allow.

Table 9.3 indicates, conversely, that part of southern China's (Shenzhen/Pear River Delta) competitive advantage itself lies in its proximity to Hong Kong where transactions costs involved in interacting with this centre were much lower than from Shanghai or Beijing. There are 24-hour information links with Hong Kong, so that Japanese and other firms can use the infrastructure of Hong Kong when manufacturing in southern China. In other words, all the advantages of Hong Kong were available in the south China area, including payments in Hong Kong dollars rather then RMB.

To illustrate the ramifications of this shift in thinking about Hong Kong's role in China (from a national to a regional service sector hub), the following section provides detailed case studies. These are based upon our interviews of local Hong Kong managers and headquarters offices of Japanese electronics firms and their own accounts of sales, finance and logistics operations for southern China (and in some cases beyond) carried. In addition, local branch managers and Japanese headquarters managers were asked specifically to describe likely implications for their Hong Kong operations following China's entry into the WTO.

Japanese MNCs in southern China: three case studies

NEC's sales and production support functions

NEC (Nippon Electronics Corporation) evolved from a firm based initially in telecommunications technologies, and which then moved into consumer electronics and computers during the 1960s. In Hong Kong (at the time of the 2001 survey) it operated three separate companies; two of which were sales operations, and one of which was production-oriented (see Table 9.1). The first, NEC Hong Kong, sold telecommunications

Table 9.3 Perceptions of the merits and demerits of post-WTO Hong Kong (based upon a survey of 15 Japanese electronics firms in Hong Kong)

A *Merits*
1   Advantages of a Hong Kong office location
   - free port
   - low taxes: 15% corporate taxes; 17% personal tax
   - little complicated customs documentation
   - no foreign exchange controls
   - fixed currency to the US dollar (HKL$7.8 to US$1)
   - superior infrastructure:
     - high watermark port
     - 24-hour airport
     - international financial centre/can conduct financial settlements in HK$
     - English speaking managers and legal staff are available
     - good telecommunications infrastructure
     - superior security for information technology and other sensitive material
   - Good airport and seaport connections with the rest of the world
2   Indirect advantages of surrounding South China area/Pearl River Delta (PRD) area
   - low-cost labour force
   - access to existing factories for OEM products and parts
   - low-value RMB currency
   - all the advantages of HK are also there across the border in the PRD; access to HK infrastructure 24 hours a day
   - payments in the PRD can be conducted in HK$
B *De-merits*
   - Hong Kong costs are very high:
     - office rents
     - apartment fees
     - costs of harbour and airport cargo facilities

Source: Authors' surveys, 2001, 2002.

products as well as computer systems into the local Hong Kong and Macao markets. The second, NEC Electronics, handled sales of integrated circuits (ICs) to mainland China and elsewhere in the region, as well as liquid crystal displays (LCDs), plasma displays and some automobile products (see Figure 9.1: sales side). Representative offices were necessary to serve its far-flung sales region (including Taiwan, South Korea, as well as the Philippines and Australia), and ICs were essentially sold on a FOB (free-on-board) basis from its Hong Kong office in US currency. The third local subsidiary, NEC Technologies HK, comprised a design team charged with IPO work, and setting up assembly factories in the Philippines, Thailand and China. All three were controlled by the International Division of NEC's head offices in Tokyo. Like other firms, NEC had to place factories in China in order to sell consumer products to the local market (e.g. audio-visual units). The Hong Kong office of the first two companies organized supplies of components and final products. Sales offices of NEC first opened in Hong Kong in 1977 and at the time of the survey the first two subsidiaries employed 82 people, including 12 Japanese.

Local NEC managers felt that it would be easier to enter China with an expanded array of sales offices in order to support NEC's customers there. They felt that a major problem with existing arrangements was that the Hong Kong office could not accept funds in the local PRC currency (RMB), only US currency; whereas clients in Shanghai or Beijing preferred to pay in the local PRC currency. Selling FOB to local Chinese agents avoided this problem, but the use of intermediaries also meant that NEC in Hong Kong never quite knew the final sales prices or sometime even the final destination or final client for its electronic products. After China joined the WTO, the local sales staff of NEC in Hong Kong hoped to develop the PRC market further and become closer to their final customers, especially those located in the central and northern regions of China.

The production side of the NEC Hong Kong operation was organized along different lines (see Figure 9.2). Thus, the third Hong Kong company, NEC Technologies Hong Kong Limited, was established in 1989 in order to purchase parts and final parts from Chinese factories, mainly those owned by Hong Kong-based producers in Shenzhen and the Pearl River Delta. Because of the special corporate links required to fulfil this role back to Japan, there were 37 Japanese staff among roughly 180 employees in 2001. Close corporate connections were maintained with the NEC Beijing RHQ, which had control over all NEC operated factories in China as well as OEM arrangements with PRC factories of Chinese and Taiwanese companies. Because Hong Kong firms controlled most OEM factories, their payments were made in Hong Kong and in US currency. Indeed, the production of many computer peripheral parts was subcontracted to Hong Kong-controlled Chinese factories by NEC, whereas previously they were made solely in Japan (products such as printers, floppy disk drives and DVD players). These OEM arrangements were necessary in order to reduce costs and maintain profits in the very competitive consumer electronics business. Local managers in Hong Kong reported that there were a number of ways of conducting OEM business, including 'payment for assembly costs', in which case NEC provided all or most of the necessary components from Japan and elsewhere. Alternatively, some sub-contractors preferred to secure their own parts, in which case NEC paid for the value of the fully-finished goods. Usually it was more advantageous for NEC to supply the parts as it is a large company and so could often negotiate lower prices than Chinese factories because of its extensive networks.

Local and head office managers thought that China's entry into the WTO was likely to encourage further Chinese OEM production. The improvement of 'soft infrastructure' in China, such as accountancy, legal systems and so on, would also allow improvement in the quality of production. Hong Kong's role as a control centre of production in southern China would likely be therefore strengthened.

SONY's international logistic centre for the PRC

Sony is perhaps Japan's most well known consumer electronic products company (although Matsushita Corporation is much bigger) and, as intimated in the previous section, it has had a long-term presence in Hong Kong extending back to the 1950s. Following Hong Kong's reversion to China in 1997, the many individual sales operations of Sony that were based in the city were amalgamated into Sony Corporation of Hong Kong Limited, and at the time of the 2001 survey this company employed about 950 people, including 80 Japanese. Due to the diverse array of products sold into Hong Kong and China the company's assigned sales territory stretched also from Pakistan to South Korea, depending upon the particular product sold. In promoting its electronic products in China, Sony Hong Kong worked closely with Sony (China) Limited, which was the regional head office in Beijing, responsible for all Sony's mainland factories. Following China's entry into the WTO, head office managers reported that Sony was likely to assign Sony China as the headquarters of a regional organization, which would have control over developing production and sales within 'Greater China', including Taiwan, Hong Kong and the mainland. As with NEC, Sony considered Hong Kong to be well placed for sales and logistical support for southern China, but other arrangements were being planned for central and northern China.

The sharp difference between the 'production culture' and 'sales culture' in Japanese electronic firms could be seen in Sony's decision to keep a separate 'production' organization for managing the purchases of OEM products and IPO operations. Consequently, Sony International (Hong Kong) Limited (known as SIH) continued as an independent company with about 60 staff and 20 Japanese. As with the NEC example, it managed a number of functions, including production management control of Sony's OEM suppliers in southern China, administering the logistics and financing of sales to Sony's overseas and Japanese sales operations, and finally the purchase and distribution overseas of parts purchase inside China. In this regard, local managers stressed the importance of Hong Kong's advantageous taxation regime compared to that in the PRC, and the ability to arrange transfer pricing of invoices through Hong Kong in order to reduce Sony's global tax bills.

Sony aggressively moved into OEM operations in the mid-1990s, and by 2001 had contracts with 48 factories operated by Hong Kong Chinese firms; 30 in southern China and the remainder in Malaysia, the Philippines and other locations in Southeast Asia. Even though Sony's head office in Tokyo organized the OEM contract, SIH provided 'on-the-ground' support for its implementation for all 48 factories. This was

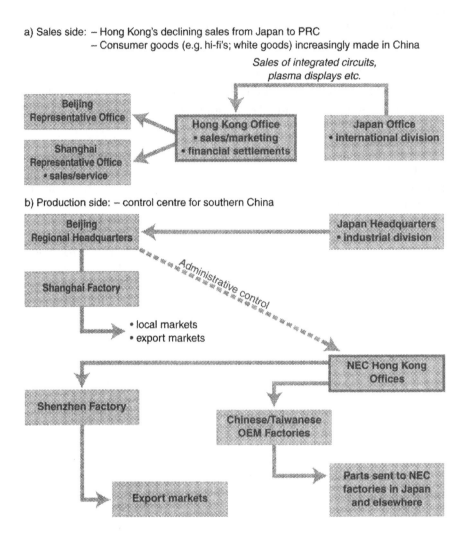

a) Sales side: – Hong Kong's declining sales from Japan to PRC
             – Consumer goods (e.g. hi-fi's; white goods) increasingly made in China

b) Production side: – control centre for southern China

Figure 9.3 Sales and production functions of NEC in Hong Kong.

carried out through an online web-based supply, logistics and financial operation called 'Dolphin', based in Hong Kong. Dolphin was a specialized internet server that ran the entire supply chain management system, and local OEM factories were expected to interface with Dolphin using a web site called 'Flipper' (see Figure 9.3). Through this essentially 'paperless' operation, Dolphin and Flipper together handled about US$1,500 million worth of products in year 2000, including audio products, home video machines, optic pickups for Playstations and CD players, and IPO parts for Sony assembly factories worldwide.

Local Sony managers in Hong Kong commented that the large investments made in the Dolphin and Flipper networks were necessary to rationalize certain complexities inherent when using an OEM system. Particularly, in rapidly changing consumer markets, a superior supply chain management was seen as an absolutely critical function in maintaining corporate competitiveness (see Mentzer, 2001). SIH managers mentioned three specific advantages of the Dolphin system. First, the computer system was important in allowing 'just-in-time' inventory control of exported products, and also avoiding bottlenecks in the supply of imported parts to OEM factories based in China and Southeast Asia. A second improvement accomplished by Dolphin involved the logistics of container ship consolidation. Smaller export 'lots' from each of 13 factories in Shenzhen, for instance, were combined and placed into a more efficient larger container ship for the haul across the Pacific to the USA or to Europe and Japan. This process had reduced the overall cost as well as the 'lead time' involved in delivery schedules. A third advance had been in upgrading the capacity to link directly to logistics systems of major customers in the USA, such as Wal-Mart and Target, as well as the ability to ship directly into their US warehouses, again saving time and money.

Matsushita's financial settlement centre for the PRC

Matsushita Electronics Industries (MEI) produced electronics products under the Panasonic brand, and has been establishing sales and factory operations in China since the 1980s, well before the arrival of most other Japanese electronics firms. In Hong Kong there were five separate operations in 2001, as follows. PSI Hong Kong sold industrial components and products into the local Hong Kong market, but with an eye for their eventual use in factories located in south China. A joint venture with a local Hong Kong firm, Shunhui Pty. Ltd., sold Panasonic consumer products into Hong Kong. Two production factories were registered in Hong Kong for taxation-avoidance purposes; and MEIL, Matsushita Electronics International Logistics Hong Kong, was set up in 1982, reporting directly to Matsushita's Beijing RHQ.

As with the Sony case study, the size and complexity of distributing parts to various Japanese-controlled factories and Hong Kong-controlled OEM factories, together with their exports around the world, led to centralization of this function in Hong Kong along with IPO activities. From the beginning of its investments into China, Matsushita considered it more efficient to have a central logistics company in Hong Kong, as the volume exported from each Chinese factory was relatively small. As a result, taking advantage of economies of scale in negotiating freight charges and financial services for each factory, such as foreign exchange, was essential. The growth of MEI factories (either joint ventures or fully owned) in China since the early 1990s has been spectacular and at the time of the survey they totalled over 35, widely distributed along the Chinese coastal region. The original strategy was to produce low- to medium-technology products for domestic PRC markets. But due to intense competition with local producers, especially in consumer products, such as televisions, camcorders and mobile phones, MEI's factories switched to export production from the mid-1990s onwards (for a review see Cheung and Wong, 2000). In addition, Matsushita now has long-term arrangements with ten OEM factories that were based in southern China. Consequently, the work conducted by the Hong Kong-based logistic operations has exploded. MEIL's staff numbered about 100 employees, including staff based at representative offices spread throughout China to monitor local production, and also to arrange negotiations with transportation and shipping through PRC ports such as Shanghai and Dalian.

Apart from minimizing transaction costs, MEIL's challenge has been how to use its centralized position and consolidated power in Hong Kong to benefit Matsushita's global operations. For instance, high transaction costs were involved in making payments to PRC factories through Chinese banks. By using an in-house 'cash settlements system' MEIL was able to pay OEM factory owners in Hong Kong directly, as well as MEI's Japanese headquarters. Conversely, by charging Matsushita's overseas network of parts makers who supplied on an OEM basis, as well as Chinese factories directly controlled by Matsushita, MEIL could bypass banks in Japan and other countries by charging their Japanese and overseas subsidiaries directly (see Figure 9.5a: cash settlements). This system was successful in achieving substantial transaction cost reductions, mainly by avoiding letters of credit for every transaction (in and out of China) and associated bank-handling charges. Of course, it relied for its success on Hong Kong's convenience in moving funds in and out of the city at will. MEIL also had the ability to consolidate parts delivery for each Chinese factory through a system called 'netting' (see Figure 9.5b: netting). Here, MEIL had taken the responsibility of supplying components that it had bought on behalf of manufacturing firms in the PRC. It then provided parts directly to the 40 or so factories in the PRC, as well as purchasing the assembled products, arranged for payment and then discounted the sales price for the cost of components supplied.

Figure 9.4 International logistics functions of Sony International (Hong Kong) Ltd.

a) Cash settlements: by-passing international banks

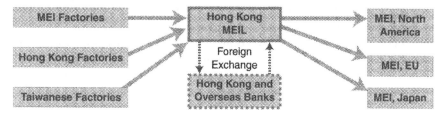

b) 'Netting' of components: reducing transaction costs

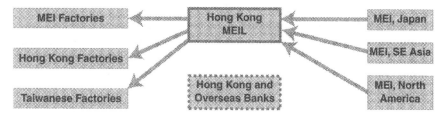

c) Consolidated loans: purchasing power

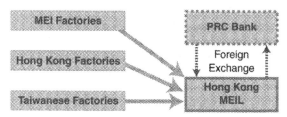

Figure 9.5 Financial settlements functions of Matsushita Electronic International Logistics (HK) Co.

Besides arranging foreign exchange and the 'netting' of components, MEIL in addition arranged loans for its Chinese factories. Here, MEIL endeavoured to consolidate each factory's loan requirements and approached mainland Chinese banks on behalf of many factories in order to reduce overall transaction costs (see Figure 9.5c: consolidated loans). In this way MEIL held short-term bank accounts for all Matsushita's Chinese factories. MEIL managers noted that should China's entry into the WTO lead to greater certainty over convertibility of the local renminbi currency, then Matsushita's logistics and financial functions might not be so concentrated in Hong Kong in the future.

Conclusions

This chapter has examined the role assigned to Hong Kong by Japanese electronics companies operating in the 'Greater China Circle', with an eye to forecasting shifts in that city's functions subsequent to China entering the WTO. The history of these companies' activities in Hong Kong, and the three detailed case studies illustrate well the commonly perceived advantages of Hong Kong's trading, logistics and financial infrastructure. However, this study also hinted at the threats to Hong Kong's existing status as a 'global city' and international service sector 'hub' subsequent to China's entry into the WTO. Specifically, if the RMB becomes freely convertible then eventually more trade will be settled in the local currency and (in theory) there will be no difference between conducting business in either Shanghai or Hong Kong. Even as the country's market and manufacturing base increases overall following WTO, the PRC will eventually generate a number of gateway cities and regional headquarters, ending Hong Kong's effective monopoly on the China trade. Moreover, higher operating costs in Hong Kong, together with the inevitable higher transaction costs of serving from southern China the remoter central and northern production clusters will exacerbate these trends. The survey results also suggested that Hong Kong was unlikely to play the role of a R & D or 'high-tech' hub in Japanese firms' strategies in the Greater China Circle.

Both local and head office Japanese managers suggested two alternative scenarios for Hong Kong. The first (optimistic) scenario involved Hong Kong maintaining its lead over Shanghai, and little diminution of its roughly 40 per cent share of mainland imports and exports. The important sales, logistics and finance operations for Japanese electronics firms (and other MNCs) would no doubt become more centralized and controlled from a Beijing RHQ. But if Guangdong were to continue to thrive as a major manufacturing base for assembly and parts components firms, as well as a domestic distribution centre, then Hong Kong's hub role in southern China would be reinforced, despite other hubs emerging around Shanghai and Beijing-Tianjin. Each hub, however, was likely to compete for future rounds of incoming foreign investment.

A second (more pessimistic) scenario involved all of the above trends, and acknowledged that Hong Kong would become the natural hub for south China. Equally, however, it recognized that the centre of gravity for production in China was likely to move inexorably away from Hong Kong, northward, especially so as the technical sophistication of Chinese manufacturing improves, both in MNC and local factories. As one Japanese manager commented:

> While the business volume is certainly higher at present in southern China, there are more information technology (IT) companies in Beijing and Shanghai, and this pattern is increasing over time. Why

the north? It seems to be a human resources driven issue. The most famous universities having electronics engineering departments are all located in northern China. Only a few engineers are interested in going south.

This evaluation indicates that Hong Kong's role as both a sourcing base and an exporting base is threatened. This concern will increase if existing Hong Kong and Taiwanese parts suppliers in Guangdong follow important IT assembly companies to the rapidly growing production clusters on the Yangtze River (Shanghai) and the PRC capital region.

The overall lesson from this Hong Kong case study is that even well-established global cities, such as Hong Kong, are sometimes under threat from shifts in the global economic and political system. WTO will be a testing time for this city of seven million people. Hong Kong commenced a strong campaign in mid-2001 to bill itself as 'Asia's World City'; a regional service hub that not only had an eye on domestic growth in China but one that allowed companies to take advantage of excellent infrastructure and a pro-business, can-do spirit (Tsang, 2001). Almost at the same time, a new report showed evidence that Shanghai was closing the gap with Hong Kong in competitiveness, and might already be ahead in its technological base and in terms of applying local research and development to commercial ventures (Seanwright and Harrison, 2001). However, the central question of to what extent Shanghai would be bound by the more conservative financial, legal and business systems of the communist regime in China, when transforming itself into a truly international marketplace post-WTO, has yet to be made clear. In other parts of the world the distinctions between domestic and international financial markets have become blurred. Yet, the Hong Kong and Shanghai comparison may be an exception, owing to very different legal, financial and business frameworks of the two cities.

The last word goes to an electronics company manager interviewed as part of this study, who recalled Japanese experience when trying to imagine Hong Kong's role in the Greater China Circle.

Even if Hong Kong is eclipsed by Shanghai in its role as service hub for industrial production, there will still be a role for Hong Kong. It is the same way that there was a role for Osaka (Japan's second largest commercial city), when business and national government connections became more important and concentrated in Tokyo after 1945. Osaka was a merchants' city before, yet it carved out a role for itself in the last 50 years by looking outward into Asia-Pacific. For that reason Hong Kong could be left with just financial operations as its overall focus. This is because the future of Shanghai as a major financial centre may be more like that of Tokyo (domestically oriented with capital allocated administratively) rather than New York or London (internationally oriented with capital allocated via market forces).

After all, Shanghai's future depends on the region's prosperity and if the region continues to prosper, Hong Kong's vibrancy, established advantages and connections and, by no means least, its location, will ensure that it too will evolve in complementary fashion (see also Leung, 1995).

Acknowledgements

The authors would like to thank the managers of Japanese electronics firms interviewed in Hong Kong (2001) and Japan (2002), and to George Lin for hosting D. W. Edgington at Hong Kong University during May 2001. Cartography was prepared by Eric Leinberger. This study was carried out with funds from a Canadian SSHRC grant # 410-2002-0272.

Notes

1 The rapid increase in Japanese DFI in China since the early 1990s has been dealt with in studies by the East Asia Analytical Unit (1996), Ernst (1997), Cheung and Wong (2000) and Nakagane (2002).
2 Hong Kong's vulnerability and future role vis-à-vis China and other Asian service hubs has been addressed by Overholt (1993), Sung (1997) and Panitchpakdi and Clifford (2002).
3 The companies that are driving China's growth, and which are potential Japanese customers, are concentrated in three areas. The first of these is the Pearl River Delta of Guangdong province, which has attracted a large number of electronics and electrical equipment firms for such labour-intensive exports as personal computers, copying machines and electrical appliances (e.g. hi-fi units). The second main area is the Yangtze River Delta, which stretches from Shanghai to Jiangsu and Zhejiang provinces. This has attracted foreign investment in areas ranging from high-tech industries such as semiconductors and mobile phones, to automobiles, steel, chemicals and clothing, mainly for the domestic Chinese market. The third area is China's 'Silicon Valley', which includes the Zhongguancun area in Beijing. Here are concentrated strongholds of software and R & D for information technology (IT) and there are many academic-industrial alliances based on links between industry and local universities. While China has large numbers of other industrial zones; these three areas stand out in that they have attracted large amounts of similar industries and enjoy the accumulative effects of concentration. These three industrial clusters, which are all quite different in nature, functionally complement each other, and at the same time, compete together for success, and are increasing their overall productive capacity. They are attracting investment from all over the world, and are continuing to spawn local industries (Kuroda, 2002).
4 As an aside, this revelation indicates there was very little chance of Hong Kong becoming a significant R & D or 'high-tech' hub for Japanese electronics firms.
5 A 'mother' factory refers to a main factory in Japan that specializes in the products transferred to 'child' production facilities overseas, and which usually exercises control over the technology transfer process. 'Focal' factories are those Japanese factories that provide a range of complex functions, including research and development and experimentation, as well as mass production (see Fruin, 1992).

References

Amano, S. (2002) 'Lights and shadows in the Chinese economy (Part I)', *Journal of Japanese Trade and Industry*, 21(5), 19–22.

Business Week (2002) Greater China, 9 December, 40–50.

Cheung, C. and Wong, K.-Y. (2000) 'Japanese investment in China: a glo-cal perspective', in Li, S.-M. and Tang, W.-S. (eds), *China's Regions, Polity, and Economy: A Study of Spatial Transformation in the Post-Reform Era*. Hong Kong: The Chinese University Press, 97–132.

East Asia Analytical Unit (1996) *Asia's Global Powers: China Japan Relations in the 21st Century*. Canberra: Commonwealth of Australia.

Economist, The (2001a) 'Singapore: death by a thousand cuts: the global downturn hits the city-state', 15 September, 38.

Economist, The (2001b) 'A panda breaks the formation: China's economic challenge to East Asia', 25 August, 57–8.

Edgington, D. W. and Haga, H. (1998) 'Japanese service sector multinationals and the hierarchy of Pacific Rim cities', *Asia Pacific Viewpoint*, 39, 161–78.

Edgington, D. W. and Hayter, R. (2000) 'Foreign direct investment and the flying geese model: Japanese electronics firms in Asia Pacific', *Environment and Planning A*, 32, 281–304.

Ernst, D. (1997) 'Partners for the China Circle? The East Asian production networks of Japanese electronics firms,' in Naugton. B. (ed.), *The China Circle: Economics and Electronics in the PRC, Taiwan, and Hong Kong*. Washington, DC: Brookings Institution Press, 210–53.

Friedman, J. (1986) 'The world city hypothesis', *Development and Change*, 17, 69–83.

Fruin, M. (1992) *The Japanese Enterprise System: Competitive Strategies and Cooperative Structures*. Oxford: Oxford University Press.

Hayter, R. and Edgington, D. W. (2004) 'Flying geese in Asia: the impacts of Japanese MNCs as a source of industrial learning', *Tijdschrift voor Economische en Sociale Geografie*, 95, 3–26.

Hayter, R. and Han, S. (1998) 'Reflections on China's open policy towards foreign direct investment', *Regional Studies*, 32, 1–16.

Hong Kong Trade Development Council (2000) *Trade Developments: China's WTO Accession and Implications for Hong Kong*. Hong Kong: HKTDC.

Hong Kong Trade Development Council (2001) *Trade Developments. The Two Cities: Shanghai, Hong Kong*. Hong Kong: HKTDC.

Kuroda, A. (2002) 'The rise of China and the changing industrial map of Asia', *Journal of Japanese Trade and Industry*, 21(5), 14–18.

Leung, M. K. (1995) 'Banking reform in the People's Republic of China and its implications for Hong Kong as an international banking centre', in Davies, H. (ed.), *China Business: Context and Issues*, Hong Kong: Longman Hong Kong, 98–116.

Lin, G. C. S. (1997) *Red Capitalism in South China: Growth and Development of the Pearl River Delta*. Vancouver: UBC Press.

Lo, C. P. (1992) *Hong Kong*. London: Belhaven Press.

Lo, F.-C. and Yeung, Y.-M. (eds) (1996) *Emerging World Cities in Pacific Asia*. Tokyo: United Nations University Press.

Mentzer, J. T. (ed.) (2001) *Supply Chain Management*. Thousand Oaks, CA: Sage Publications.

Meyer, D. R. (2000) *Hong Kong as a Global Metropolis.* Cambridge: Cambridge University Press.

Meyer, D. R. (2002) 'Hong Kong: global capital exchange', in Sassen, S. (ed.), *Global Networks, Linked Cities.* New York, NY: Routledge, 249–72.

Mukoyama, H. (2002) 'Growing economic interdependence between Taiwan and China', *Rim: Pacific Business and Industries,* II(5), 28–48.

Mukoyama, H. (2003) 'How globalization will change Japan-China trading patterns', *Rim: Pacific Business and Industries,* III(9), 2–24.

Nakagane, K. (2002) 'Japanese direct investment in China: its effects on China's economic development', in Hilpert, H. G. and Haak, R. (eds), *Japan and China: Cooperation, Competition and Conflict.* Houndmills, Basingstoke: Palgrave, 52–71.

Naughton, B. (ed.) (1997a) *The China Circle: Economics and Electronics in the PRC, Taiwan, and Hong Kong.* Washington, DC: Brookings Institution Press.

Naughton, B. (1997b) 'The emergence of the China Circle', in Naugton, B. (ed.), *The China Circle: Economics and Electronics in the PRC, Taiwan, and Hong Kong.* Washington, DC: Brookings Institution Press, 3–37.

Overholt, W. H. (1993) *China: The Next Economic Superpower.* London: Weidenfeld and Nicholson.

Panitchpakdi, S. and Clifford, M. L. (2002) *China and the WTO: Changing China, Changing World Trade.* Singapore: John Wiley and Sons (Asia) Pte. Ltd.

Sassen, S. (2000) *Cities in the World Economy* (2nd edn), *Sociology for a New Century.* Thousand Oaks, CA: Pine Forge Press.

Sassen, S. (2001) *The Global City: New York, London, Tokyo* (2nd edn). Princeton, NJ: Princeton University Press.

Sassen, S. (ed.) (2002) *Global Networks, Linked Cities.* New York, NY: Routledge.

Scott, A. J. (ed.) (2001) *Global City-regions: Trends, Theory, Policy.* Cambridge, MA: Oxford University Press.

Seanwright, S. and Harrison, S. (2001) 'Shanghai closes gap in competitiveness survey', *South China Morning Post (Business Section),* 15 May, 1.

Skeldon, R. (1997) 'Hong Kong: colonial city to global city to provincial city?', *Cities,* 14, 265–71.

So, A.-Y. (1999) 'Hong Kong's pathway to a global city: a regional analysis', forthcoming in Gugler. J. (ed.). *World Cities Beyond the West: Globalization, Development and Inequality.* Cambridge: Cambridge University Press.

Sung, Y.-W. (1997) 'Hong Kong and the economic integration of the China Circle', in Naugton, B. (ed.), *The China Circle: Economics and Electronics in the PRC, Taiwan, and Hong Kong.* Washington, DC: Brookings Institution Press, 41–80.

Taylor, R. (1996) *Greater China.* London: Routledge.

Tsang, D. (2001) 'Our dynamism and openness', *Sunday Morning Post* (Hong Kong), 13 May, 9.

Vogel, E. F. (1989) *One Step Ahead in China: Guangdong Under Reform.* Cambridge, MA: Harvard University Press.

Yeung, H. W. and Olds, K. (eds) (2000) *Globalization of Chinese Business Firms.* Houndmills: MacMillan Press Ltd.

Yeung, Y.-M. (1997) 'Planning for Pearl City: Hong Kong's future, 1997 and beyond', *Cities,* 14, 249–56.

10 Assembling the 'global schoolhouse' in Pacific Asia

The case of Singapore[1]

Kris Olds and Nigel Thrift

Introduction

It is a near constant in the history of capitalism that what there is to know about the conduct of business is surrounded by a garland of service-sector institutions which do not only impart that knowledge but attempt to codify and improve upon it, so producing new forms of conduct. But since the 1960s, this roundelay has accelerated as the institutions of business knowledge have joined up to form a fully functioning 'cultural circuit of capital' (Thrift, 1997, 1998, 1999, 2002). This cultural circuit of capital is able to produce constant discursive-cum-practical change with considerable power to mould the content of people's working lives and, it might be added, to produce more general cultural models that affect the rest of people's lives as well.

But we cannot stop there. For the discursive and practical tenets of this world have increasingly become entangled with state action, producing new practices of government that are also redefining who counts as a worthy citizen. In other words, the kind of subject positions that are deemed worthy managers and workers are increasingly similar to the kinds of subject positions that define the worth of the citizenry (and, it might be added, other actors like migrant workers). This is particularly true of that network of global cities where these tenets are most likely to be put into action (Olds, 2001; Sassen, 2001; Ong, 2004).

The different centres of 'calculation' (if calculation is quite the right word) that make up the cultural circuit of capitalism can perhaps best be thought of as shifting 'assemblages' of governmental power, made more powerful by their strictly temporary descriptions and attributions. Assemblages are:

> secondary matrices from within which apparatuses emerge and become stabilized or transformed. Assemblages stand in a dependent but contingent and unpredictable relationship to the grander problematizations. In terms of scale, they fall between problematization and apparatuses, and function differently from either one. They are a

distinctive type of experimental matrix of heterogeneous elements, techniques and concepts. They are not yet an experimental system in which controlled variation can be produced, measured, and observed. They are comparatively effervescent, disappearing in years or decades rather than centuries. Consequently, the temporality of assemblages is qualitatively different from that of either problematizations or apparatuses.

(Rabinow, 2003: 56)

In turn, the denatured notion of assemblage makes much more room for *space*. Assemblages will function quite differently, according to local circumstance, not because they are an overarching structure adapting its rules to the particular situation but because these manifestations are what the assemblage consists of. Indeed, the cultural circuit of capital allows the knowledges of very different situations to circulate much more freely (and rapidly) and to have a relatively greater say within a space that is precisely tailored to that circulation, consisting of numerous sites and specialized route ways.

In this chapter, we want to look at one of these spaces, a space which is attempting to recast itself as a 'global schoolhouse' for business and management knowledge. Singapore, a Pacific Asian city-state with a population of 3.9 million (of whom about 600,000 are foreigners) is a rapidly evolving laboratory for the corporate interests of both the cultural circuit of capital and the state. But while Singapore is a very intense example, we would argue that the trajectory it has set out to follow – towards a kind of *kinetic utopia* – is one which many western and some Asian states (e.g. India or Malaysia) would like to emulate to a significant degree. This is a space where accumulation becomes the very stuff of life, through persuading the population to become its own prime asset – a kind of people mine (in a mineral sense) of reflexive knowledgeability.

This chapter therefore consists of four parts, including these introductory comments. In the second part, we go on to consider the cultural circuit of capital, concentrating especially on the role of business schools as the key nodes in this circuit. In the third part, we will consider the practices of the Singaporean state, as it seeks to engender socio-economic and discursive change via translating the internationalization/globalization agendas of a number of elite western universities (including three 'world-class' business schools: INSEAD, Wharton and the Chicago Graduate School of Business). Finally, we offer a few speculative comments and about the future direction of the Singaporean management experiment, for the phase that we focus on in this chapter is designed to lay the groundwork for a much more ambitious goal of transforming Singapore into an 'enterprise ecosystem', not just for Singaporeans but for the entire Pacific Asian region (ERC, 2002b).

The cultural circuit of capital

The world may consist of a constantly moving horizon of situated actions, learning experiments and makeshift institutional responses, but that does not mean that it cannot be held together. Since the 1960s, one of the more impressive of these holdings together has been the link up of a series of institutions to produce and disseminate business knowledge. In particular, this circuit arises from the concentration of three different institutions – management consultants, management gurus and especially business schools – all surrounded by the constant presence of the media, which in itself constitutes a purposeful part of the circuit.

Management consultancies date from the late nineteenth and early twentieth centuries. But their heyday has been since the 1960s, when companies like Bain and Co. and McKinsey began to gel into vast consulting combines. Consultancies subsequently became the important producers and disseminators of business knowledge through their ability to take up ideas and translate them into practice – and to feed practice back into ideas.

Management consultancies were helped in these ambitions by the oracles of business knowledge, *management gurus*, nearly all of whom were (or are) consultants. Gurus packaged business ideas as aspects of themselves (Micklethwait and Wooldridge, 1996). Though they existed before the 1980s, gurus have become particularly prevalent since the phenomenal success of Peters and Waterman's *In Search of Excellence* (1982), 'a Zen gun that was fired 20 years ago' (Peters, 2001). Gurus tend to embody particular approaches to business knowledge through performances that are meant to both impart new knowledge while also confirming what their audiences may already know (but need bringing out or confirming). Increasingly, gurus come replete with moral codings, 'they do not only tell managers how to manage their organizations, they also tell them what kind of people they should become in order to be happy and morally conscious citizens with fulfilling lives' (ten Bos, 2000: 22).

But the primer for the system of producing and disseminating management knowledge is now the *business school*. Though a small elite of business schools was formed in the late nineteenth and early twentieth century in the United States, the main phase of expansion took place much later – from the late 1940s on – on the back of the Masters in Business Administration (MBA) degree. In the rest of the world, business schools only slowly came into existence until, in the 1950 and 1960s, they began to open and expand in Europe and then in Asia. They now form the most visible tips of a vast global business education iceberg, one that turns over billions of dollars per year.

Producers of business knowledge necessarily have to have a voracious appetite for new knowledge since it is the continuous conveyor of new knowledge that keeps the system going. In particular, this means a central bank of knowledge that can be stripped of many of its local contingencies and can therefore be made mobile across the globe. So for example ideas like 'complexity theory' (Thrift, 1999) or 'community of practice' (Vann and Bowker, 2001) can be made into ready-made resources that give up a hold on certain aspects of the world for the sake of portability. But while the universalizing nature of much business knowledge is evident, business schools also produce rich case studies of practical corporate strategy that more often than not recognize the complex socio-spatial embeddedness of firms and market processes. The case study method is a prominent one in many business schools, with upper tier schools such as Harvard Business School, the Richard Ivey School, Darden and INSEAD producing the bulk of the 15,000 plus cases that now circulate through business school class rooms and corporate education centres. Given the interdependencies between business schools and corporations, business school academics have relatively deeper access to the primary 'movers and shapers' (Dicken, 1998) of the global economy than the vast majority of social scientists.

The kinds of knowledge that are pursued in business schools necessarily range widely. So there is functional knowledge of all kinds – from principles of accounting and finance to logistics. Then, there is knowledge that is organizational and strategic. And, finally, there is knowledge that is especially concerned with subjectification, how to be a 'global leader', for example (Roberts, 2003). But, whatever the case, what is effectively being pursued is a constant process of adaptation through continuous critique of the status quo (Boltanski and Chiapello, 1999). The critical feedback loop produced by the cultural circuit of capital is meant to produce a kind of dynamic equilibrium in which the brink (the 'edge of chaos') is the place to be.

Weaving in and out of this set of actions and ideas are the global *media*, key means of transport, amplifiers and generators of business knowledge in their own right. Through the vast range of different general and special media outlets which now exist, and through the vast range of general and special media intermediaries which vie to get their ideas circulated in these outlets, the media acts to force the production of ideas. Newspapers such as *The Financial Times* also shape institutional conduct at a wide variety of levels via their regular surveys and ranking exercises; a form of audit culture that competitive states are increasingly cognizant (and terrified) of. In addition, business knowledge is also circulated via the continual production of conferences, seminars, workshops, and the like as the meeting has increasingly been turned into a means of dissemination, which is itself sold as a product.

Though these different sets of institutions which make up the cultural circuit of capital, dispersed knowledges can be gathered up and centred/concentrated, practical knowledges and skills (including soft skills like leadership) can be codified, the miasma of too-much-information can be cut down and simplified, and large numbers can be made into small and manageable numbers. But three points need to be made here. First, we are **not** claiming that the knowledge being produced is somehow false, for example because it is caught up in 'fashion'. The hard and fast lines between the kind of studied objectivity which, in its various forms, academic knowledge still strives for, and the mutable contingencies of management knowledge were long ago broken down by superficial models of relativist or quasi-relativist approaches to knowledge, and the continuous process of osmosis between academic and management knowledge. But, second, that does not mean that we consider management knowledge to be neutral. The process of instrumentalized commodification which calls it into being brings with it a set of highly politicized values which cannot be denied (Vann and Bowker, 2001; Dezalay and Garth, 2002); values that underlie the influential spread of neoliberal policies through much of the world. Still, and third, both academic and management knowledges increasingly share certain values: a commitment to conceiving the world as continuously rolling over, continually on the brink; a commitment to fantasy as a vital element of how knowledge is constructed; and a commitment to tapping the fruitfulness of the contingency of the event.

One element of management knowledge which we want to foreground here is the constant attempt to produce new, more appropriate kinds of *subjects*, what we might call 'souls', that fit contemporary and especially future systems of accumulation. In pursuit of high performance, both workers and managers must be refigured. Of course, this kind of explicit engineering is hardly new. F. W. Taylor and others plotted bodily configurations which they believed would produce better workers at the end of the nineteenth century. Similarly, by the middle of the twentieth century, managers were beginning to be expected to embody themselves in ways that would make them better leaders. But the emergence of the subject as a quite explicit focus of management knowledge has taken on a new urgency of late, boosted by the growing power of human resources departments and the growing body of knowledge and practice devoted to such practices. In particular, we can see much greater attention being paid to attempts to produce 'knowledgeable' subjects – by harnessing tacit knowledge, by producing communities of practice within which learning is a continuous activity, by working with and making more of affect, by understanding the minutiae of embodied time and space, and so on. In other words, a partially coherent set of practices of 'government of the soul' (Rose, 1999) is starting to be produced by the cultural circuit of capital, a kind of instrumental phenomenology which can produce subjects that disclose the world as one which is uncertain and risky, but a world that can also be stabilized (in profitable ways) by the application of particular kinds of intense agency that are creative, entrepreneurial and business-like.

The Singaporean state, the global schoolhouse, and the cultural circuit of capital

Of course, other organizations have interests in producing pliant but enterprising subjects, not least the state. And, as has been shown many times now, a considerable part of this interest has come about as states have become more and more aligned with global corporate interests, redescribing themselves as competitive guarantors of economic growth through their ability to produce subjects attuned to this objective. Enterprise becomes both a characteristic and a goal of the new supply-side state. Nearly all western states nowadays subscribe to a rhetoric and metric of modernization based upon fashioning citizens who can become an actively seeking factor of production, rather like a mineral resource with attitude. And that rhetoric, in turn, has been based upon a few key management tropes – globalization, knowledge, learning, network, flexibility, information technology, urgency – which are meant to come together in a new kind of self-willed subject whose industry will boost the powers of the state to compete economically, and will also produce a more dynamic citizenry.

Many of the states of Asia have bought into this rhetoric of a knowledge economy, often with· good reason. Thus, beyond the purely economic advantage that is seen to arise from it, there is also its ability to both respect and minimize ethnic difference, and to provide an unthreatening (or difficult to critique) national narrative (c.f. Bunnell, 2002). Of these states, perhaps the most enthusiastic participant has been the paternalist but ultimately pragmatic State of Singapore, an independent city-state since 1965. Indeed, it would not be entirely unfair to say that Singapore has become a kind of management primer come true, with the fantasies of the serried rows of management texts in its main bookshops embodied in the person of its citizens and its 'professional' migrant workers. In Singapore accumulation often seems to have become the work of life, a passion of production (and consumption – Singaporeans are expected to be 'prosumers') in its own right.

Periodically, prompted by circumstance, Singapore re-focuses its economy. In the process this 'modern day garrison state' reworks a post-independence discourse of survivalism. Frequent tropes include both real and manufactured concerns about the country's small size and its resultant openness to competition from Malaysia, Hong Kong and most recently China; the gradual rundown of its traditional long-term geographic advantages (such as the port); and its lack of natural resources and consequent dependence upon its people. This concern seemed to be confirmed by the Asian economic crisis of 1997–1998 which meant that Singapore, though on the edge of events, saw its growth rate fall from 8 per cent in 1997 to 1.5 per cent in 1998. Singapore reacted predictably, with a 15 per cent wage cut, a 30 per cent reduction in rentals on industrial properties, and the liberalization of its financial sector (allowing for more foreign bank presence in the domestic banking sector). But the crisis also hastened a longer-term strategic shift fuelled especially by the later downturn in information technology industries, as well as more general concerns about a sluggish world economy, and the heightened geo-economic power of China.

The Government of Singapore, a technocratic 'soft-authoritarian' government that has been controlled by the People's Action Party (PAP) since 1959, is responsible for reshaping the economy. The Ministry of Trade and Industry (MTI) is the most important formal institutional mechanism for economic governance. While the MTI has only one functional department – the Singapore Department of Statistics – nine statutory boards (semi-independent and well resourced agencies) under the MTI jurisdiction carry out policy and programme work. The most significant MTI statutory boards are the:

- Economic Development Board (EDB);
- Singapore Productivity and Standards Board (PSB);
- Singapore Trade Development Board (TDB).

The Singapore EDB (http://www.sedb.com/) was founded in 1961 to formulate and implement economic development strategy for Singapore (Schein, 1996; Low, 1999; Chan, 2002). While relatively well resourced and staffed by Singaporeans, the EDB is open to the cultural circuit of capital through regular visits by management gurus and consultants – figures like Tom Peters, Gary Hamel and Michael Porter (the latter having worked with the EDB since 1986, anointed as a 'Business Friend of Singapore' in 2001), and most recently Richard Florida.

While the EDB is the shaper and mediator of most economic change within Singaporean territory, a powerful guidance role is played by select committees that report on a one-off or ad-hoc basis. An example of the former is the Committee on Singapore's Economic Competitiveness that reported on Asian crisis related matters in 1998. An example of the latter is the Economic Review Committee (ERC) (http://www.erc.gov.sg/), a Singapore-based network of state and private sector representatives responsible for making recommendations to generate structural shifts in economy and society. The most recent ERC was set up by Prime Minister Goh Chok Tong in October 2001 with a mandate 'to fundamentally review our development strategy and formulate a blueprint to restructure the economy, even as we work to ride out the current recession'. The Committee's composition is revealing: nine members of the government or government functionaries (including the President of the National University of Singapore), two Union representatives, and nine private sector representatives (including Arnoud De Meyer, the Dean of INSEAD's Singapore campus). Arnoud De Meyer also serves on the Sub-Committee on Service Industries in the ERC.

While the current ERC was given a relatively new mandate in 2001, it is building upon initiatives first established in the mid-1980s to promote the services sector as actively as manufacturing, thereby firing up 'twin engines' in a city-state drive for more diversified economic growth (ERC, 2002a). This service-oriented agenda subsequently merged with the trope of the 'knowledge-based economy' (KBE) that began circulating on a global scale in the 1990s. As Coe and Kelly (2000) demonstrate in the Singaporean case, this phrase first surfaced in a speech by the Prime Minister in 1994. By 1998 the phrase was gaining some currency. By 1999 it was in wholesale circulation, having 'seemingly entered the common vocabulary of all Ministers, bureaucrats and media commentators in Singapore' (Coe and Kelly, 2000: 418; see also Coe and Kelly, 2002).

In line with the goal of transforming Singapore into 'a vibrant and robust global hub for knowledge-driven industries', the EDB accordingly announced its detailed 'Industry 21' strategy, a strategy whose product would be a Singapore capable of developing:

> manufacturing and service industries with a strong emphasis on technology, innovation and capabilities. We also want to leverage on other hubs for ideas, talents, resources, capital and markets. To be a global hub and to compete globally, we require world-class capabilities and global reach. The goal is for Singapore to be a leading centre of competence in knowledge-driven activities and a choice location for company headquarters, with responsibilities for product and capability charters.
>
> The knowledge-based economy will rely more on technology, innovation and capabilities to create wealth and raise the standard of living. For our knowledge-based economy to flourish, we will need a culture which encourages creativity and entrepreneurship, as well as an appetite for change and risk-taking.
>
> (http://www.sedb.com: accessed 20 May, 2001)

More bluntly, Senior Minister Lee Kuan Yew (quoted in Hamlin, 2002) put it this way:

> Where do you produce your entrepreneurs from? Out of a top hat?
>   There is a dearth of entrepreneurial talent.
>   We have to start experimenting. The easy things – just getting a blank mind to take in knowledge and become trainable – we have done. Now comes the difficult part. To get literate and numerate minds to be more innovative, to be more productive, that's not easy. It requires a mind-set change, a different set of values.

Such comments crystallize the connection between structural reform (in a sectoral sense) and the need to construct new subject-citizens. Hence the shift from 'I21' to 'E21' – from industrial development to educational reform – and the emergence of the global assemblage that is the focus of this chapter.

The development of E21, in the mid-1990s, led the state to focus on the development of a 'world-class' education sector. Significantly, from our perspective, E21 led to the import of 'foreign talent', both to expose Singaporean educational institutions to competition (thereby forcing them to upgrade), and to produce (in discursive and institutional senses) a 'global education hub' that would be attractive to students from around the Asia-Pacific region. In theory, this cluster of educational institutions would produce and disseminate knowledge at a range of scales, supporting local and foreign firms in Singapore, state institutions in Singapore, and firms and states in the Southeast, East and South Asian regions.

Significantly, much of this educational strategy was concerned with those key institutions of the cultural circuit of capital, business schools. In turn, this hub would hypothetically act as the core of a series of industrial clusters, through spin-offs and the like in industries like medicine, engineering and applied sciences.

This education upgrade strategy hinged on attracting ten 'world-class' educational institutions to set up independently or in collaboration with Singaporean partners by the year 2008, plus a series of large corporate training concerns. In fact, by mid-2003 that target had been surpassed (see Table 10.1).

This striking development trend is likely to continue: in August 2003 Singapore's Trade and Industry Minister (George Yeo) stated that a foreign university is likely to be permitted (within the next year) to establish a large university campus in Singapore to offer a comprehensive curriculum from liberal arts to engineering. This is without taking note of the numerous corporate organizations which have set up training facilities in Singapore, including the New York Institute of Finance (set up in 1997) which trains senior financial executives and professionals, Motorola University South East Asia, Cable and Wireless, Citibank, ABN Amro, St Microelectronics, Lucent Technologies, and so on.

The 'grounding' of these foreign universities, including elite western business schools, was, and is, far from guaranteed. Policies do not beget the stabilization, even if only temporary, of the heterogeneous elements that make up this evolving global assemblage. What also mattered was government support via the powers and capacities of a Pacific Asian developmental state (e.g. large scale targeted financial subsidies), along with doses of bureaucratic persistence and persuasion (on the Singaporean developmental state see Olds and Yeung, 2004). For example, the EDB played an important role in courting select universities in R & D rich contexts (e.g. those in Boston). However, universities are less hierarchical than the transnational corporations Singapore is used to dealing with. As Tan Chek Ming, Director EDB Services Development, put it, 'Every faculty

Table 10.1 Substantial Singapore-foreign university initiatives (as at August 2003)

	Initiatives (by date of establishment)
Johns Hopkins University (JHU)	Three medical divisions of JHU were established in January 1998: Johns Hopkins Singapore Biomedical Center; Johns Hopkins Singapore Affiliated Programs and Johns Hopkins – National University Hospital International Medical Centre. These institutions facilitate collaborative research and education with Singapore's academic and medical communities. Web link: http://www.jhs.com.sg/. JHU's Peabody Institute is also collaborating with the National University of Singapore (NUS) to create the Singapore Conservatory of Music (now known as the Yong Siew Toh Conservatory of Music). An agreement was established in November 2001. Web link: http://music.nus.edu.sg/index.htm.
Massachusetts Institute of Technology (MIT)	The Singapore – MIT Alliance (SMA) was established in November 1998. Local alliance partners include the National University of Singapore (NUS) and Nanyang Technological University (NTU). The focus is on advanced engineering and applied computing. SMA-1 runs until 2005, and involves approximately 100 professors and 250 graduate students (who receive MIT certificates). A new phase (known as SMA-2) will run from July 2005–2010, with deeper MIT-degree granting capacity. Web link: http://web.mit.edu/sma/.
Georgia Institute of Technology (GIT)	The Logistics Institute – Asia-Pacific (TLI-AP) was established in February 1999. TLI-AP is a collaboration between NUS and the Georgia Institute of Technology. TLI-AP trains engineers in specialized areas of global logistics, with emphasis on information and decision technologies. TLI-AP facilitates research, and the acquisition of dual degrees and professional education. Web link: http://www.tliap.nus.edu.sg/.
University of Pennsylvania (Penn)	Singapore Management University (SMU) was officially incorporated in January 2000. Wharton School faculty from the University of Pennsylvania (Penn) provided intellectual leadership in the formation of SMU's organizational structure and curriculum. The Wharton-SMU Research Center was also established at SMU. 306 students were enrolled at SMU in 2000, 800 in 2001, 1,600 in 2003, with eventual enrolment levels expected to top out at 9,000 (6,000 undergraduates and 3,000 graduate students). A US$650 million campus is currently being built in Singapore's downtown. Web link: http://www.smu.edu.sg/.
INSEAD	INSEAD, the prominent French business school, established its second campus in Singapore in January 2000. A US$40 million building was built to enable Singapore-based faculty, and European campus visiting faculty, to offer full- and part-time courses, as well as executive seminars. As of February 2003 there were 255 MBA students in Singapore. In February 2003 INSEAD formally decided to launch phase two which will involve doubling the size of the Singapore campus. Web link: http://www.insead.edu/campuses/asia_campus/index.htm.
University of Chicago	The University of Chicago Graduate School of Business (GSB) established a dedicated Singapore campus in July 2000 to offer the Executive MBA Program Asia to a maximum of 84 students per programme. The curriculum is identical to the Executive MBA Programmes in Chicago and Barcelona, and faculty are flown in from Chicago to teach on it. Web link: http://gsb.uchicago.edu/.

Technische Universiteit Eindhoven (TU/e)	The Design Technology Institute (DTI), jointly administered by National University of Singapore (NUS) and Technische Universiteit Eindhoven (TU/e), was established in May 2001. The courses and projects offered by DTI aim to provide a balance between basic engineering concepts, and product design and development. TU/e has strong links to Philips, both in the Netherlands and in Singapore. Web link: http://www.dti.nus.edu.sg/.
Technische Universität München (TUM)	The National University of Singapore (NUS) and the Technische Universität München (TUM) established a Joint Master's degree in Industrial Chemistry programme in January 2002, and a Joint Master of Science in Industrial and Financial Mathematics in late 2003. The German Institute of Science and Technology (GIST) in Singapore coordinates these education programmes, as well as executive training and contract research. A significant proportion of specialists from industry are involved. Web link: http://www.gist-singapore.com/.
Carnegie Mellon University (CMU)	Carnegie Mellon University (CMU) signed a Memorandum of Understanding with Singapore Management University in January 2003 to collaborate on the development of a School of Information Systems (SIS). The MOU runs from 2003–2007. The School will be SMU's fourth since it was established in 2000. Web link: http://www.smu.edu.sg/sections/schools/information.asp.
Stanford University	Stanford University and Nanyang Technological University (NTU) signed an official Memorandum of Understanding, in February 2003, to offer joint graduate programmes in environmental engineering. The Stanford Singapore Partnership education programme began in June 2003. The programmes will be mix of distance education with student and faculty exchanges. Web link: http://www.ntu.edu.sg/CEE/ssp/Index.htm.
Cornell University	A Memorandum of Understanding was signed in February 2003, between Nanyang Business School (NBS), Cornell University's School of Hotel Administration, and the International Hotel Management School (a Singaporean entity), to set up a joint Cornell-Nanyang Business School of Hospitality Management. The School will be established by 2004 and offer joint graduate degrees while facilitating research on the Asian hospitality industry. Web link: http://www.hotelschool.cornell.edu/.
Duke University	A Memorandum of Understanding was signed in June 2003, between Duke University Medical Center and the National University of Singapore, to establish a graduate medical school in Singapore by 2006. Web link: http://medschool.duke.edu/.
Karolinska Institutet (KI)	A Memorandum of Understanding was signed in July 2003 between the Stockholm-based Karolinska Institutet (KI) and the National University of Singapore to operate joint postgraduate programmes in the areas of stem-cell research, tissue engineering and bio-engineering. Web link: http://info.ki.se/index_en.html.
Indian Institute of Technology (IIT)	The multi-sited Indian Institute of Technology (IIT) gave in principle approval, in May 2003, to the idea of establishing a Singapore campus in 2004. The IIT campuses involved include Bombay, Chennai, New Delhi, Kharagpur, Kanpur, and Roorkie. The exact modality of presence is currently being worked out in conjunction with the Singapore Economic Development Board.

member has to agree. All you need is one person to disagree and the whole deal will be thrown out of alignment.' In this context, EDB:

> team members act as tour guides, flying in faculty staff for a look-see trip to Singapore. The usual highlight is a meeting between the dons and senior Cabinet Ministers, namely Deputy Prime Minister Tony Tan, who oversees university education, Education Minister Teo Chee Hean, and Trade and Industry Minister George Yeo.
>
> These meetings are important, stresses Mr Tan, as they send a strong signal to the visitors of the political will and commitment in drawing reputable universities to Singapore. Team members also double up as property agents, scouting around for suitable premises in Singapore to locate the foreign university. They also help look into the legal and financial aspects of setting up shop in Singapore.
>
> (Nirmala, 2001)

In order to tempt business schools, the EDB played up Singapore's cosmopolitan nature, and then used tangible material resources in the form of financial and other incentives. For example, INSEAD received $10 million in research funding over four years, plus soft loans, reduced land values (about one-third of the commercial price), easier-to-get work permits, housing access, and so on. The University of Chicago GSB received several million dollars worth of subsidy via the renovation of the historic House of Tan Yeok Nee building they now use as their 'campus'. Finally, the Government of Singapore effectively funds the Wharton-SMU Research Center (http://www.smu.edu.sg/research/) at SMU, providing monetary and in-kind support for research projects, seminars, scholarships and the like.

These forms of material support are clearly important, and short- and long-term financial opportunities needed to be viewed favourably by the three business schools before they would commit the necessary intellectual and material resources required to stretch complicated institutional fabric across space. But there were some additional factors that led the cultural circuit of capital into Singapore space: the city-state's strategic geographical position within Asia (boosted by Changi Airport, an efficient award winning airport 20–30 minutes taxi ride from all three campuses), 'quality of life' for expatriates, the fact that many alumni were Singaporean, and the large number of transnational corporations with presences in Singapore. All of these factors were often put together as 'international feel' or a genuinely 'cosmopolitan nature'; characteristics associated with global cities (Olds, 2001; Sassen, 2001).

In summary, elite institutions of higher education are recognized by the Singaporean state as playing a fundamental role in restructuring the economy via the refashioning of the local citizenry, while simultaneously providing retooling opportunities for the 75,000–100,000 professional migrants who use Singapore as a temporary base. The key idea is the creation of a virtuous circle: draw in the 'best universities' with global talent; this talent then creates knowledge and knowledgeable subjects; these knowledgeable subjects, through their actions and networks, then create the professional jobs that drive a vibrant KBE. As Tharman Shanmugaratnam (Senior Minister of State for Trade and Industry) puts it, the Government seeks to create 'a new breed of Singaporean':

> We have strong institutions and a highly credible government. We start from a position of strength, both financially and socially. All we want to have now is a stronger individual, more adaptable to the business world with a global mindset and concrete experience.
>
> (*Straits Times*, March 17, 2002: 18)

And, again, elite business schools are perceived by the State to support (and attract to Singapore) the highly prized 'global talent' associated with TNCs.

Conclusions

This chapter has sought to describe the way in which the cultural circuit of capital has become aligned with the Singaporean state, and has thereby become involved in global geoeconomic and geopolitical interventions. These interventions are producing new forms of governmentality that privilege the mass production of knowledgeable and enterprising subjects, subjects who can simultaneously optimize their relationship to themselves and to work. We paid particular attention to the case of Singapore as a story of how what are still relatively loose functions that bring into play populations, territories, affects and events appear to be finding common cause in particular places, at particular times, and can co-evolve new strategies of government which are intended to re-code Singapore's citizens.

In our view, the injection of new knowledges into Singapore space is designed to diversify and enhance Singapore's functional role, and discursive identity, as a global services hub. These interventions, so it is hoped, will generate synergy between the new foreign educational institutions, the TNCs that have bases in Singapore, and Government of Singapore bodies (including indigenous universities). Clearly there is a direct investment aspect to the development process. However, the real value and total employment impacts of the foreign universities setting up in Singapore cannot help but be relatively limited given the limited size of the city-state. Indeed, the Government of Singapore is heavily subsidizing the foreign universities, Singaporean ventures (to the tune of $5–20 million per university over the first five years). Apart from the import of new services agents (the foreign universities), we feel that the overriding objectives are discursive and subject-making in nature. The interventions we have written about are designed to discursively engineer a 'global education hub', especially in the discursive fields associated with the cultural circuit of capital. And these interventions are designed to create 'a new breed of Singaporean', one that will be more entrepreneurial, connected to the world, yet (so the state hopes) still committed to 'our best home'. On a related note, these new strategies of government are designed to enable the local and regionally-based professional migrants (expatriates) to discipline themselves through a continual 'upgrading' process, while simultaneously 'branding' Singapore as a suitable hub for 'global talent'.

So far as Singapore is concerned, the strategy of bringing the cultural circuit of capital and the state together as a relatively loose and opportunistic assemblage is clearly intended to be a critical element of 'Remaking Singapore', one which if successful may lift Singapore further out of the Southeast Asian region, flinging it into an orbit where its region can be the globe itself.

Of course no strategy is without risk. One risk is that the strategy of attracting the cultural circuit of capital will be too successful and that the pile-up of new educational institutions of one sort or another will grow beyond what Singapore can deliver. Indeed in September 2002 the ERC recommended that Singapore become a 'Global Schoolhouse' for an 'additional 100,000 international fee-paying students and 100,000 international corporate executives for training' (ERC, 2002b), a challenging policy goal for both the state and the cultural circuit of capital, to put it but mildly. The first public voicing of such concerns were recently levelled in *The Financial Times'* highly influential 'Business Education Report' (January 2004 version), with a 400,000 plus paper-copy circulation. A long article (Bala, 2004) in this report cites various senior foreign academics who express support but also some concerns about the speed and management of the education hub development process. For example, Ronald Frank, the American president of Singapore Management University, puts it this way:

> 'Unlike in Australia and in Hong Kong, two role models for education, the market developed gradually over time. But Singapore's market is being artificially created,' says Prof Frank.
>
> 'The good thing is the country has the best record in social engineering than any other society.'
>
> 'However, the danger in higher education is that speed and the building of quality education are antithetical to each other.'
>
> Prof Frank, who has taught at Harvard and Stanford and has held various posts at Wharton, adds Singapore's policymakers need to 'carefully work out the rate of growth they want as over-capacity can damage not just the new entrants but the entire education system'.
>
> (Bala, 2004: 11)

A second risk is that the informal agreement on academic freedom for these foreign universities will be tested, just as foreign media freedoms in Singapore are tested from time to time.

The third risk is that contradictions may emerge between growing economic sectors in this small island nation: a services sector, and services employees, that demand high quality of life, versus a fast-growing chemicals sector, one that is injecting increasing volumes of noxious emissions into the atmosphere of the coastal zones (especially in the Jurong area which is close to NUS, INSEAD, the Singapore Science Parks, the 200 hectare One-North biotechnology-oriented development project, and several international schools).

The final risk, for a control-oriented Singaporean state, is that the Singaporean subjectivities being engineered in this denationalized context may unhinge 'rootedness' and facilitate permanent mobility and/or emigration.

What is clear, then, is that the future shape and effectiveness of the set of assemblages that are associated with making 'literate and numerate minds to be more innovative, to be more productive' has yet to be fully worked through. This is clearly a case of development in the making.

Acknowledgements

This chapter is derived, with kind permission, out of a chapter for the forthcoming book edited by Aihwa Ong and Stephen Collier titled *Global Assemblages: Technology, Politics and Ethics as Anthropological Problems*, Malden, MA: Blackwell. We would also like to thank the National University of Singapore (especially the Department of Geography) for its generous support. Kris Olds would also like to acknowledge the assistance of all interviewees, some anonymous Singapore-based professionals, and important financial support from the US Department of Education via the UW-Madison Center for International Business Education and Research (CIBER).

Note

1   This chapter is based upon a range of primary and secondary sources. While no direct quotes are utilized in this chapter, the text has been shaped by the insights derived from formal interviews and correspondence with key actors. These interviews took place in June 2001 and October 2002, and include the following people: Janice Bellace, President, Singapore Management University. Professor Bellace was the first President of SMU, and she stepped down in September 2001, becoming Vice Chairman (Academic Affairs) of SMU's Board of Trustees. Professor Bellace is Samuel Blank Professor of Legal Studies, The Wharton School, University of Pennsylvania; Lily Kong, Dean, Faculty of Arts and Social Sciences, National University of Singapore; Arnoud De Meyer, Dean, INSEAD Singapore Campus; Beth Bader, Managing Director, Executive MBA Program Asia, University of Chicago Graduate School of Business. Kris Olds

also worked in Singapore from 1997 to June 2001 at the Department of Geography, National University of Singapore, and conducted field research in Singapore in November 2001 and January 2003. It is also important to note that Nigel Thrift acted as an external examiner for Nanyang Technological University, and he was a Distinguished Visiting Professor at NUS from January to April 2002.

References

Bala, S. (2004) 'The city-state makes its point as a regional hub: Singapore: which schools will stand at Asia's pinnacle', *Financial Times*, FT Report – Business Education, 11.

Boltanski, L. and Chiapello, E. (1999) *Le Nouvel Esprit du Capitalisme*. Paris: Gallimard.

Bunnell, T. (2002) '(Re)positioning Malaysia: high-tech networks and the multicultural rescripting of national identity', *Political Geography*, 21, 105–24.

Chan, C. B. (ed.) (2002) *Heart Work: Stories of How EDB Steered the Singapore Economy from 1961 into the 21st Century*. Singapore: EDB.

Coe, N. and Kelly, P. (2000) 'Distance and discourse in the local labour market: the case of Singapore', *Area*, 32(4), 413–22.

Coe, N. and Kelly, P. (2002) 'Languages of labour: representational strategies in Singapore's labour control regime', *Political Geography*, 21, 341–71.

Dezalay, Y. and Garth, B. (2002) *The Internationalization of Palace Wars: Lawyers, Economists, and the Contest to Transform Latin American States*. Chicago, IL: University of Chicago Press.

Dicken, P. (1998) *Global Shift: Transforming the World Economy*, 3rd edn. London: Paul Chapman.

ERC (2002a) *Report of the ERC Subcommittee on Services Industries: Part I*. Singapore: ERC, available at: http://www.erc.gov.sg/frm_ERC_ErcReports.htm (accessed 1 December 2003).

ERC (2002b) *Developing Singapore's Education Industry*. Singapore: ERC, available at: http://www.erc.gov.sg/frm_ERC_ErcReports.htm (accessed 1 December 2003).

Hamlin, K. (2002) 'Remaking Singapore', *Institutional Investor*, May.

Low, L. (ed.) (1999) *Singapore: Towards a Developed Status*. Singapore: Oxford University Press.

Micklethwait, J. and Wooldridge, A. (1996) *The Witch Doctors: Making Sense of the Management Gurus*. New York, NY: Times Books.

Nirmala, M. (2001) 'Campus courtships', *Straits Times*, 24 June, R1.

Olds, K. (2001) *Globalization and Urban Change. Capital, Culture and Pacific Rim Mega-Projects*. Oxford: Oxford University Press.

Ong, A. (2004) 'Ecologies of expertise: technology and citizenship in Asian knowledge society', in Ong, A. and Collier, S. (eds), *Global Assemblages: Technology, Politics and Ethics as Anthropological Problems*. Malden, MA: Blackwell.

Peters, T. (2001) 'True confessions', *Fast Company*, 53, 78.

Peters, T. and Waterman, R. (1982) *In Search of Excellence*. New York, NY: Warner Books.

Rabinow, P. (2003) *Anthropos Today: Reflections on Modern Equipment*. Princeton, NJ: Princeton University Press.

Roberts, S. (2003) 'Global strategic vision: managing the world', in Perry, R. and

Maurer, B. (eds), *Globalization Under Construction: Governmentality, Law and Identity.* Minneapolis, MN: University of Minnesota Press.

Rose, N. (1999) *Powers of Freedom: Reframing Political Thought.* Cambridge: Cambridge University Press.

Sassen, S. (2001) *The Global City: New York, London, Tokyo,* 2nd edn. Princeton, NJ: Princeton University Press.

Schien, E. (1996) *Strategic Pragmatism: The Culture of Singapore's Economic Development Board.* Cambridge, MA: MIT Press.

Straits Times (2002) 17 March, 18.

ten Bos, R. (2000) *Fashion and Utopia in Management Thinking.* Amsterdam and Philadelphia, PA: John Benjamins Publishers.

Thrift, N. (1999) 'The place of complexity', *Theory, Culture and Society,* 16(3), 31–70.

Thrift, N. J. (1997) 'The rise of soft capitalism', *Cultural Values,* 1, 29–57.

Thrift, N. J. (1998) 'Virtual capitalism: some proposals', in Carrier, J. and Miller, D. (eds), *Virtualism: The New Political Economy.* Oxford: Berg.

Thrift, N. J. (2002) 'Think and act like revolutionaries: episodes from the global triumph of management discourse', *Critical Quarterly,* 44, 19–26.

Vann, K. and Bowker, G. (2001) 'Instrumentalizing the truth of practice', *Social Epistemology,* 15(3), 247–62.

11 Air cargo services, global production networks and competitive advantage in Asian city-regions

Thomas R. Leinbach and John T. Bowen

Introduction

The importance of services in the contemporary global economy is beyond dispute. Clearly services facilitate economic transactions and play a vital role in development (Daniels, 1993; Daniels and Moulaert, 1991; Illeris, 1996; Begg, 1994). But in addition, services may be used to secure competitive advantage in the global economy (Porter, 1990; Porter, 1996; Dicken, 2003). National economies are much less insular than even a decade ago and more interdependent, a process that is inextricably linked to, in part, the presence of 'enabling or space-shrinking technologies' – transportation, communication and organizational innovations that aid in the internationalization process (Dicken, 2003: 89–93). In this process, forward and backward linkages in global production networks make the service sector essential to manufacturing and development planning. Remaining competitive means that firms require, as they develop value chains, efficient logistics services applied to their procurement and distribution networks. Yet the service sector remains relatively understudied, leaving large gaps in our understanding of services especially in global context (Daniels, 1987; Martinelli, 1991).

This is particularly true of air cargo. The thousands of routes operated by the world's airlines have become vital arteries of trade. Between 1980 and 2000, the volume of international air cargo traffic, measured in freight ton-kilometres, grew fivefold (*ICAO Journal*, 2001). With its growing importance, air cargo has taken a prominent place among the services to which states, regions and firms must have access in order to survive in global markets. Air cargo has become the principal mode of international transport used for a wide variety of (especially knowledge-intensive) goods.

This chapter argues that air cargo services are a major instrument in the ability of firms and states to internationalize and maintain competitive advantage in the global economy (Porter, 1990; Nilsson *et al.*, 1996). Moreover, air cargo services are critical because they enhance and may even induce regional development (Bowen *et al.*, 2002; Begg, 1994;

Hansen, 1990). Our major aim is to show how the firms that use air cargo services, the airlines[1] that carry air cargo, and the forwarders that act as intermediaries among these other firms interact in ways that engender the development of firms and regions. More specifically, we attempt to show how manufacturing firms, and in particular those from the electronics sector, heighten their ability to internationalize and compete by using air cargo services and at the same time how logistics operators operate to deliver such services amidst competition from other providers. In turn, the rise of a handful of global forwarders, the emergence of cargo airline alliances, and the new importance of FedEx and UPS in the region affect the performance of Asian city-regions in the global economy.

The chapter, which is based on interviews and data gathered over the last three years with both the users and providers of air cargo services in Malaysia, the Philippines and Singapore, uses case study evidence to achieve a greater understanding of the demand and supply side of air services (Martinelli, 1991). The focus of our analysis is the Southeast Asian electronics industry (e.g. Borrus, 2000; Ernst and Guerrieri, 1998). Our aim is to situate our work in the growing body of work on global production networks (GPNs) (e.g. Borrus *et al.*, 2000; Coe, *et al.*, 2003; Henderson *et al.*, 2002; Yeung, 2000; Gereffi, 1994; Gereffi and Kornzeniewicz, 1994) and advanced producer services. Especially important here is to show how producer services (of which air cargo services are one type) and competitive advantage fit into GPNs. In this effort we show that, in part, competitive advantage is afforded by the flexibility, diversity and differentiation of air cargo services in various markets. More specifically we place our work against existing research on regional development and the competitive advantage of firms within producer services industries (O'Farrell *et al.*, 1993; Lindahl and Beyers, 1999).

We address first the nature of air cargo services and the growth of these services in Asia. This section is followed by our conceptual approach to air cargo services and competitive advantage within the context of GPNs. Next we examine the interaction between firms that provide and firms that use air cargo services to show how competitive advantage can be 'pulled through' to foster competitive advantage in structural and geographical context. Subsequently, and in light of the theme of this volume, we examine the differences among air cargo services in and their impact upon the several city regions under study. A concluding section draws some generalizations and provides a comment on further research directions.

Air cargo services in Asia

The increased importance of the service sector, or tertiarization, in Asian economies has differed from that of other regions in the greater importance of transportation and logistics services (O'Connor and Hutton,

1998). The region's dependence on export-driven industrialization, its distance from other major world markets, and the elaboration of production linkages across Asia have made services related to the movement of goods more important to Asia than other regions. Rapid industrialization in Asia has been accompanied not only by the vast expansion of traffic by land, at sea, and in the air, but also by qualitative changes in the sophistication and diversity of transportation and logistics services (Bowen and Leinbach, 1995). Together, these changes have made such services critical in shaping patterns of development at the city-region level.

The importance of transportation services to the region is manifest in Asia's disproportionate share of the world's air cargo and related air logistics services (Rimmer, 1997; Bowen, 2004). Routes to, from, and within Asia accounted for 43 per cent of air freight ton-kilometres and just under 30 per cent of air freight tons in 2000 (*Boeing World Air Cargo Forecast 2000/2001*, 2001). Asia hubs rank among the most important with regard to the volume of air cargo handled.

The rapid expansion of air cargo flows globally, and in Asia more specifically, reflects the greater emphasis placed on time as an element of competitive strategy (Schoenberger, 1994). The complex interplay between economic globalization and time-space compression has fostered a business environment in which firms in a wide array of manufacturing industries now compete on time. The unprecedented pressure to compress the elapsed time from product innovation to product consumption has been felt with particular keenness by the electronics manufacturers (broadly defined) that have been both leading agents in Asian industrialization and chief customers for the air cargo industry. Major electronics manufacturers have developed elaborate production networks that traverse much of the world, especially Asia, in search of competitively priced resources (e.g. components, labour); the use of air cargo services on links of those networks minimizes the time-cost of internationalization.

As the air cargo industry has grown, its services have been transformed by two forces that have affected other services industries as well: advances in information technology (IT) and the deeper externalization of services. First, IT has been used to accelerate the flow of information among shippers, freight forwarders, airlines, customs authorities and consignees regarding the movement of goods (Button and Owens, 1991). That information, particularly when delivered through electronic cargo tracking systems, has been critical to maintaining the precision of just-in-time production systems. Second, the externalization of service functions, which has been a powerful factor propelling the growth of advanced producer services more generally (Bagchi-Sen and Sen, 1997), has fostered the proliferation of air logistics services that are offered in conjunction with conventional air freight transportation (Bowen and Leinbach, 2004). Major freight forwarders now offer a variety of value-added services that together are termed third party logistics (3PL) services (Zhu *et al.*, 2002).

Such services are, unsurprisingly, unevenly developed across the global economy and are more developed in and around major hub airports. Within Asia, in particular, major hub airports are surrounded by logistics parks within which 3PL services are offered by major multinational freight forwarders. It is also in major hubs that the widest range of air cargo transportation services are offered, both with respect to the destinations directly served and in the classes of services offered (e.g. different levels of time-definite services guaranteeing delivery within 1 day, 2 days etc.). These knowledge-intensive services fit squarely within the rubric of advanced producer services (Bowen and Leinbach, 2003).

Producer services, global production networks and competitive advantage

The importance of advanced producer services in internationalized production networks has received too little attention. Although the enabling role of services, particularly transportation and communication services, has been amply noted (Dicken, 2003), the manner in which services shape both the geography of such networks and the competitiveness of firms and places within those networks has not. In this section, we propound a conceptual framework for analysing the manner in which the competitive advantage of services firms is captured by the firms that use their services and by the places in which those services are situated. In pursuit of this objective, we synthesize recent work on global production networks (Coe *et al.*, 2003) and the conceptual treatment of competitive advantage articulated by Porter (1985, 1990). We turn first to the former.

Networks have gained currency as descriptive, analytical, and conceptual tools in geography and related disciplines (Yeung, 1994; Yeung, 2000; Dicken *et al.*, 2001; Smith *et al.*, 2002). The network metaphor has been applied at a variety of scales: from the internal structure of transnational corporations (TNCs), to the relationships among multiple firms linked through the production of a single commodity, to the connections among national economies in global trade.

The global production network (GPN) holds the potential to integrate these overlapping scales of analysis (Coe *et al.*, 2003) by incorporating intrafirm and interfirm linkages with the institutional and economic settings of the regions across which those linkages are elaborated. The GPN is 'the globally organized nexus of interconnected functions and operations through which goods and services are produced and distributed' (Coe *et al.*, 2003: 18). The linkages comprising a GPN carry not only the commodities of international trade and the raw materials and components with which they are made, but also money, information and ideas and, ultimately, value. As described below, value is the conceptual bridge between GPNs and competitive advantage.

Networks provide an incisive approach to understanding the manner in which places are affected by the globalization of economic activity.

> A network methodology forces us to address the direct and indirect connections between economic activities stretched across geographical spaces but embedded in particular places. Thus, we have a mutually constitutive process: while networks are embedded within territories, territories are, at the same time, embedded into networks.
>
> (Dicken *et al.*, 2001: 97)

GPNs offer a theoretically rich framework to assess the success of regions in creating, enhancing, and capturing value. The degree to which a region is successful in this regard depends upon the 'coupling' of the strategic needs of transnational corporations *and* region-specific assets (Coe *et al.*, 2003). Those assets comprise knowledge, skills and expertise, but also a cooperative atmosphere conducive to learning.

How do services fit into GPNs? Producer services integrate and coordinate the dispersed functions and operations of a GPN, tying together production and consumption. Indeed, the geographic disaggregation of production has been made possible by the growing capabilities of the services sector, and conversely the disaggregation of the production has spurred the rapid growth of services (Rabach and Kim, 1994). Services shape what is produced (e.g. through market research), how it is produced (e.g. through process engineering), and, most importantly in the current context, where it is produced.

Producer services, by lowering the cost (including the time-cost) of transacting business on an international basis, enlarge the scope over which a firm can feasibly leverage the assets of distant regions. Producer services are, to a varying degree depending on the specific industry, provided on a worldwide, integrated basis by a shrinking number of services TNCs [see Thrift, 1987; Agnes, 2000 and Dymski, 2002 on finance and banking; Daniels *et al.*, 1989 on accountancy; Leslie, 1995 and Grabher, 2001 on advertising; Warf, 2001 on legal services; Heaver, 2001 on international sea freight; and Bowen and Leinbach, 2003 on international air freight]. These services TNCs mitigate the inherent complexity of operating an internationally dispersed production network by, for example, enabling a global merchandiser to deal with one or a handful of advertising agencies in most markets worldwide.

Furthermore, services are a part of the constellation of territorially-specific assets with which the networks of transnational corporations intersect. As described above, sophisticated air logistics services, for example, are unevenly developed (Bowen and Leinbach, 2004) and consequently distinguish some regions (particularly global city-regions where advanced producer services are concentrated) from other competitors for investment. It is partly such resource disparities that give rise to the 'differentiated networks' operated by TNCs (Nohria and Ghoshal, 1997). In places with dense, local interfirm knowledge networks, TNCs are better able to develop more complex operations (e.g. product development instead of routine assembly); but such places also require a higher level of TNC commitment in order to compete effectively. In other words, regions that are rich in sophisticated knowledge-based firms drawn together through a skein of overlapping networks both facilitate and necessitate broader and more complex TNC investments (c.f. Coe *et al.*, 2003). Those investments in turn sustain and augment the density of knowledge networks in favoured regions, creating a kind of virtuous cycle through which the further development of regions that best 'couple' with TNC needs is propelled. Such regions will better create, sustain, and capture value in GPNs.

The creation of value links GPNs and the earlier work of Porter (1990) on competitive advantage. Porter (1990: 40) claimed:

> To gain competitive advantage over its rivals, a firm must either provide comparable buyer value but perform its activities more efficiently than its competitors (lower cost), or perform activities in a unique way that creates greater buyer value and commands a premium price (differentiation).

The value chain (or, from a GPN perspective, network of value) linking supplier and user firms represents an avenue via which the competitive advantage of the former can be conveyed to the latter. Porter's 'value system' allows for only a one-way flow of value from supplier value chains to the firm value chain and then to distributor and retailer value chains. One merit of the GPN approach is that it does not privilege any particular direction in value creation and appropriation (Smith *et al.*, 2002). A TNC manufacturer can, through its focal position in one or more GPN, capture part of the value produced both by upstream services (e.g. product design) and downstream services (e.g. distribution and retailing).

Services comprise part of the 'related and supporting industries' that Porter (1990) identified as one of the four determinants of competitive advantage. In explaining the significance of this determinant, Porter pointed to several advantages that accrue to a firm that has internationally-competitive domestic supplier industries: freer information flow (attributable to proximity and reduced cultural distance), reduced transaction costs, greater prospects for joint product development and other forms of 'deep interchange', and higher likelihood of new domestic entry in the firm's industry (given the ease of securing relevant inputs), fostering greater dynamism.

Porter (1990: 106, emphasis added) argued, 'Services industries *pull through* sales of linked manufactured goods from that nation and vice versa'. An example is the Swedish specialty shipping industry whose success in the automobile shipping business was partly the result of that nation's internationally successful car and truck exporters (Saab and Volvo) and which, in turn, gave impetus to sales of the country's specialty shipbuilding industry. We suggest that the concept of pull-through can be applied more broadly to competitive advantage such that, via the value chain linking supplier and buyer, a firm can *pull through* competitive advantage from other firms to which it is linked through a GPN. A firm's cost or differentiation advantage is necessarily transmitted through the value system. The ability of a firm, and by implication a place or region, to capture the advantage (and value) derived by a linked supplier or buyer in a related industry will depend on the firm's relative power in a GPN. A focal firm, such as a large manufacturing TNC in a producer-driven GPN, will be better able to do so than a more peripheral firm.

Porter contended that the pull-through effect (in his more narrowly circumscribed use of the concept) would be dependent upon the technological interdependencies between the products concerned and that, consequently, the effect would be greater earlier in the product cycle. The same conditions likely mediate the pull-through of competitive advantage. Further, the pull-through effect will be muted or amplified depending upon the degree to which a particular link between a supplier and buyer is common or unique. A world-class manufacturer of a particular component that supplies that product to most major manufacturers in an industry will have little impact on the competitive advantage of any of the latter. Conversely, if a supplier has a singular relationship with a buyer, the prospects for pull-through are enlarged. We are especially interested in the manner in which a services firm with a competitive advantage in its industry can convey that advantage to a manufacturer using those services.

Because air cargo services are part of the rapidly changing transportation and logistics industries (Bowen and Leinbach, 2004; Leinbach and Bowen, 2004), there is greater scope for a pull-through effect upon competitive advantage in this industry than in services industries that lack such dynamism because firms can more easily gain a 'first-mover advantage' (Porter, 1990) (Figure 11.1). Moreover, because transportation and logistics services, and the distributive infrastructure (Martinelli, 1991) more generally, are so directly related to production, in contrast to some other producer services (e.g. legal services, advertising), their potential impact upon the competitive advantage of manufacturing firms is augmented.

We further argue that, more broadly, the concentration of such services in particular regions positively affects the articulation of those places in GPNs because such places offer a richer and deeper variety of region-specific assets with which the needs of TNCs are more likely to be favourably coupled. In making this case, we extend the theorization of competitive advantage in advanced producer services. Previous work has examined the competitive advantage of firms within producer services industries (O'Farrell *et al.*, 1993; Lindahl and Beyers, 1999), but in this

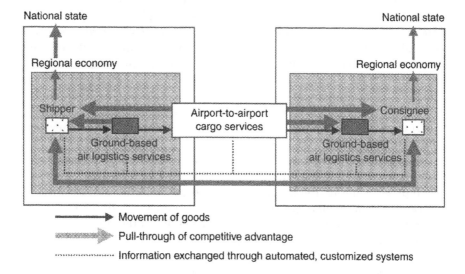

Figure 11.1 Air freight services and the pull-through of competitive advantage.

work we posit a conceptualization of the pull-through of competitive advantage between services firms and user firms and between services firms and places. We illustrate that conceptualization through an examination of the linkages between the electronics and air cargo industries in the context of several Southeast Asian economies.

Competitive advantage and air cargo services: a firm-level perspective

Based on more than 200 in-person interviews with managers of electronics manufacturers, freight forwarders, airlines and relevant government agencies in the Philippines, Malaysia and Singapore, we have analysed both variation in the demand for air cargo services (Bowen and Leinbach, 2003; Leinbach and Bowen, 2004) and in the supply of services provided by forwarders across the same region (Bowen and Leinbach, 2004). Here we employ case studies of selected firms from our larger samples (the sampling methodology is fully described in the earlier publications) to further explore the connections between the firms that provide and the firms that use air cargo services and between those firms and the places in which they are situated.

Case 1: Intel and FedEx

Intel, creator of the first microprocessor and today the largest manufacturer of microchips, operates facilities in several Asian countries, including Malaysia (commencing 1972), the Philippines (1974) and China (1998). Here we focus on the firm's Philippine operations where Intel has its largest assembly and test facilities in the world. Two plants and one sales and marketing headquarters together employ 6,000 people, mainly in labour-intensive operations although Intel has gradually expanded its back-end and engineering processes in the country (Perez, 2002). Intel is enormously important to the economy of the Philippines. Since 1997, Intel's exports (mainly of Pentium 3 and Pentium 4 chips in 2002) from the Philippines have been greater than those of any other firm in the country.

Intel was one of the first American manufacturers to invest in the Philippines and was dependent on air cargo services from the beginning. Its chips were among the first manufactured goods in the world for which air cargo was a regular (as opposed to emergency) mode of transport. With the acceleration of electronics product cycles, Intel's use of air cargo services has evolved towards a greater dependence on express cargo. From the Philippines, for example, 50 per cent of shipments to the USA move by express service, as do 15 per cent of shipments within Asia. Intel's standard throughput time to the USA is three to five days, but for expedite (i.e. express) shipments elapsed time is less than three days. At the time of our interview with Intel-Philippines in 2001, the firm relied mainly on FedEx for outbound express shipments. FedEx established its principal Asian hub at Subic Bay in 1995 (Bowen *et al.*, 2002). That decision gave Intel-Philippines direct, express access to the United States, its principal export market and major source of silicon wafers (the raw material for microchips). Via FedEx and its Subic Bay hub, Intel also has superior access to other Intel production sites and rapidly growing markets in Asia (particularly China which is increasingly important in both respects).

Intel's association with both UPS (through its hub at the former Clark air force base, also near Manila) and FedEx is expected to expand with the shift towards e-business, though the additional traffic will not necessarily be express. Rather, Intel expects to make greater use of the integrators' slower time-definite products as part of a just-in-time style production system tied into electronic ordering. By 2003, more than 60 per cent of Intel's material transactions worldwide were processed electronically (www.intel.com). E-business is associated with more frequent, smaller shipments, raising the logistics complexity. One of the important implications of Intel's transition to e-commerce is that the firms (including air cargo services firms) to which it is linked in a GPN must have the ability to connect electronically to Intel's track-and-trace system; for Intel this is a 'contractual mainstay'. The integrators, especially FedEx, have been pioneers in the development of electronic shipment tracking, a competitive advantage that Intel is able to tap (i.e. to pull-through) through its more aggressive use of the integrators' services compared to other electronics manufacturers.

As one of the two pillars of the 'Wintelist' duopoly (Microsoft and Intel) in the computer industry (Borrus, 2000), Intel can sustain its competitive advantage as long as it delivers innovations to the market faster than its rivals (differentiation) and at a competitive price (cost). Air cargo services are important in both respects. Intel uses its dominant position in both the semiconductor industry and the Philippine economy to pressure the air cargo industry there to accelerate its services and to do so at lower prices. At the time of our interview, Intel-Philippines was trying to shift its shipments away from FedEx towards UPS in order to foster greater competitiveness. In so doing, Intel places itself in a stronger position to pull-through competitive advantages from each integrator.

Case 2: Adaptec and GeoLogistics

GeoLogistics Corporation is one of the largest non-asset based providers of logistics services and the only privately-held company of its type. The current firm was established in 1997 but in fact it has its roots in the Lep Corporation founded in the UK in the mid-nineteenth century. With over 6,000 employees, its greatest presence is in Europe, followed by the Asia-Pacific regions and the Americas. Within the Asian region, the firm handled 148,000 tons of air freight in 2001 and had over two million square feet of facilities.

The Singapore office was opened in 1975, employs 180 staff and as elsewhere emphasizes supply chain management (responsibility for design, optimization and management of operations with customized information systems), integrated logistics (multi-service, integrated solutions with specific information systems) and traditional freight forwarding (traditional transportation, freight forwarding and storage). Major customers in Singapore include IBM, Flextronics, Toshiba, Epson and JVC.

One interesting example of the relationships that have developed between manufacturing firms and logistics firms is that between Adaptec and GeoLogistics. The former focuses on taking complex storage technologies mainstream. The company pioneered, and is the world's largest producer of, SCSIs (small computer system interfaces) that allow end-users to easily and affordably connect their personal computers (PCs), servers and workstations to storage devices and peripherals. Their products are marketed to PC and server OEMs (original equipment manufacturers) as well as end-users through more than 115 distributors and value-added resellers worldwide.

Adaptec's major inputs are printed circuit boards and customized integrated circuits from Singapore, Hong Kong and South Korea. All products and raw materials are carried by air. Further, Adaptec makes intensive use of externalized air logistics. For example, 20 per cent of its output is fed into a remote inventory management system (RIMS) in Europe and the United States where a 3PL operates warehouses containing Adaptec products that can be quickly delivered to nearby customers.

Similarly, in Singapore, Adaptec has appointed GeoLogistics to manage and run its entire supplier vendor managed inventory (VMI) and assembly programme. VMI is not a new application, but its role in today's supply chain is attracting more and more interest (Malone, 2001). Driven by a need for greater competence in supply chain processes, the roles of some supply chain partners are shifting away from traditional inventory management. Now, the vendor tracks the quantity of products transported to distributors or other outlets and has the ability to know when more products need to be sent. Replenishment is automatic and only occurs when there is a need. This system requires the electronic transfer of pertinent inventory data over a network. VMI has become a necessary planning tool that focuses on both the replenishment and fulfilment processes.

More broadly, GeoLogistics offers services dealing with just-in-time inventory, merge-in-transit, pick and pack, product testing, assembly, warranty and return services, and customs capabilities. A key service is the Purchase Order Management System, which features the tracking of a product from the inception of purchase order to final delivery to end-user. The status is logged at every stage, with purchase order and transport activities. Searches may be carried out by product code, purchase order number, commercial invoice number, carrier reference number, etc. The service is accessible via the Internet, using any browser.

Another variant of the enhanced visibility of supply chains services is a product labelled as GeoVista, which touts remote access in order to increase the visibility of supply and distribution changes. This service provides shipment, tracking, order/product tracking and virtual inventory reports. It allows shippers to answer the questions; for example, what is the availability of product x in the hub, what is the cycle time for shipment x in the warehouse and what is the total transaction? Thus a customer has Internet portal visibility to orders, inventory, product status, tracking and compliance.

Case 3: Jabil Circuit and FedEx/Danzas-AEI

Our next case concerns a new model of outsourced manufacturing that has emerged as a centrepiece of some GPNs: contract manufacturing. This form of network-based mass production is being linked to the erosion of the value chain and the emergence of the 'Wintelist' (the duopoly of Microsoft and Intel noted above) model of competition and the rise of 'fabless' product design companies in key sectors of the economy. This latter term refers to a company that does not manufacture its own silicon wafers and concentrates instead on the design and development of semiconductor chips.

The role of contract manufacturing in the supply chains of the IT industry is captured clearly in the following paraphrased quote:

The current supply chain is part of a time based segment of the marketplace. People want what they want, when they want it, and where they want it. At the same time product cycles in the supply chain are getting shorter and shorter. Putting these together means that companies are forced to outsource. They can only focus on so many things. The more sophisticated companies work on wealth creation and demand creation. And they let somebody else do everything in between.

(quoted in Luthje, 2002: 230)

The point here is that increasingly there is a move away from 'focal' corporations coordinating their value chain through their own operations. The acceleration of technology and product development has produced instability across the value chain.

Jabil Circuit is one of a half dozen major contract manufacturers around the globe. The firm was founded in Detroit in 1966 and its earliest work involved manufacturing replacement circuit board assemblies for a computer manufacturer. A 1976 contract with General Motors catapulted Jabil into a major player in the electronic manufacturing services industry. It established a high-volume manufacturing partnership with the automotive giant in which Jabil purchased all parts and provided engineering services. To meet those demands, it invested in advanced assembly technology and automated manufacturing equipment. Headquartered in St Petersburg, Florida, the firm has approximately 36 worldwide locations, including four plants in China and one each in Penang and Singapore. There are about 17,000 total employees and the firm had revenues of $3.55 billion in 2002.

First established in 1995 in the Bayan Lepas Industrial Park, Jabil's Penang campus includes three manufacturing buildings and a Jabil Global Services facility. A major operation here, and one of their basic functions worldwide, is the assembly of circuit or 'mother boards'. Currently the firm has 2,200 employees locally engaged in various assembly and service operations. While several other products are assembled in Penang, the circuit board is the only one which uses air freight dominantly. The major inputs, the integrated circuit (chip) and basic circuit board, are obtained from suppliers in Taiwan. Approximately 70 per cent of the completed circuit boards are shipped to other sub-contractors in the US, while roughly 20 to 30 per cent are air lifted to the UK. The remaining product is sent to China. But in addition Jabil Global Services in Penang serves the Asia-Pacific region and offers component level repair and logistics services to support its various customer-specific programmes in the region. Product types serviced include communications hubs and routers, desktop and laptop computers, PCB board rework and PCB component level repair.

The firm is somewhat typical in its use of both a traditional forwarder (Danzas-AEI) and an integrator (FedEx). Smaller parcels (under 15 kilograms) are usually sent by FedEx and as much as 60 per cent of Jabil's business – especially to the US – is handled by this US based firm, and the remaining 40 per cent is dealt with by Danzas-AEI. Goods handled by Danzas are normally destined for Europe via Malaysian Airlines or British Airways. Cargo within Asia, especially China, is uplifted by Cathay Pacific.

The nature of 'pull through' crystallized by air cargo services takes several forms in this case study. First, air cargo services allow the firm to heighten its competitive advantage by facilitating the delivery of state of the art services and logistical support to its various customer-specific programmes in the region. The 'pull through' of competitive advantage is also expressed as the firm may offer its supply chain management (SCM) and IT competence underpinned by air cargo services. The articulation of e-business with air cargo services provides a more seamless and flexible operation as information is exchanged between customers and suppliers. The firm's industry knowledge and ability to deliver state of the art solutions is recognized by a customer base of world-leading technology companies, including Cisco Systems and Hewlett-Packard. The integration of the SCM and IT organizations of the firm is geared towards ensuring that business processes are strongly supported with the correct systems and tools in order to remain responsive to their customers and their supply bases.

Air cargo services in Asian city-regions

As the global production networks within which electronics manufacturers are embedded extend further with internationalization, the services firms that interconnect production processes in those networks internationalize in response as well (Rabach and Kim, 1994). The result is potentially denser clusters of interrelated firms in overseas production sites and thicker networks of production and related services linking those clusters. We examine how that potential is realized by several city-regions across the industrializing economies of Southeast Asia.

Singapore

Singapore is the ninth ranked air cargo hub in the world and is the principal Southeast Asian hub for both cargo and passenger air traffic (O'Connor, 1995; Bowen, 2000; Bowen, 2004). Several aspects of Singapore's stature in the airline networks of the region (which in turn helps to funnel the elaboration of GPNs) merit further elaboration because they set the stage for an evaluation of the manner in which the city-region economy of Singapore pulls-through the competitive advantage of the air cargo services firms situated within it.

First, the completion of Changi Airport in 1980 and its subsequent repeated expansion has given Singapore an unrivalled air freight capacity, although the developments at Kuala Lumpur are intended specifically to challenge the city-state's primacy. The completion of an eighth multi-level terminal in 2002 raised the total air freight capacity to 2.5 million tons per year (Feller, 2002), well ahead of the actual throughput of 1.7 million tons and the 0.7 million tons capacity of the new Kuala Lumpur International Airport. The building of infrastructure in advance of demand conveys an advantage to firms based in Singapore versus rivals operating from more congestion-prone Asian city-regions.

Second, Singapore's Open Skies policy reflects a broader commitment to the principles of free trade in a state whose economy is critically dependent on exports (Raguraman, 1997). Those policies have not only facilitated the emergence of Singapore as an air transport hub (served by 65 airlines with direct flights to 146 cities worldwide) but also as an air logistics hub. Goods can enter and leave Singapore with relative ease, making it an attractive setting for regional distribution centres. The concentration of the city-state's manufacturing sector in very high value-added, technologically advanced goods (creating strong local demand for inbound and outbound sophisticated air logistics services) and the government's use of diverse tax incentives to attract regional services centres (Raguraman, 1997) have further enhanced Singapore's importance in this regard.

Third, SIA is an especially important cargo airline, ranking fifth in the world in freight ton-kilometres (behind FedEx, Lufthansa, Korean Air and UPS). By mid-2003, with its fleet of 100 passenger aircraft and 12 Boeing 747 freighters, SIA is one of three founding members of the New Global Cargo alliance that seeks to develop a 'one-stop shop', relatively seamless global network more commensurate with the scale of operations of major air cargo shippers, especially in the electronics industry (Taverna, 2001). It is not simply the scale and scope of SIA's cargo services that are important but also their quality. In our surveys of electronics manufacturers and freight forwarders, the quality of Singapore's air freight services were given markedly higher ratings versus the other three markets sampled.

In sum, Singapore provides an environment in which air cargo services and related air logistics services have flourished. In turn, those services have contributed to the vitality of the city-state's economy. Logistics, both sea and air, account for about 8 per cent of Singapore's economy (Feller, 2002). The expanded role of air logistics services in the economy of Singapore is manifest in the massive multi-storey warehouses that most large multinational freight forwarders have built near Changi since the mid-1990s.

Singapore has sought to amplify its advantages with yet another state-led initiative. In 2003, the government opened the Airport Logistics Park of Singapore (ALPS) to further leverage Singapore's connectivity, efficiency, information technology and infrastructure advantages (Urquhart, 2003). Unlike other logistics parks in Singapore, ALPS is located inside the free trade zone, enabling firms operating within the park to save as many as 18 hours in handling time (Lee, 2003). ALPS will augment the time-cost advantage of logistics firms operating in Singapore, conveying a competitive advantage to user firms and to the broader Singapore economy.

It is worth noting, however, that Singapore does face daunting challenges in maintaining its stature as an air cargo services hub. The cost of labour and land in the city-state are a constraint on its competitiveness. These problems are not new and Singapore has dealt with them in the past through a combination of aggressive automation, land reclamation and intensive use of land (e.g. multi-storey warehouses). More seriously, the shift of electronics manufacturing from Southeast Asia to China threatens the longer-term outlook for air cargo and air logistics services, particularly because the previously discussed importance of time as a competitive factor in electronics manufacturing has precipitated a cascade of relocation decisions as suppliers follow lead firms and vice versa to China. Moreover, the shift of manufacturers (especially lead or focal firms in GPNs) to China weakens the favourable cluster of manufacturing and related services (including air cargo and air logistics services) that have been important to Singapore's competitiveness.

Kuala Lumpur

Competitive advantage in global production networks plays out somewhat differently in the case of Kuala Lumpur. Partly to counter the dominance of nearby Changi, Kuala Lumpur International Airport (KLIA) at Sepang, about 50 kilometres from the KL metro area, was designed as a world-class hub airport for the Asia-Pacific region. Completed in 1998 at a cost of approximately US$3.5 billion, the air facility is part of a long-term development strategy wherein both IT and infrastructure are major features. It is part of the infrastructure expansion in the Klang Valley core that includes the Shah Alam industrial zone and particularly the MultiMedia Super Corridor (MSC). The nation has developed this massive corridor – larger than the entire state of Singapore – in order to create a desirable environment for firms wanting to create, distribute and employ multimedia products and services. The MSC brings together three key elements: a high-capacity global telecommunications and logistics infrastructure built upon the 2.5–10 gigabit digital optical fibre backbone; the KLIA airport; and new policies and cyberlaws designed to enable and encourage electronic commerce and facilitate the development of multimedia applications. The success of the MSC, which has yet to be realized, and other industrial zones clearly feed into the growth of the airport and vice versa.

Malaysia Airlines System Berhad (MAS), with its main base at KLIA, has experienced a variety of management problems which have interfered with its ability to expand its cargo uplift in both a regional and global perspective. In 2003 it reported its first annual profit in six years – a net profit of RM339 million ($89.2 million), reversing an RM835 million ($219.7 million) net loss the year before. Better than expected demand in passenger and cargo services, and reduction in expenditure through a 'Widespread Asset Unbundling' (WAU) exercise the previous year, helped the airline return to the black. Revenue rose 1.9 per cent to RM8.9 billion ($2.3 billion) versus RM8.7 billion ($2.2 billion) previously, while operating expenses were reduced to RM8.7 billion ($2.2 billion) from RM9.1 billion ($2.4 billion) before. Cargo traffic increased by 17.7 per cent to 2,073 ton-kilometres.

An intent is to have KLIA serve as a major developmental anchor and provide the basic uplift capacity for firms in close proximity. In addition, major incentives designed to attract investors are separate free commercial and industrial zones. Here, as elsewhere, the attraction is a location in which goods and services may be brought into, produced, manufactured or provided without any customs duty, excise duty, sales tax or service tax. An additional strategy has been to create added-value services to the activities in the free zone whilst enhancing the efficiency and proficiency of related operations. The value-added activities include trading, breakbulking, grading and sorting, repackaging and relabelling, partial clearance of cargo, combined cargo acceptance and transshipment.

The entire zone is managed by Malaysia Airport (Sepang) Sdn. Bhd., a wholly owned subsidiary of Malaysia Airports Holdings Berhad. Currently there are two main operators in the FCZ namely, MASKargo and KLAS Cargo. MASKargo's Advanced Cargo Centre (ACC) covers an area of 108 acres and houses 92,900 square metres of processing area. Currently, the ACC has the capacity to handle 650,000 tons of cargo annually, with the potential to expend to 3 million tons of cargo throughput per annum.

The operational environment of the zone has not yet measured up to Singapore's standard and therefore the value-added services noted above have been slow to gain strength. In an attempt to create further efficiencies, a customized system has been put in place by Dagang Net Technology. Called the KLIA Community System (KLIACS), it requires all agents and operators to link up to it. Currently, it allows fast processing of Customs and Free Zone declarations electronically. It is also able to monitor and track consignment status and movement within the FCZ, balance the inbound and outbound cargo manifest and capture trade statistics. Although Malaysia does have an advantage in lower land and labour costs, as with Singapore the shift of electronics to China poses a threat to its development plans.

Penang

An interesting contrast to KLIA, and a case in which the electronics industry still does clearly provide considerable demand for air cargo, is Penang International Airport. The PIA, located approximately 12 kilometres from

Georgetown, is the second largest airport in Malaysia. One of the key reasons for Penang's success as a high-tech manufacturing centre and home for world-class companies is the availability of an efficient air freight industry providing essential supporting services. Especially important are the free trade zones of Bayan Lepas and Butterworth in Penang as well as Kulim in the state of Kedah. The free trade zones have attracted a large number of major electronics manufacturers including Intel, Dell, Sony, Seagate, Bosch, Agilent, Acer and the contract manufacturer Jabil to name only a few. The primary actors affecting cargo at PIA are Malaysian Airlines (MAS), Malaysia Airport Berhad (MAB), KL Airport Services and the Penang Freight Forwarders Association (PFFA). Sixteen (16) air carriers currently deal with cargo. The most recent cargo entrants are China Air and Cathay Pacific.

The infrastructure consists of the MAS complex (80 metric tons) and a newer complex, MAB, opened in November 2000 at a cost of US$18 million (equivalent to RM$72 million) which can accommodate 360 metric tons per year. The MAB Complex is to complement the present MAS Complex by making available much needed additional handling capacity. The MAS cargo complex is handling well in excess of its capacity of 80,000 tons a year. The cargo volume has exceeded the complex's handling capacity just after three years since the start of its operations in 1992 and since then, has been on the uptrend. In 1995 it handled 87,470 tons and 171,000 tons in 2000.

The State Government has a 40 per cent stake through Penang Development Corporation (PDC), a joint venture company to construct the Airfreight Forwarders Warehousing Complex comprising 74 units (part of the Malaysia Airports Cargo Complex). However, there has been considerable controversy regarding the integrated operations of both facilities and their ability to handle cargo efficiently. Air freight forwarders in Penang have complained that the airport should have an integrated and seamless system in place for its cargo operations yet does not. The Penang Freight Forwarders Association (PFFA) has argued that a more efficient system will allow improved cargo movements, delivery, collection and accessibility between the Malaysia Airlines Cargo Complex and the second complex owned by Malaysia Airport Bhd (MAB). Along with the management problems associated with the cargo handling facilities, development expansion has been constrained by other factors. Foremost among these is a limited runway which reduces direct flight opportunities.

The ability of Penang to attract and retain high technology firms in the commercial zones is clearly related to the provision of efficient air cargo services. Yet it is not unusual for electronics firms to truck goods to KLIA from the Penang free trade zones for export given the greater availability of uplift capacity and especially given the wider set of destinations from that facility. Management problems between competitive operators and the inability to modernize and extend runways along with the federal government's primary interest in insuring the success of KLIA have all contributed to the inability of PIA to develop further. Given the relatively short trucking distance to KLIA from anywhere on the peninsula, it will be of interest to see whether the expanding Kulim FTZ and other developing zones in Kedah swing their traffic to the capital district's air hub.

Metro Manila

The air cargo services requirements of Metro Manila are served by three airports. South of the city, and close to the important electronics export processing zones (EPZs) in Cavite and Laguna, lies Ninoy Aquino International Airport (NAIA), the principal airport for the Philippines' capital and largest city. NAIA handled 0.4 million tons of cargo in 2001, a total that was down sharply from the pre-Asian financial crisis peak in 1997. North of Manila lie the two former American bases that now serve as the major intra-Asian hubs for the two largest integrators (FedEx at Subic and UPS at Clark). The establishment of the integrators' hubs in the Philippines and the development of large 3PL facilities close to NAIA by leading multinational freight forwarders (Bowen and Leinbach, 2004) have brought more sophisticated air cargo services and air logistics services to this city-region, though those services remain somewhat weak compared to Manila's rivals in the region.

The development of the air cargo industry in Metro Manila has been hampered by several factors. First, the Philippine government's protectionist policy towards the airline industry has gradually eroded Manila's centrality in airline networks, a problem which has been exacerbated by the country's persistent poverty. Second and related to the first point, PAL, the oldest airline in the region, has been a relatively weak competitor in the international airline industry and that weakness has in turn adversely affected PAL's principal hub. PAL operates no freighter aircraft and its cargo services (with cargo carried in the bellyhold of passenger aircraft) are not well-regarded in the industry. The paucity of capacity offered by the flag carrier means that shippers must depend on foreign airlines but because yields (the revenue per kilogram an airline earns) are low in the Philippines (related in part to the country's position in electronics GPNs), capacity is not abundant.

Third, corruption, cronyism and bureaucratic inefficiencies substantially raise the cost (including the time-cost) of air cargo services. In the interviews we conducted with freight forwarders in 2002, for instance, the need to bribe airport handlers in order to secure the release of cargo was repeatedly mentioned as an endemic problem. More generally, these practices warp the nature of the articulation of the Metro Manila regional economy within GPNs because they interfere with the 'coupling' of TNC needs and the region's resources. The foreign firms that we interviewed, especially large multinational forwarders, were especially unhappy. The representative of one large Swiss forwarder in Manila disparaged the 'revolving funds' that are required to do business in this market. Each of these three problems undermines the pull-through of competitive advantage from air cargo services firms to user firms.

Nevertheless, since about 1998, multinational freight forwarders have expanded their presence in the Philippines with the development of 3PL warehouses near Manila. These investments have been client-following. Japanese forwarders have been especially prominent in this regard, reflecting the very strong presence of Japanese (especially electronics) manufacturers. One foreign forwarder stated that his firm had little choice but to establish a presence in Manila because 'for worldwide business, there's always a little bit of the Philippines in every deal'.

Moreover, the electronics industry, by virtue of its contribution to the Philippine economy and the very real threat of investments being shifted to China and other foreign markets, has successfully exerted pressure on the Philippine government to adopt some policies to suit the requirements of the industry. For example, the Semiconductor and Electronics Industries in the Philippines Inc (SEIPI), a trade group, was able to get permission for a 'back-door' customs channel permitting goods to be cleared into the Philippines more rapidly and at a times of the day when the regular customs channels are closed. The 'back-door' channel, of course, facilitates the development of just-in-time type logistics management strategies across the GPNs in which Metro Manila is incorporated.

With the elaboration of GPNs across the Philippines, multinational forwarders and the integrators have come to increasingly dominate the air cargo services industry. That domination is resented by Philippine local firms but can be expected to force a harmonization of practices in the country towards those of other major centres of electronics production. This foreign-led change in practices is likely to mitigate the competitive disadvantage of the Metro Manila region.

Conclusions

Our intent in this chapter has been to point up the critical role of air cargo services in the continuing process of globalization and especially with reference to a set of Asian city-regions. We argue that these services are a major instrument in the ability of firms and states to internationalize and maintain competitive advantage. While competitive advantage for both firms and regions can be measured and examined from a variety of perspectives, we suggest that an elucidation of the concept of 'pull through' in the context of Porter's discussions on this topic is useful. Particularly germane to this analysis is the seminal contribution of air cargo services within the broad theme of global production networks.

In this perspective on GPNs our previous research has shown that air cargo usage is strongly associated with the degree to which a firm has internationalized, not only its production sites and final markets but also its material procurement sites (Leinbach and Bowen, 2004). But in addition, cycle time, an indication of a firm's material management strategy, and the role of the customer as an influence on logistics choices illustrate the complexity of air cargo usage. Our findings also suggest that the use of advanced producer services is likely to increase with the attenuation of GPNs as firms engage more fully in the process of the externalization of services. Time is also critical, as meeting market demand is critically important. In this connection firms are increasingly outsourcing production in order to minimize the time between product genesis and consumption but at the same time to hold down costs.

The case studies presented in this chapter offer further insights by examining the way in which air cargo enhances the value system embedded in GPNs. By analysing the specific needs of particular electronics firms and their use of services provided by integrators and logistics firms some additional knowledge is gained relevant to competitive advantage.

Intel, both in Manila and also in Penang, is a firm that has an extremely time sensitive product and thus relies almost exclusively on 'higher order' forms of air freight. In this situation the need for fast secure shipment outweighs the heavier costs associated with tailored services such as those provided by FedEx. In Manila, Intel's association with both UPS and FedEx is expected to expand with the shift towards e-business, though the additional traffic will not necessarily be express. Rather, Intel expects to make greater use of the integrators' slower time-definite products as part of a just-in-time style production system tied into electronic ordering. Furthermore, Intel-Philippines appears to be shifting its shipments away from FedEx towards UPS in order to foster greater competitiveness. In so doing, Intel places itself in a stronger position to pull-through competitive advantages from each integrator. Finally part of the 'pull through' effect in creating value is captured in other ways. In 1994 Intel opened an IC design centre in Penang and in carrying out its responsibilities, which include manufacturing, marketing and customer service, air freight is also quite critical (Linden, 2000: 216).

Competitive advantage is increasingly being expressed through supply chain processes and the roles of some supply chain partners are shifting away from traditional inventory management. The relationship between GeoLogistics and Adaptec is especially interesting and perhaps typical of growing symbiotic relationships between shipper and firms offering increasingly sophisticated services related to supply chain management. In this particular case GeoLogistics operates the entire vendor managed inventory of Adaptec which provides a more efficient supply chain management approach for this firm and hence allows it to remain extremely competitive against other related sellers.

Finally, Jabil Circuit, the example of a contract manufacturer, uses both an integrator and multi-functional forwarder. The air freighted product, in this case, circuit boards, is somewhat less time and handling sensitive than the integrated circuits of a manufacturer such as Intel. In this light Jabil, as with many other electronics manufacturers, will utilize the highest forms of air freight (FedEx) to meet peak season demand and in emergency situations. Routine air freight, as well as other supply chain management functions such as pick and pack assembly and a vendor hub, is carried out by Danzas-AEI. The firm felt that competitive advantage accrues from the ability to respond faster to supplier and market situations given shorter lead times. 'Pull through' comes about as air cargo services allow the firm to heighten its competitive advantage by facilitating the delivery of state of the art services and logistical support to its various customer-specific programmes in the region. In addition, heightened competitive advantage and 'pull through' is also expressed as the firm may offer its supply chain management (SCM) and IT competence underpinned by air cargo services. The articulation of e-business with air cargo services provides a more seamless and flexible operation as information is exchanged between customers and suppliers.

The nature of competitive advantage which we have discussed above is directly related to the efficiency and cost of particular interactions between individual firms and forwarders. Yet the development of 'pull through' advantage is not independent of the nature of air hub capacity and operations in the individual city region contexts. First and foremost the efficient management of operations and labour practices in each situation is critical. There are important differences among the air hubs in the region in regards to these measures. Singapore's Changi is an example of a hub where innovative technological applications are almost routine and efficiency is unparalleled. Capital investment for upgrading and expansion do not constrain its competitive posture. Moreover, labour, management and rent-seeking issues are not problematic. In contrast, in both Kuala Lumpur and Penang management and operational inefficiency have been deterrents. In Manila, despite the nearby presence of FedEx and UPS and the development of 3PL facilities, the city-region suffers from several lingering problems. Philippine Airlines, like Malaysia Airlines, has been a weak competitor in the air cargo industry. In part this is due to the Philippine government's protectionist policies toward the airline industry but in addition rent-seeking behaviour has introduced considerable inefficiencies in the cargo system.

To a considerable extent the 'pull through' advantage which is transmitted through the application of air cargo services will also be increasingly determined by the efficient application of e-commerce. The example of the new internet portal, Air Cargo Exchange (ACX), offered by Singapore Airlines which allows customers to book space online, track cargo in transit, and review the flight schedules of multiple carriers, all on a single website, is evidence of this (Sakran, 2003). In order to remain competitive and to prevent the leakage of cargo traffic air hubs such as Penang must modernize or face the prospects of decline.

Finally it is clear that changes in industry structure are altering the form and spatiality of global production networks and are producing a variety of complex outsourcing relationships. Such networks are being created in response to more conspicuous and sometimes urgent market demands. The result is a quest for shorter and shorter product cycles. The complexity of arranging outsourcing to create cost and temporal efficiency means that logistics and air cargo services must increasingly rely on new innovations in information technology. Vertical specialization in the IT industry will continue to produce changes in the manufacturing process. The separation of product design and manufacturing, for example, is only one way in which producer services such as air cargo must respond. Research on transportation services which recognizes the dynamics of production structures and competition is essential.

Note

1 Including the integrators, which are firms like FedEx and UPS that both fly the cargo among airports and handle ground pick up from and delivery to customers.

References

Agnes, P. (2000) 'The "end of geography" in financial services? Local embeddedness and territorialization in the interest rate swaps industry', *Economic Geography*, 76(4), 347–66.

Bagchi-Sen, S. and Sen, J. (1997) 'The current state of knowledge in international business in producer services', *Environment and Planning A*, 20, 1153–74.

Begg, I. (1994) 'The service sector in regional development', *Regional Studies*, 27(8), 817–25.

Boeing Commercial Airplane Group (2004) *World Air Cargo Forecast 2004/2005*. Available online at www.boeing.com/commercial/cargo/sitemap.html.

Borrus, M. (2000) 'The resurgence of U.S. electronics: Asian production networks and the rise of Wintelism', in Borrus, M., Ernst, D. and Haggard, S. (eds), *International Production Networks in Asia: Rivalry or Riches?* London: Routledge, 57–79.

Borrus, M., Ernst, D. and Haggard, S. (eds) (2000) *International Production Networks in Asia: Rivalry or Riches?* London: Routledge.

Bowen, J. T. (2000) 'Airline hubs in Southeast Asia: national economic development and nodal accessibility', *Journal of Transport Geography*, 8(1), 25–41.

Bowen, J. T. (2004) 'The geography of freighter aircraft operations in the Pacific Basin', *Journal of Transport Geography*, 12(1), 1–11.

Bowen, J. T. and Leinbach, T. R. (1995) 'The state and liberalization: the airline industry in the East Asian NICs', *Annals, Association of American Geographers*, 85(3), 468–93.

Bowen, J. T. and Leinbach, T. R. (2003) 'Air cargo services in Asian industrializing economies: electronics manufacturers and the strategic use of advanced producer services', *Papers of the Regional Science Association*, 82(3), 303–23.

Bowen, J. T. and Leinbach, T. R. (2004) 'Market concentration in the air freight forwarding industry', *Tijdschrift voor Economische en Sociale Geografie*, 95(2), 174–88.

Bowen, J. T., Leinbach, T. R. and Mabazza, D. (2002) 'Air cargo services, the state and industrialization strategies in the Philippines: the redevelopment of Subic Bay', *Regional Studies*, 36(5), 451–67.

Button, K. and Owens, C. A. (1991) 'Transport and information systems: a case study of EDI deployment by the air cargo industry', *International Journal of Transport Economics*, 26(1), 3–21.

Coe, N. M., Hess, M., Yeung, H. W. C., Dicken, P. and Henderson, J. (2003) '"Globalizing" regional development: a global production networks perspective', Paper presented at the 99th Annual Meeting of the Association of American Geographers, New Orleans, 5 March.

Daniels, P. W. (1987) 'Producer-services research: a lengthening agenda', *Environment and Planning A*, 19, 569–74.

Daniels, P. W. (1993) *Service Industries in the World Economy*. Oxford: Blackwell Publishers.

Daniels, P. W. and Moulaert, F. (eds) (1991) *The Changing Geography of Advanced Producer Services: Theoretical and Empirical Perspectives*. London: Belhaven.

Daniels, P. W., Thrift, N. and Leyshon, A. (1989) 'Internationalisation of professional producer services: accountancy conglomerates', in Enderwick, P. (ed.), *Multinational Service Firms*. London: Routledge, 79–106.

Dicken, P. (2003) *Global Shift: Reshaping the Global Economic Map in the 21st Century*, 4th edn. New York, NY: Guilford.

Dicken, P., Kelly, P. F., Olds, K. and Yeung, H. W. C. (2001) 'Chains and networks, territories and scales: towards a relational framework for analysing the global economy', *Global Networks*, 1, 89–123.

Dymski, G. A. (2002) 'The global bank merger wave: Implications for developing countries', *The Developing Economies*, 40(4), 435–66.

Ernst, D. and Guerrieri, P. (1998) 'International production networks and changing trade patterns in East Asia: the case of the electronics industry', *Oxford Development Studies*, 26(2), 191–212.

Feller, G. (2002) 'Can Singapore become the preferred choice for logistics management?', *Payload Asia*, November. Available online at www.payloadasia.com.

Gereffi, G. (1994) 'The organization of buyer-driven global commodity chains: how U.S. retailers shape overseas production networks', in Gereffi, G. and Kornzeniewicz, M. (eds), *Commodity Chains and Global Capitalism*. Westport, CA: Greenwood Press, 95–122.

Gereffi, G. and Kornzeniewicz, M. (eds) (1994) *Commodity Chains and Global Capitalism*. Westport, CA: Greenwood Press.

Grabher, G. (2001) 'Ecologies of creativity: the village, the group, and the hierarchic organisation of the British advertising industry', *Environment and Planning A*, 33, 351–74.

Hansen, N. (1990) 'Do producer services induce regional economic development', *Journal of Regional Science*, 30(4), 465–76.

Heaver, T. D. (2001) 'The evolving roles of shipping lines in international logistics', *International Journal of Maritime Economics*, 4, 210–30.

Henderson, J., Dicken, P., Hess, M., Coe, N. and Yeung, H. W. C. (2002) *Review of International Political Economy*, 9(3), 436–64.

ICAO Journal (2001) Annual Civil Aviation Report. 56(6), July/August, 13.

Illeris, S. (1996) *The Service Economy: A Geographical Approach.* Chichester: John Wiley.

Lee, R. (2003) 'Logistics giant's S'pore address helps it to bag deals', *The Straits Times* (Singapore). March 19, Money.

Leinbach, T. R. and Bowen, J. (2004) 'Air cargo services and the electronics industry in Southeast Asia', *Journal of Economic Geography*, 4(2), 1–24.

Leslie, D. A. (1995) 'Global scan: the globalization of advertising agencies, concepts, and campaigns', *Economic Geography*, 71(4), 402–26.

Lindahl, D. P. and Beyers, W. B. (1999) 'The creation of competitive advantage by producer service establishments', *Economic Geography*, 75(1), 1–20.

Linden, G. (2000) 'Japan and the United States in the Malaysian electronics sector', in Borrus, M., Ernst, D. and Haggard, S. (eds), *International Production Networks in Asia: Rivalry or Riches?* London: Routledge, 198–225.

Luthje, B. (2002) 'Electronics contract manufacturing: global production and the international division of labor in the age of the Internet', *Industry and Innovation*, 9(3), 227–48.

Malone, R. (2001) 'VMI: managing supply based upon demand', *Inbound Logistics*, October. Available at www.inboundlogistics.com.

Martinelli, F. (1991) 'A demand-oriented approach to understanding producer services', in Daniels, P. W. and Moulaert, F. (eds), *The Changing Geography of Advanced Producer Services: Theoretical and Empirical Perspectives.* London: Belhaven, 15–29.

Nilsson, J., Dicken, P. and Peck, J. (eds) (1996) *The Internationalization Process: European Firms in Global Competition.* London: Paul Chapman.

Nohria, N. and Ghoshal, S. (1997) *The Differentiated Network: Organizing Multinational Corporations for Value Creation.* San Francisco: Jossey-Bass Publishers.

O'Connor, K. (1995) 'Airport development in Southeast Asia', *Journal of Transportation Geography*, 3(4), 269–79.

O'Connor, K. and Hutton, T. (1998) 'Producer services in the Asia Pacific region: an overview of research issues', *Asia Pacific Viewpoint*, 39(2), 139–43.

O'Farrell, P., Moffat, L. and Hitches, D. (1993) 'Manufacturing demand for business services in a core and peripheral region: does flexible production imply vertical disintegration of business services?', *Regional Studies*, 27, 385–400.

Perez, B. (2002) 'Intel widens spread of Asia production; Philippines to get largest chip-test site in the world', *Business Post*, January 23, 10.

Porter, M. E. (1985) *Competitive Advantage: Creating and Sustaining Superior Performance.* New York, NY: The Free Press.

Porter, M. E. (1990) *The Competitive Advantage of Nations.* New York, NY: The Free Press.

Porter, M. E. (1996) 'Competitive advantage, agglomeration economies and regional policy', *International Regional Science Review*, 19(1 and 2), 85–94.

Rabach, E. and Kim, E. M. (1994) 'Where is the chain in commodity chains? The services sector nexus', in Gereffi, G. and Kornzeniewicz, M. (eds), *Commodity Chains and Global Capitalism.* Westport, CA: Greenwood Press, 123–44.

Raguraman, K. (1997) 'International air cargo hobbling: the case of Singapore', *Asia Pacific Viewpoint*, 38(1), 55–70.

Rimmer, P. J. (1997) 'Trans-Pacific oceanic economy revisited', *Tijdschrift voor Economische en Sociale Geographie*, 88(5), 439–56.

Sakran, S. (2003) 'Role of websites in freight forwarding', *Malaysian Business*, May 1: http://www.maskargo.com/technology/?nav=features.

Schoenberger, E. (1994) 'Competition, time and space in industrial change', in Gereffi, G. and Kornzeniewicz, M. (eds), *Commodity Chains and Global Capitalism*. Westport, CA: Greenwood Press, 51–66.

Smith, A., Rainnie, R., Dunford, M., Hardy, J., Hudson, R. and Sadler, D. (2002) 'Networks of value, commodities, and regions: reworking divisions of labour in macro-regional economies', *Progress in Human Geography*, 26(1), 41–63.

Taverna, M. A. (2001) 'Lufthansa spearheads alliances in freight, express and logistics', *Aviation Week & Space Technology*, 155(9), 58–60.

Thrift, N. (1987) 'The fixers: the urban geography of international commercial capital', in Henderson, J. and Castells, M. (eds), *Global Restructuring and Territorial Development*. London: Sage, 203–33.

Urquhart, D. (2003) 'Logistics FTZ boosts Changi as cargo hub; $35m facility will help Republic stay ahead in logistics, SCM industries', *Business Times* (Singapore). Shipping Times section, March 21.

Warf, B. (2001) 'Global dimensions of U.S. legal services', *Professional Geographer*, 53(3), 398–406.

Yeung, H. W. C. (1994) 'Critical reviews of geographical perspectives on business organizations and the organization of production: towards a network approach', *Progress in Human Geography*, 18, 460–90.

Yeung, H. W. C. (2000) 'Organising "the firm" in industrial geography I: networks, institutions and regional development', *Progress in Human Geography*, 24, 301–15.

Zhu, J., Lean, H. S. and Ying, S. K. (2002) 'The third-party logistics services and globalization of manufacturing', *International Planning Studies*, 7(1), 89–104.

Part III

Services and urban development in the Asia-Pacific

City case studies

12 Understanding mega city-region development

A case study of Melbourne

Kevin O'Connor

One of the distinguishing characteristics of the urban settlement of the Asia-Pacific region is its very large cities. They are found in circumstances as diverse as the eastern Pacific Rim at Los Angeles and San Francisco, in the west in Bangkok, Jakarta and Hong Kong (and its adjacent Pearl River delta), in the north around Shanghai and Tokyo, and in the south around Melbourne and Sydney. These urban regions are intensively settled and spatially integrated units that spread discontinuously over long distances across local statistical or administrative borders. They create a major public policy challenge in terms of management. They are variously labelled 'mega city', 'megalopolis', 'extended metropolitan region', 'mega city-region' and 'global city-region'. These terms are applied to a series of large cities, such as the Pearl River Delta (Sit and Yang 1997) and the Tokyo-Osaka Corridor (Yeung 2000) and other axes of urban development where cities are clustered in narrow corridors, often along coastal strips (Rimmer 1991) and across national borders (Heikila 2002; Artibise 1995). These terms are also applied to the large agglomeration associated with individual cities like Shanghai, Bangkok, Tokyo and Jakarta. It is the latter, geographically smaller, unit that is the focus of the current research and the label to be used here for these areas is mega city-region, acknowledging their large physical and economic size, as well as the urban character of a region of a nation. This chapter will explore ways of describing and understanding the development of these mega city-regions.

The chapter will explore ways of describing and understanding the development of a mega city-region around Melbourne, with special attention to the role played by employment in producer services. It is the focus of the research for a number of reasons. First, Melbourne's historic low density development, an economic structure that favours manufacturing, transport and logistics and its lack of physical barriers to urban development to the west, north and south-east make it an ideal candidate to observe the forces of mega city development at work. Second it is entering a new era of urban management; recent policies have established an urban growth boundary to limit the physical extent of the metropolitan area while at the same time policy has been enacted to strengthen the links between the metropolitan area and a series of surrounding regional centres within its mega city-region. These apparently contradictory policy stances deserve closer scrutiny, which can be built upon an understanding of the evolution of the mega city-region.

Describing and understanding the mega city-region

Mega cities have been explored in considerable detail over an extended period of time. The initial interest centred on their sheer population size, which alone made them distinctive. That can be seen in a foundation book on these and other cities by Dogan and Kasarda (1988). Closer study in the Asian context quickly focused attention upon the spatial dimension as seen in an early study of *extended metropolitan regions* edited by McGee and Robinson (1995). Here the word 'extended' captured the notion that these cities were not only big in terms of population but big in a spatial sense. That thinking evolved into a long term project carried out by researchers at the UN University which produced a number of books (for example Lo and Yeung 1996, 1998; Lo and Marcotullio 2001).

These early descriptions of mega cities have found resonance in recent US and European research on what have come to be called 'global city-regions' (Scott 2001). The underlying geography of these regions has been well expressed in diagrammatic form by Healey who (in Graham and Marvin 2001: 205) produced a diagram of what she calls a *multiplex urban region*. That shows a CBD, along with suburban business park, retail and leisure complex and industrial zone, as well as a small old centre on the fringe. Her diagram is consistent with Hall's (2001: 73) description of these areas as 'quintessentially polycentric'; his description includes the CBD, a new business district usually on the CBD fringe, an internal and external 'edge city' as well as 'outermost edge city complexes and specialised subcentres'. Simmons and Hack (2000: 185) provide detailed examples of these ideas in case studies of 11 mega city-regions. They found a particular infrastructure framework was central to the emergence of what they termed the 'polynucleated spread city'. That framework included the CBD, the road system, sea ports, airports and suburban industrial zones.

These efforts at description have established that mega city-regions have a number of characteristics. The research challenge is to move beyond description and attempt to develop explanations of these special parts of the Asia-Pacific's urban settlement system.

The process of mega city-region development

Initial work that tried to account for the geographic dispersal of these cities tended to emphasise the spread of residential development. The well-spring of much of this thinking was the concept of the *urban field* developed by Friedmann and Miller (1965). That showed metropolitan areas in the US began to extend and become mega city-regions as residential location choices favoured the outer edges of cities and smaller rural places when the interstate highway system stretched deep into surrounding rural areas. The heritage of this thinking can be seen in Leaf's (1994) work on Jakarta, which traces the residential spread associated with the construction of freeway systems in that city. In the case of Seoul, Jun and Hur (2001) have shown how this spread effect was shaped by planned new town development. The focus upon residential patterns is an important one as the successive rings of settlement account for much of the population change in these city-regions. As an example, Jones (2001) has shown that what he identifies as an 'inner zone' (an area extending 30–50 kilometres from the city centre) is where population growth was most rapid in the extended metropolitan regions of Asia in the 1990s.

However, the modern city-region is shaped by forces more complex than the dispersal of residential settlement. Demattais and Governa's (1999: 547) work on the *periurbain* in France and the *citta diffusa* in Italy suggested that the outcome was really due to new arrangements in the spatial division of labour within a network of cities. Ideas of changing economic structure as well as new demographic trends and expanding transport infrastructure lay behind McGee's (1994) idea of *desakota* which implied a set of forces that spread employment, particularly in manufacturing, into the outer edges of a city. McGee believed this produced a mixed pattern of urban and rural uses, a process that in Europe has been seen to follow 'interurban and rural road grids in a discontinuous pattern, including wide rural open spaces' (Demattais and Governa 1999: 548). That pattern meant the settlers in the outer areas of the mega city-region are not fringe dwellers forced out by the cost of housing as the case studies carried out by Browder *et al.* (1995) illustrated. Rather these settlers are there because manufacturing plants associated with foreign direct investment provide employment opportunities, a process summarised in a model of peri-urban development developed by Webster (2002b) and illustrated in a case study of the development of the outer areas of Bangkok (Webster 2002a).

These descriptions and initial efforts to establish the process shaping mega city-regions confirm that they are large in population and geographic area, often manufacturing based, multi-nodal in character and draped around a particular infrastructural frame. Their distinctive feature however, and perhaps the fundamental underlying force shaping their development, is that their various parts are 'integrated to form a regional urban network connected by an efficient transport system ... (that) ... transcends administrative boundaries' (Tang and Chung 2000: 277). In this arrangement a central core provides the services that facilitates the production of goods in suburban industrial parks, which move along the infrastructural framework to other producers in the region, or out to global markets through the port and airport. This functional integration and connectivity between various areas across great distances and across local, regional and, (in some cases) even national borders is the really distinctive feature of a mega city-region, and the key to understanding its vitality.

It is easy to envisage the integration of the region expressed in the movement of labour between socially-sorted residential sites to industrial and commercial centres. However, the integration of these regions draws upon more than the commuting of labour. Analysis of Hong Kong and the Pearl River Delta by Sit and Yang (1997) provides a good lead here. They argue that Hong Kong has developed a front-of-shop office service cluster to deal with the back-room factories that are spread across the delta region. In these terms the integration of the core of the region and its far flung parts is based on the flow of knowledge through the provision of services and reflects, in some cases, control and ownership. In a sense this is an extension of the first waves of industrial suburbanisation in the 1950s in US and European cities where it was common for manufacturing plants to relocate to the suburbs (to take advantage of larger land sites and easier road access) while the head office remained in the city. However, in today's flexible and specialised manufacturing the linkages between plants and across industries involve much greater integration than that implied by head office-production plant separation. Leung (1993) working on the Pearl River delta, and Marton (2000: 45) working on the Yangtze delta inland from Shanghai, show 'spatial economic restructuring ... (is) best understood and explained in terms of the complex interactions and interrelationships which constitute the transactional environment'. McGee (1994: 75) has acknowledged this idea with reference to an 'intense transactional environment' as one of the characteristics of his concept of *deskota*. The focus upon transactions implies a role for producer services in the evolution and growth of these regions.

Rodrigue (1994: 72) connected these industrial linkage perspectives with an understanding of transport systems in research that identified the extended metropolitan region of Singapore. He outlined an area that stretched more than 100 kilometres into southern Malaysia and incorporated the Indonesian islands of Batam and Bintan which reflected the conscious decision by Singapore firms to 'densify' some activities within the city and 'disseminate' others into the extended zone. He has followed the Singapore study with research on the logistical integration of the New York-Boston corridor (2003: 6). He finds most of the region's commercial development is within 25 miles of a sea- or rail container-handling facility, with some very substantial clusters and nodes of activity, notably the Port of New York. Hence container traffic within a region could be a useful indicator of the extent of a mega city-region.

With that in mind, data on the origins and destinations of containers at the Port of Melbourne was assembled as a first step toward mega city-region identification. Recent survey work (Sinclair, Knight, Merz 2003) shows, however, that the main origins and destinations of containers are suburban and inner industrial zones and large distant agricultural regions; the edges of the metropolitan area were not significant sites on this indicator. In fact the geography of container movement could provide a misleading impression of the physical integration of the Melbourne mega city-region, implying that it lies across several Australian states, consistent with an average length of domestic truck haul in the US of 550 miles (Rodrigue 2003: 10).

Studying logistical networks could also be misleading as they can span inter-continental distances in a short period of time. At an extreme Bowen *et al.*'s (2002) work on electronic products made in Southeast Asia shows that the logistical systems in that industry, even counting daily movements, can span continents. If data was available, the daily geography of urban freight movements would be a useful source to show the integration between each part of a mega city-region. Sheard (1983) used this understanding to show that the networks within and between urban areas associated with Japanese automobile manufacturers reflected the frequency of demand for car parts. If analysed for more industries in a mega city-region this approach would expose the forces behind the integration of parts of the region as Leung's (1993) detailed case study of the Pearl River Delta has done.

Apart from the difficulty of getting the data for such a study, the particular challenge here is to look beyond the movements of physical goods to create an insight on the linkages and connections expressed through the transactions of producer services. The latter are not necessarily just the traditional finance and banking activities found in the business centres of the core of these mega city-regions (important though they are) but can involve a range of activities like design, labour market services, project management, construction and engineering, as well as logistics planning associated with just-in-time imports and exports. It may be that the functional integration that makes the mega city-region so distinctive might emanate from the movement of these services between suppliers and firms in different parts of a region. This would mean the provision of technical and design skills associated with modern manufacturing from research and development centres could be part of the functional integration of a mega city-region in the same way as the movement of goods in supply chains. The frustration is that information on these aspects is not collected in any convenient form; expensive and time consuming case studies are needed. Hence, although the provision of an array of services provides a conceptually powerful way of understanding the integration of the various parts of a mega city-region, the lack of easy-to-collect information in an appropriate spatial framework makes it difficult to explore in the current circumstances.

That brings the focus back to commuting as a reflection of the integration of a city-region. Commuting data has the advantage that it provides a focus not only on the travel in a region, but also on the location of jobs (of various kinds) and homes, and can be collected separately for producer service type workers, so exposing the role they play in mega city-region development. With that in mind, this chapter identifies the way that patterns of commuting of producer service employees are consistent with the pattern of development of a mega city-region.

Melbourne and its mega city-region

Background to the case study

The research reported below has been carried out in the mega city-region that spreads beyond the metropolitan area of Melbourne, in the core of the State of Victoria, around Port Phillip Bay. The metropolitan area at the core of the mega city-region is the second largest in population terms in Australia. It has undergone major economic re-structuring in the past 30 years. That change was expressed in three ways. The first shook the structure of manufacturing industry as Australian markets were opened to international competition, and many Melbourne firms, especially in clothing and textiles, were unable to compete. In a number of sectors, especially in the auto industry, the change eventually produced some internationally-focused firms, and export-oriented manufacturing is now an important part of the metropolitan area's economy. Second, the refinement of national transport networks allowed many Melbourne firms to expand and win market share at the expense of smaller plants in other parts of the country (O'Connor *et al.* 2001). Taken together these changes have consolidated Melbourne's position as the nation's major manufacturing location. Third, change in the global links of the finance industry worked in favour of Sydney at the expense of Melbourne (O'Connor and Edgington 1991) which changed the commercial role of the core of the metropolitan area away from finance and banking toward other producer service activities, as shown in O'Connor (2002). These changes have had important consequences for the development of the mega city-region as two of them tended to favour job growth in middle and outer suburban locations, broadening and deepening the job base of these areas, creating the suburban business parks and industrial zones that figure in Healey and Hall's descriptions cited earlier and so providing employment opportunities for residents commuting across the broader city-region.

The central part of the State of Victoria has other features that have fostered mega city-region growth. These include the lack of major physical restrictions on urban development to its west, north and south-east around Port Phillip Bay. In addition it has a freeway and rail system that stretches out to state and national markets along major corridors to the south-west, west, north and south-east. Apart from the movement of goods this network enables commuters to live outside the metropolitan area. In addition, Melbourne is the nation's major container terminal, and its airport's share of air freight has been increasing steadily (O'Connor 2002). Hence it has the infrastructure that has been shown to be critical to mega city-region development.

The analysis below has been carried out on a number of spatial units. The first is the official statistical definition of Melbourne, called the Melbourne Statistical Division. This includes an urban area and a fringe that spreads 10–15 kilometres beyond the urban area, including low-density urban settlements and some small towns. This is a similar unit to the Functional Urban Region used in Europe and the Standard Metropolitan Statistical Area used in the US. The first step in the analysis is to establish the extent to which a mega city-region extends beyond that already generous definition of the metropolitan area into what has been labelled the 'Rest of Victoria'. This extension is explored separately for producer service jobs and for other employment.

The second set of spatial units are municipalities within the statistical division. The research focuses upon three particular municipalities. The first is the City of Melbourne, at the heart of the metropolitan area, and the largest in employment terms and recent employment growth, as can be seen in Table 12.1. Investigating commuting to this location from the rest of Victoria will establish the extent to which jobs in and near the CBD (biased in favour of producer service employment as displayed in Table 12.2) act as a driver of mega city-region development. The second and

Table 12.1 Rank of case study municipalities in jobs and job growth in the Melbourne statistical division 1996–2001

Rank	Municipality	Jobs 1996	Municipality	Jobs 2001	Municipality	Change in jobs 1996–2001
1	*Melbourne*	246,482	*Melbourne*	275,700	*Melbourne*	28,218
2	*Monash*	77,493	*Monash*	83,275	Casey	8,436
3	Kingston	63,602	Greater Dandenong	66,979	*Hume*	8,110
4	Greater Dandenong	61,745	Kingston	66,855	Whitehorse	8,091
5	Port Phillip	56,432	Port Phillip	63,923	Port Phillip	7,491
6	Boroondara	53,314	Whitehorse	57,416	Wyndham	6,999
7	Yarra	53,285	*Hume*	57,205	Brimbank	6,688
8	Knox	49,616	Boroondara	55,765	Whittlesea	6,456
9	Whitehorse	49,325	Yarra	55,179	Mornington Peninsula	6,275
10	*Hume*	49,095	Knox	53,444	Yarra Ranges	6,206
11	Darebin	44,106	Stonnington	42,516	*Monash*	5,782

Source: ABS: Census 1996, 2001. Journey to Work data.

Table 12.2 Employment structure in selected municipalities

	Melbourne	Hume	Monash
Share of jobs in administrative-managerial and professional categories (%)	42	21	31

Source: ABS Population Census 2001.

third municipalities are suburban. One, Hume, is near the outer edge of the metropolitan area while the other, Monash, is more centrally located within the suburbs. Table 12.1 confirms that these two municipalities are among the largest outside of the City of Melbourne in employment (Monash) or in recent growth (Hume ranked at number three). They differ in terms of job mix, with Monash having a larger share of producer service type jobs within it, as can be seen in Table 12.2. Investigating commuting to these suburban locations will identify the role that suburban nodes or 'business centres' play in mega city-region development. Given that they are closer to the edge of the metropolitan area it is possible that they have a greater effect than the centrally-located City of Melbourne, although the scale of employment in the latter could still give it a strong influence.

Data to quantify the journey to work links reported in the following analysis was derived from answers to questions in the national Census of 2001. This information is a tabulation of the 'principal residence' of the worker. In the main this includes people commuting on a daily basis. Some individuals however have temporary accommodation in the metropolitan area and a principal residence in a distant non-metropolitan location; they are in effect weekly commuters rather than daily commuters. Taken together the two groups provide a comprehensive picture of the way that the metropolitan labour market impacts upon non-metropolitan locations, and shows how the mega city-region has spreads its influence into a surrounding area.

Commuting from the rest of the State of Victoria into the Melbourne Statistical Division involves just 55,000 workers in a metropolitan workforce of 1.4 million. Hence the commuting element of mega city-region development is small. However those 55,000 workers represent 10 per cent of the workforce of non-metropolitan Victoria. Of the commuters, 30 per cent are in what can be broadly classified as producer service type activities (defined here as administrative, managerial and professional categories of occupations). This is equivalent to their share of the Victorian workforce overall, suggesting their role in mega city-region commuting is consistent with their role in the workforce generally. The other 70 per cent of the commuters have been put in a category labelled 'other workers'.

Commuting to the Melbourne Statistical Division from the rest of Victoria

Insight on the pattern of commuting and in turn the character of the mega city-region around Port Phillip Bay has been developed by displaying aggregate figures for journey to work flows into the Melbourne Statistical Division in Figures 12.1 and 12.2. These show that a mega city-region has emerged in the core of the state with commuting flows from immediately adjoining municipalities, particularly to the north and west. The main section of the mega city-region (those areas supplying at least 1,000 workers) stretches some 250 kilometres from its furthest western border to that in the south-east. There are few differences between the pattern of commuting of management, administration and professional workers compared to other workers, although the larger number in the second category has produced a bigger western and northern ring of municipalities (Figure 12.2). In both maps the core area on the north, west and south-west is prominent, as is the extension to the west, which corresponds with a major regional city, Ballarat.

The information displayed in Figures 12.3 and 12.4 provides an insight on the relative size of the two categories of workers travelling from municipalities arrayed by distance from Melbourne. These two graphs share a common feature: labour commutes from very long distances within the

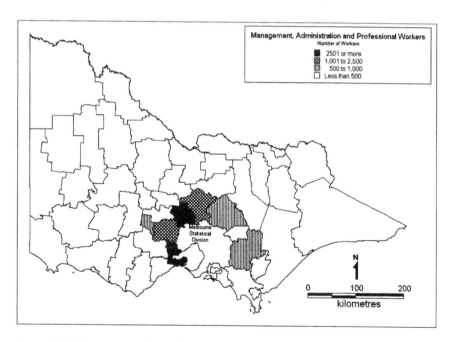

Figure 12.1 Commuting from the rest of Victoria to the Melbourne Statistical Division: management, administration and professional workers.

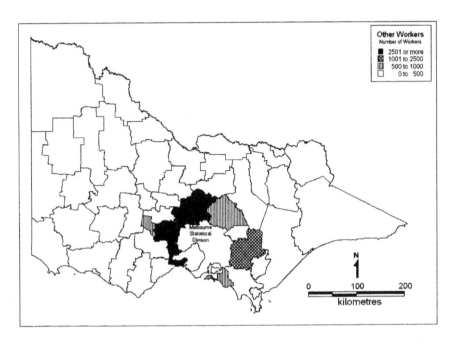

Figure 12.2 Commuting from the rest of Victoria to the Melbourne Statistical Division: other workers.

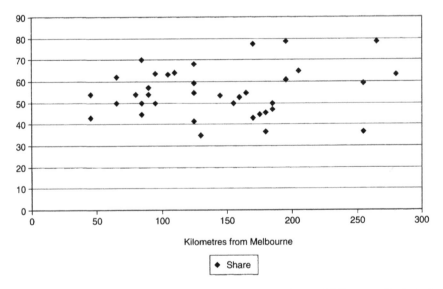

Kilometres from Melbourne

◆ Share

Figure 12.3 Share of commuters from each municipality to Melbourne Statistical Division by distance from Melbourne: management, administration and professional workers.

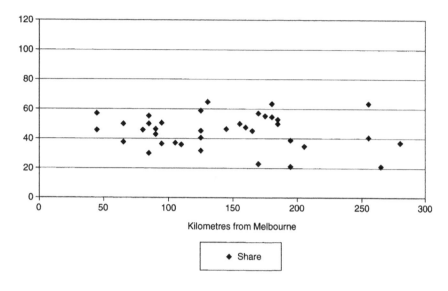

Figure 12.4 Share of commuters from each municipality to Melbourne Statistical Division by distance from Melbourne: other workers.

state to jobs in Melbourne. However the graphs are very different in that the producer service jobs account for a much larger share of the workers commuting from each municipality (around 60 per cent and up to 80 per cent of those travelling from some municipalities). For other employees at each distance from Melbourne the shares are commonly around 40 per cent and rarely over 60 per cent. This suggests that producer service employment, even though it may involve a small share of the total workforce, has a major impact on mega city development by fostering more long distance commuting.

That effect is geographically selective however, as can be seen in the 11 municipalities where the share of the local workforce commuting into the Melbourne Statistical Division is over 10 per cent, which are displayed in Figure 12.5. That shows the proportion of the local labour force commuting into Melbourne, alongside the share of the commuters who have producer service jobs. These municipalities form a ring around the metropolitan area and probably all involve the day-to-day trip to and from work. It is apparent that there are four places (Macedon, Moorabool, Mitchell and Murrindindi) where commuting is very significant, but producer service commuters are not prominent. These four places adjoin Melbourne to the west and north. In a number of cases they provide lower cost housing opportunities, so attract a wide array of the workforce. At the

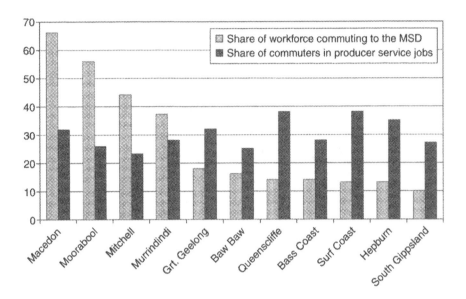

Figure 12.5 Significance of producer service workers amongst commuters from main municipalities.

other end of the scale are four municipalities where commuting shares are just above the state average of 10 per cent, but producer service commuters account for up to 40 per cent of the movers in two cases. These include more distant coastal locations where house prices have been high and rising rapidly over the past few years (Surf Coast and Queenscliffe), and an inland location (Hepburn) that has attracted very substantial housing investment in the past decade. For these places it seems the higher level of income of the producer service employee has made it possible to purchase housing a longer distance away from Melbourne. Hence the part that producer service employees play in mega city-region commuting is shaped by the working of the housing market as well as the willingness of some employees to gain local residential amenity through longer distance commuting.

Commuting to selected municipalities within the Melbourne Statistical Division

The commuting flows into Melbourne City (shown in Figures 12.6 and 12.7) illustrates some of these themes in more detail. The broad area supplying 50 or more workers is similar to the core at the centre of the state that was identified in Figures 12.2 and 12.3. The differences in type of

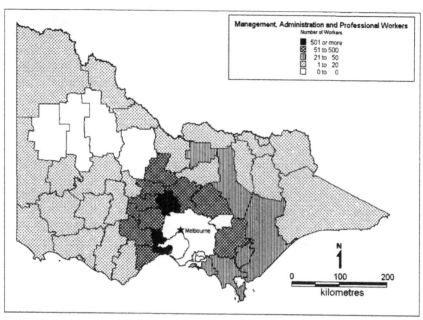

Figure 12.6 Commuting from the rest of Victoria to the City of Melbourne: management, administration and professional workers.

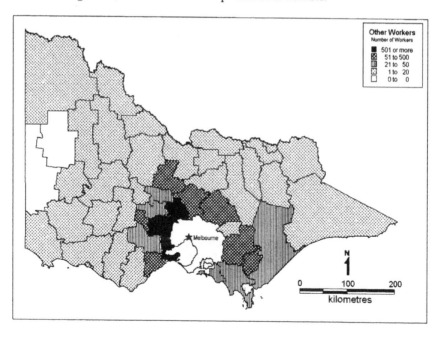

Figure 12.7 Commuting from the rest of Victoria to the City of Melbourne: other workers.

work have only a small effect upon the movement of labour. Management, administration and professional workers are drawn from a larger area to the west, while the catchment for other workers is stronger in the immediately adjoining western areas. This small difference may reflect the effect detected in Figures 12.2 and 12.3 where the producer service workers seem to travel further. For both types of workers the very large central Melbourne City labour market exerts an influence across much of the state, although numbers travelling from the distant parts are very small.

The next four maps provide information to identify the impact of middle and outer suburban employment on the shape and growth of a mega city-region. The impact of employment at the City of Monash (shown in Figures 12.8 and 12.9) can be seen in its northern, eastern and southern catchment; it also draws labour from the other side of the metropolitan area. Commuting patterns are broadly similar for both categories of work although the producer service workers are drawn from a wider set of municipalities, especially to the north. This suggests that suburban industrial and service development has an impact well beyond its nearby corridor especially for jobs in producer services. That is confirmed by the data displayed in Figures 12.10 and 12.11 for Hume. Here the work location is very close to the statistical division border, and the job market is biased in favour of other workers. Commuting connections with two nearby municipalities are quite strong for other workers, while the catch-

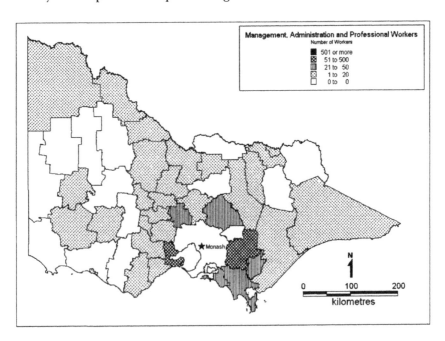

Figure 12.8 Commuting from rest of Victoria to the City of Monash: management, administration and professional workers.

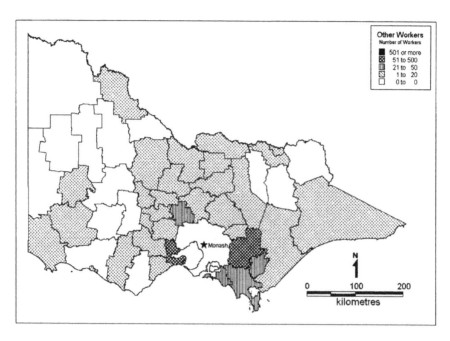

Figure 12.9 Commuting from rest of Victoria to the City of Monash: other workers.

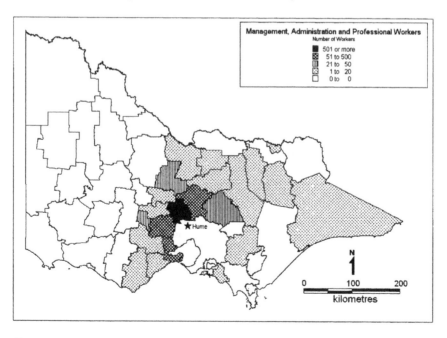

Figure 12.10 Commuting from rest of Victoria to City of Hume: management, administration and professional workers.

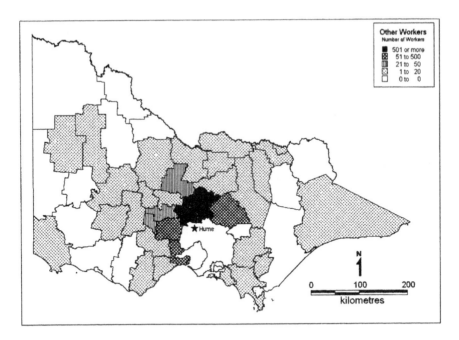

Figure 12.11 Commuting from rest of Victoria to City of Hume: other workers.

ment area for producer service workers is spread a little further, especially to the north.

The availability of jobs in suburban locations has contributed to mega city-region development around Melbourne although the outcome is not any different from the pattern created by the flow of workers into the major labour market in the centre of the metropolitan area. The patterns of flows of commuters seems to depend more upon the numbers of jobs than their location. The very big central city labour market associated with the City of Melbourne has a similar impact to that recorded for destinations like Monash and Hume with smaller numbers of jobs, even though the latter are closer to the current fringe of the metropolitan area. In short it seems the Melbourne mega city-region experience is still largely shaped more by the residential location choice of workers wherever those jobs are located, rather than by the dispersal of employment that brings jobs closer to the fringe.

Conclusion and implications

The data analysis carried out here shows that mega city-region development is occurring around Melbourne and has been felt primarily in a ring of adjoining municipalities, stretching out some 50–75 kilometres beyond the metropolitan area. Beyond that a weaker effect is felt across much of the State of Victoria. That outcome can be seen in two types of jobs, and from central as well as suburban work places. Some differences do exist in the pattern of commuting of producer service workers, as the data show they select particular home locations and travel further, so play a significant part in shaping mega city development.

The results confirm that residential dispersal with long distance commuting associated with availability of housing and facilitated by road and rail infrastructure is central to the growth of a mega city-region. Suburban and outer area job concentrations, especially in manufacturing and producer services, have contributed to this development, but their effects have not been locally distinctive. An interesting issue is whether residential dispersal will eventually stimulate more complex job concentrations in outer areas and so stimulate an even stronger local labour market effect. This conjecture calls for more detailed analysis of the commuter and other links that occur in the two broad models of mega city-region development.

These results have a number of other implications. The first is the way that broader strategic planning policy can accommodate or direct them and so shape the long term growth of the mega city-region. In the Melbourne case that strategic perspective has two components. The first component is the imposition of an urban growth boundary around the suburban parts of the metropolitan area, with designated corridors for development that correspond to the corridors that have shaped the mega city development identified here. This has been done in a policy context that favours urban consolidation and higher suburban residential densities. That policy is expected to achieve a set of objectives in urban sustainability especially to do with energy consumption and public transport utilisation, as outlined in the State Government's management strategy for the Metropolitan area (State of Victoria 2003). That approach will have some effect upon the development of the mega city-region as the restriction on residential development opportunities may stimulate some more of the commuting that has been displayed in the figures analysed above. That could be helped by a second strategic thrust which is a decision to provide higher-standard rail transport services following the motorways along the main corridors of the mega city-region. This component is displayed in Figure 12.12, which is drawn from the state government strategy. It shows plans to provide improved rail services along four corridors connecting Melbourne to major regional centres. The provision of improved rail services in these corridors could stimulate more commuting (although the proportions of the workforce using rail transport to get to work is very low) and improve access to the service base of the central city, which together could stimulate more residential dispersal.

These two components of policy would seem to ensure that the mega city-region of Melbourne will become a larger and more substantial area

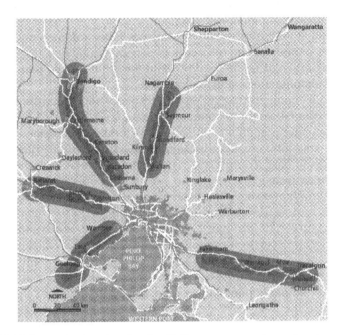

Figure 12.12 Strategic planning for the Melbourne mega city-region (source: State of Victoria 2003).

in the long term. These two strategic planning components, one channelling growth into corridors and restricting it elsewhere, and a second providing opportunities for growth in the larger region, could provide a framework for mega city-regions elsewhere in the world. In this way the Melbourne experience could be a useful model for those dealing with long term strategic planning in other Asian mega cities.

A broader issue emanating from this research is whether commuting is the best indicator of mega city-region development. At one level it is a very good indicator, as it provides an understanding of how metropolitan area jobs can be expressed in residential development and the urban services in a mega city-region. By taking a broad definition of that link, and counting weekly as well as daily commuters, the current project has displayed the spread of mega city-region influence. It also makes it possible to expose the extent to which suburban or fringe locations have mega city-region impacts; in the Melbourne case it showed residential location and long distance commuting even to central city jobs was still an important influence upon mega city-region development. However, it is a limited indicator in that it accounts just for labour movement and provides none of the insight on functional ties and linkages that have formed the basis of research by Marton (2000) and Leung (1993). A second stage of the current research could explore the location and use of producer services by firms in the outer parts of the mega city-region. One particular focus would be on the suburban clusters of producer services in a municipality like Monash for example to establish if firms in this location have customers in the wider city-region.

Another issue is the possibility of translating the patterns and processes of an Australian city to the broader Pacific Asian experience. That is part of a long-running debate about whether European and US urban experiences have any relevance to Asian development outcomes. Parai and Dutt (1994) explored this issue with some comparative statistics, noting that there are major underlying differences in the social and demographic contexts of the rapidly growing Asian cities of today compared to their European counterparts in an earlier era. That was consistent with the foundation laid by McGee's (1971) work on the distinctive characteristics of the Asian city. However, that sense of difference may be changing. Just as Webster (2002b) is reporting results based upon a research framework using the notion of employment sub-centres in the peri-urban spaces beyond the fringes of several big cities in Asia, a group in Europe is embarking upon an analysis of the poly-nucleation of the very large urban region of north-west Europe (Hall *et al.* 2003). Although the real drivers (manufacturing and foreign direct investment in the Asian case and producer services and knowledge-based activities in Europe) are different, the underlying structures of the local labour markets, and the significance of backbone infrastructure in the form of roads and rail networks is common to both approaches. The Melbourne research would seem to sit between these two, suggesting that journey-to-work data could provide good insight on the underlying processes, but warning at the same time that suburban work clusters may need to be very large to have distinctive mega city-region effects over and above those emanating from the core of the metropolitan area.

A final consideration here is the management of these complex spatial units. This has attracted considerable attention to date, with a particular focus upon the environmental pressures that the rapid spread of urban settlement has created (Brennan 1995) and the institutional and management issues that are associated with urban management structures (Ruland 1997). Webster's (2002b) review of several mega city-regions in Asia illustrates some very difficult land and service supply management issues, often involving lower skill and poorly resourced rural area governments. His work shows that the key role in planning and management of these areas needs to lie at the national or major provincial level, as can be seen with the plans for the improved rail networks in the Melbourne case. However, those approaches need to be matched by a broad array of actions directed at social and personal services; how that can be done in a form and speed to match the rates of industrial and land use change in the outer parts of these mega city-regions by governments with limited funds and expertise is a major concern.

Acknowledgements

The author thanks the Faculty of Architecture, Building and Planning for financial assistance to purchase Australian Bureau of Statistics journey to work data, and Bill Callagan's assistance with the preparation of that data. The maps were drawn by Peter Elliott.

References

Artibise, A. (1995) 'Achieving sustainability in Cascadia: an emerging model of urban growth management in the Vancouver-Seattle-Portland corridor', in Kresl, P. K. and Gappert, G. (eds), *North American Cities and the Global Economy: Challenges and Opportunities.* Thousand Oaks, CA: Sage, 221–85.

Bowen, J. T., Leinbach, T. R. and Mabazza, D. (2002) 'Air cargo services, the state and industrialisation strategies in the Philippines: the redevelopment of Subic Bay', *Regional Studies*, 36, 451–67.

Brennan, E. M. (1995) 'Developing management responses for mega-urban regions', in McGee, T. G. and Robinson, I. M. (eds), *The Mega Urban Regions of South East Asia.* Vancouver: UBC Press, 242–65.

Browder, J. O., Bohland, J. R. and Scarpaci, J. L. (1995) 'Patterns of development on the metropolitan fringe', *Journal of the American Planning Association*, 61, 310–27.

Demattais, G. and Governa, F. (1999) 'From urban field to continuous settlement networks: European examples', in Aguilar, A. G. and Escamilla, I. (eds), *Problems of Mega Cities: Social Inequalities, Environmental Risks and Urban Governance.* Mexico City: Universidad Nacional Autonoma de Mexico, 543–56.

Dogan, M. and Kasarda, J. (eds) (1988) *The Metropolis Era.* Newbury Park, CA: Sage.

Friedmann, J. and Miller, J. (1965) 'The urban field', *Journal of the American Institute of Planners*, 31, 312–19.

Graham, S. and Marvin, S. (2001) *Splintering Urbanism.* London: Routledge.

Hall, P. (2001) 'Global city regions in the twenty-first century', in Scott, A. (ed.), *Global City-Regions: Trends, Theory, Policy.* Oxford: Oxford University Press, 59–77.

Hall, P., Walker, O. and Evans, O. (2003) *POLYNET: Sustainable Management of European Polycentric Urban Regions.* Global and World Cities Research Project Number 33. Department of Geography, University of Loughborough.

Heikila, E. (2002) 'Seoul: regional realities and global ambitions', Paper presented to the Annual Conference of the Association of Collegiate Schools of Planning. Baltimore. November.

Jones, G. (2001) 'Studying extended metropolitan regions in South East Asia', Paper presented to the XXIV General Conference of the IUSSP. Salvador, Brazil.

Jun, M.-J. and Hur, J.-W. (2001) 'Commuting costs of "leap frog" new town development in Seoul', *Cities*, 18, 151–8.

Leaf, M. (1994) 'The suburbanization of Jakarta: a concurrence of economics and ideology', *Third World Planning Review*, 16, 341–56.

Leung, C. K. (1993) 'Personal contacts, sub-contracting linkages and development in the Hong Kong-Zhujiang Delta region', *Annals, American Association of Geographers*, 3, 297–302.

Lo, F. and Marcotullio, P. J. (2001) *Globalisation and the Sustainability of Cities in the Asia Pacific Region.* Tokyo: United Nations University.

Lo, F. and Yeung, Y. (1996) *Emerging World Cities in Pacific Asia.* Tokyo: United Nations University.

Lo, F. and Yeung, Y. (1998) *Globalisation and the World of Large Cities.* Tokyo: United Nations University.

Marton, A. (2000) *China's Spatial Economic Development: Restless Landscapes in the Lower Yangste Delta.* London: Routledge.

McGee, T. G. (1971) *The South East Asian City.* London: Bell.

McGee, T. G. (1994) 'Labour force change and mobility in the extended metropolitan regions of Asia', in Fuchs, R., Brennan, E., Chamie, J., Lo, F.-C. and Uitto, J. (eds), *Mega City Growth and the Future.* Tokyo: United Nations University Press, 62–102.

McGee, T. G. and Robinson, I. M. (eds) (1995) *The Mega Urban Regions of South East Asia.* Vancouver: UBC Press.

O'Connor, K. (2002) *Monitoring Greater Melbourne.* Melbourne: State of Victoria Department of Infrastructure.

O'Connor, K. and Edgington, D. (1991) 'Producer services and metropolitan development in Australia', in Daniels, P. (ed.), *Services and Metropolitan Development: International Perspectives.* London: Routledge, 204–25.

O'Connor, K., Stimson, R. and Daly, M. (2001) *Australia's Changing Economic Geography: A Society Divided.* Melbourne: Oxford University Press.

Parai, A. and Dutt, A. K. (1994) 'Perspectives on Asian urbanisation: an East Asian comparison, in Dutt, A. K., Costa, F. J., Aggarwal, S. and Noble, A. G. (eds) *The Asian City: Processes of Development Characteristics and Planning.* Dordrecht: Kluwer Academic, 369–89.

Rodrigue, J.-P. (1994) 'Transportation and territorial development in the Singapore extended metropolitan region', *Singapore Journal of Tropical Geography,* 15, 56–74.

Rodrigue, J.-P. (2003) 'Freight, gateways and mega urban regions: the logistical integration of the Bostwash corridor', *Tijdscrift voor Social en Economische Geographie,* 95(2), 147–61.

Ruland, J. (ed.) (1997) *The Dynamics of Metropolitan Management in Southeast Asia.* Singapore: Institute for Southeast Asian Studies.

Scott, A. (ed.) (2001) *Global City-Regions: Trends, Theory, Policy.* Oxford: Oxford University Press.

Sheard, P. (1983) 'Auto production systems in Japan: organizational and locational features', *Australian Geographical Studies,* 21, 49–68.

Simmons, R. and Hack, G. (eds) (2000) *Global City Regions: Their Emerging Forms.* London: Spon Press.

Sinclair, Knight, Merz (2003) *Melbourne Port Container Origin Destination Study.* Consultants Report to State of Victoria Department of Infrastructure. Melbourne: Sinclair Knight Merz Pty Ltd.

Sit, V. F. S. and Yang, C. (1997) 'Foreign-investment-induced exo-urbanisation in the Pearl River Delta, China', *Urban Studies,* 34, 647–77.

State of Victoria (2003) *Melbourne 2030: Planning for Sustainable Growth.* Melbourne: Department of Infrastructure.

Tang, W. S. and Chung, H. (2000) 'Urban rural transition in China: beyond the *Desakota* model', in Li, S.-M. and Tang, W.-S. (eds), *China's Regions, Polity and*

Economy: A Study of Spatial Transformation in the Post-Reform Era. Hong Kong: Chinese University Press, 275–307.

Webster, D. (2002a) 'Bangkok: evolution and adaptation under stress', in Gugler, J. (ed.), *World Cities Beyond the West.* Cambridge: Cambridge University Press, 82–118.

Webster, D. (2002b) *On the Edge: Shaping the Future of Peri-Urban East Asia.* Urban Dynamics of East Asia Program. Asia Pacific Research Center, Stanford University.

Yeung, Y. (2000) *Globalisation and Networked Societies: Urban and Regional Change in Pacific Asia.* Honolulu: University of Hawaii Press.

13 Service industries and development trajectories in the Seattle metropolitan region

William B. Beyers

Introduction and overview

Like many regions in advanced economies today, the Seattle metropolitan region has the bulk of its employment accounted for by various service industries. This is not something new, for even a half century ago service employment accounted for most jobs in the Seattle metropolitan region. The share of services jobs has continued to rise, and the composition of employment within the local service sector has changed.

This chapter documents trends in employment in the Seattle metropolitan area, and places the structure of employment in Seattle in a national US context in order to provide some basis for interpreting changes observed in our local economy. The chapter begins with a description of employment trends in the Seattle area, and then the local experience is contrasted with trends in the US economy as a whole. Next, I raise the question of the structural mechanisms at work explaining the large ongoing shift to services employment observed in the Seattle metropolitan region. I conclude with some research questions that would help sharpen our understanding of the evolution of the service economy in Seattle and in other advanced urban regions.

For the purposes of this chapter I am considering the Seattle metropolitan region to be King, Pierce, Kitsap and Snohomish counties. These counties are collectively members of the Puget Sound Regional Council (PSRC), a metropolitan planning organization for the region. The contiguous urban area within the central Puget Sound region extends beyond these four counties. To the south, Thurston County has employment and residential interdependencies with the counties to the north. Activity in Kitsap County has spillover impacts on Jefferson County to the west. Snohomish County has people working in it who live to the north in Island, Skagit and Whatcom counties. And there are some people who commute across the Cascade Mountains, over Snoqualmie Pass, to work in King County. So, the four-county region that I have used for the purposes of this chapter is not a 'clean' definition. An alternative regionalization might be the US Bureau of Economic Analysis (BEA) Economic Area set of counties, which includes the central urban counties as well as more peripheral counties (such as San Juan or Clallam) that are considered by BEA to be in the commuter shed of the Central Puget Sound region urban areas. However, the four counties included in the tables in this chapter account for the bulk of the population and employment within the larger BEA economic area, and are also joined into a regional planning organization (the PSRC) that provides key statistics regarding the Seattle area economy.

Trends in employment in the Seattle metropolitan region

In approaching the development of a database that could be used to characterize the evolution of the structure of the Seattle metropolitan regional economy, I considered several alternative sources. The PSRC provides data series extending back to 1958 on an annual basis, while the Employment Security Department (ESD) of the State of Washington also provides measures of covered employment extending back to the 1950s. These ESD data exclude proprietors, and in the early days did not include some services and agriculture. The federal government in the United States provides data through the economic census statistics on employment, but these do not provide coverage of the entire economy. The Bureau of Labour Statistics (BLS) also provides regional estimates of employment, but this is primarily covered employment similar to that estimated by the Washington State Employment Security Department. The US Bureau of Economic Analysis also provides estimates of employment by industry, and includes employment covered by the unemployment compensation programmes as well as proprietors in the Regional Economic Information System (REIS). Table 13.1 provides a comparison of total employment for the year 1996 from these various sources for King and Snohomish counties, and it is evident in this table that the REIS estimates of employment are above the other three sources.

A major reason for the higher levels of employment in the REIS statistical series is the inclusion of a larger number of proprietors in this statistical series than in the PSRC series. The ESD and BLS series exclude proprietors. REIS utilizes information from income tax returns to estimate

Table 13.1 Alternative measures of King–Snohomish county employment, 1996

Source	Employment
WA State Dept. of Employment Security	1,179,004
Puget Sound Regional Council	1,411,400
US Bureau of Economic Analysis (REIS)	1,516,192
US Dept. of Labor (BLS)	1,209,000

income from proprietorships. As this is the broadest measure of employment, I have chosen to use it for the purposes of this chapter, but will also make use of earnings data developed by the BEA, as they provide greater detail by industry.

The magnitude of employment in the Central Puget Sound region between 1969 and 1999 is presented in Table 13.2. Over this 30-year time period employment has expanded from 900 thousand to 2.1 million persons. Some 86 per cent of this growth in jobs occurred in services (including government as services). The bulk of the non-service employment growth occurred in manufacturing and construction; farm, forestry, fishing and mining, which are very small employers in the Central Puget Sound region. Figure 13.1 depicts the changing shares of employment for the industrial categories included in Table 13.2. There are a number of

Table 13.2 Central Puget Sound region employment 1969–1999 ('000s)

	1969	1974	1979	1984	1989	1994
Total employment	925	919	1,190	1,319	1,692	1,851
Wage and salary employment[1]	827	804	1,043	1,120	1,426	1,544
Proprietors' employment[1]	98	115	147	199	267	307
Farm proprietors' employment	3	3	5	5	5	5
Non-farm proprietors' employment	95	112	142	193	262	302
Farm employment[2]	6	8	10	10	8	7
Non-farm employment[2]	919	911	1,180	1,309	1,684	1,844
Private employment[3]	696	700	961	1,071	1,417	1,555
Ag. services, forestry, fishing	6	7	13	16	20	23
Mining	1	1	1	2	2	2
Construction	45	37	71	69	94	100
Manufacturing	188	154	197	183	250	231
Transportation and public utilities	50	49	63	67	83	90
Wholesale trade	46	49	68	72	90	102
Retail trade	129	136	190	216	272	310
Finance, insurance, and real estate	77	81	103	116	151	150
Services	155	186	256	330	454	546
Government and government enterprises[3]	222	211	219	239	267	289
Federal, civilian	44	43	46	50	51	50
Military	78	51	47	55	57	53
State and local	100	117	126	134	159	186
State	ND	0	42	45	52	60
Local	ND	0	84	89	106	127

Source: BEA Regional Economic Information System. ND – no data.

Notes
1 'Total employment' subdivided into 'wage and salary employment' and 'proprietors' employment'.
2 'Total employment' subdivided into 'farm employment' and 'non-farm employment'.
3 'Private employment' and 'Government and government enterprises = 'Total employment'.

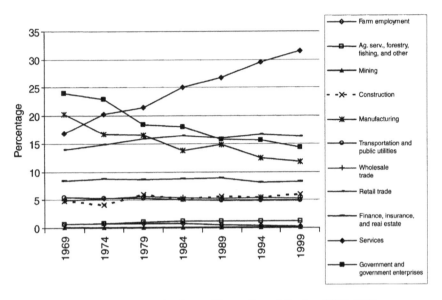

Figure 13.1 Employment trends, Central Puget Sound Region 1969–1999.

sectors whose relative position in the regional economy has been comparatively stable over this 30-year time period, including retail trade, transportation and public utilities, construction and wholesaling. In contrast, government and manufacturing have accounted for a declining share of jobs, while the broad services category has soared upwards in its share of jobs. Within this service group are the non-financial services components of producer services, health services and consumer services.

The broad geographical distribution of employment within the four Central Puget Sound region counties is shown in Figure 13.2. It is clear that King County (in which Seattle is located) dominates employment in the region, and that its share has remained relatively stable over the past 30 years. Pierce County has grown less rapidly than Snohomish County, such that the share of regional employment found in Snohomish County has grown somewhat over time.

One feature of the changing level of employment in the Central Puget Sound region is the rising share of employment by proprietors. Table 13.3 summarizes the absolute measures of wage and salary versus proprietors jobs reported in Table 13.2 as shares of total employment. This table indicates that the percentage share of people working as proprietors has increased by about 50 per cent over the 1969–1999 time period. In Table 13.2 the employment by industry is a combination of wage and salary as well as proprietary workers; the BEA does not publish the share within particular industries that are wage and salary as opposed to proprietorships.

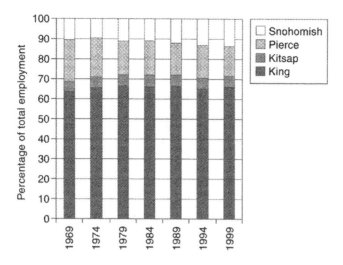

Figure 13.2 Share of regional employment by county.

The employment series included in the REIS does not disaggregate the very important services sector into detailed categories. However, BEA does supply a much more detailed file of earnings by industry. I have developed a decomposition of the earnings components within the services category into producer services (excluding the financial services component), health services, and the balance of this category, which are components of consumer services. Figure 13.3 presents a graph indicating the changing relative importance of these categories over the last 30 years. This graph makes it quite clear that the producer services have grown in the Central Puget Sound relative to the other components within the non-financial services category included in Table 13.2. The spike in values in 1999 is particularly related to stock options exercised at the Microsoft Corpora-

Table 13.3 Shares of total employment, by type of work (%)

Year	% wage and salary	% proprietors
1969	89.37	10.63
1974	87.48	12.52
1979	87.67	12.33
1984	84.92	15.08
1989	84.23	15.77
1994	83.43	16.57
1999	84.31	15.69

Source: Calculated by author from BEA Regional Economic Information System.

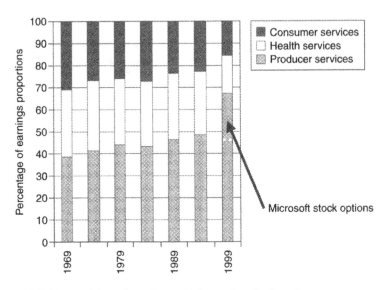

Figure 13.3 Composition of earnings within services, by broad category.

tion, a leading employer in the computer services industry located in the Central Puget Sound region. I have identified an approximate value of earnings associated with a particular portion of earnings in this industry in 1999, the value of exercised stock options. I have valued these at 7 billion dollars in 1999, which is within a billion or two of their real value. If these values were extracted from Figure 13.3, the jump in the value of producer services shown in 1999 would not be there; the value would be similar to 1994. That said, the clear long run trend in the Central Puget Sound region is for producer services to expand relative to consumer services, and for the health services sector to essentially hold its own as a share of the local services sector.

Contextualizing the Seattle metro area experience

How has the experience of the Central Puget Sound region differed from that of the US as a whole in recent years? I have chosen to view this question through the use of location quotients that compare trends between the two. A sharper distinction might have been with urbanized areas in the US, but knowing that at least 85 per cent of total employment in the country is located in areas designated as urban, I have chosen for the purposes of this chapter to use the US as whole as the benchmark for the analyses that follow.

One way of contrasting the Central Puget Sound region with the US as a whole is through the use of location quotients, measures that compare

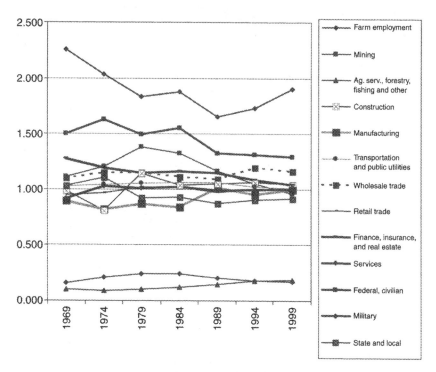

Figure 13.4 Location quotients, Puget Sound industries.

the structure of a region with that of a benchmark region. In this case, I have chosen to use the data in Table 13.2 along with national employment statistics to calculate the values graphed in Figure 13.4, and shown in Tables 13.4 and 13.5. Figure 13.4 and Table 13.5 are based on REIS employment data, while Table 13.4 is based on REIS personal income statistics (REIS does not provide as detailed estimates of employment as it provides of earnings and components of personal income). These data clearly indicate the importance of military employment to the Central Puget Sound region, and the surge in the importance of producer services

Table 13.4 Location quotients, Seattle metropolitan area, based on earnings (values over 1.1 are boldened)

	1969	*1974*	*1979*	*1984*	*1989*	*1994*	*1999*
Producer services	**1.188**	**1.390**	**1.466**	**1.295**	1.045	**1.201**	**1.866**
Health services	0.979	1.018	0.953	0.948	0.964	0.946	0.863
Consumer services	0.616	0.629	0.638	0.724	0.884	0.938	0.841

Source: Calculated by author from BEA REIS.

Table 13.5 Location quotients, Central Puget Sound region 1969–1999 (values greater than 1.1 boldened)

	1969	1974	1979	1984	1989	1994	1999
Wage and salary employment	1.034	1.017	1.018	1.002	0.995	0.996	1.011
Proprietors' employment	0.785	0.894	0.887	0.990	1.028	1.021	0.947
Farm proprietors' employment	0.107	0.122	0.176	0.200	0.179	0.159	0.154
Non-farm proprietors' employment	0.979	1.070	1.021	**1.114**	**1.130**	**1.112**	1.018
Farm employment	0.158	0.210	0.242	0.244	0.205	0.176	0.170
Non-farm employment	1.038	1.033	1.026	1.024	1.019	1.018	1.016
Private employment	0.962	0.965	1.004	0.997	1.014	1.012	1.010
Ag. services, forestry, fishing, and other	**1.113**	**1.201**	**1.378**	**1.326**	**1.167**	1.052	0.961
Mining	0.103	0.089	0.100	0.120	0.145	0.178	0.183
Construction	0.985	0.807	**1.145**	1.035	1.047	1.063	1.044
Manufacturing	0.900	0.823	0.873	0.843	1.015	0.957	1.001
Transportation and public utilities	1.032	1.043	1.058	1.064	1.064	1.026	1.013
Wholesale trade	**1.101**	**1.149**	**1.150**	**1.110**	1.094	**1.194**	**1.164**
Retail trade	0.945	0.968	1.015	1.016	0.972	1.001	0.991
Finance, insurance and real estate	**1.278**	**1.197**	**1.149**	**1.162**	**1.149**	1.091	1.043
Services	0.914	1.030	1.010	1.024	0.990	0.999	1.002
Government and government enterprises	**1.382**	**1.343**	**1.135**	**1.161**	1.045	1.053	1.054
Federal, civilian	**1.499**	**1.625**	**1.494**	**1.551**	**1.325**	**1.311**	**1.292**
Military	**2.255**	**2.035**	**1.830**	**1.876**	**1.653**	**1.728**	**1.902**
State and local	1.032	**1.107**	0.924	0.929	0.871	0.905	0.914
State	ND	ND	**1.105**	1.072	0.995	0.996	1.018
Local	ND	ND	0.855	0.870	0.820	0.867	0.874

Source: Calculated by author from BEA Regional Economic Information System.

Note
ND = no data.

in the past few years. The producer service value for 1999 is influenced by the Microsoft stock options that are included; their exclusion would move the index more to the 1.2 range for producer services. With regard to the balance of the service sector, these data indicate that the Central Puget Sound region has no striking specialization compared to the US economy, and if anything, its industrial structure has converged towards the national structure over time.

Why have services grown so rapidly? An argument for trade in services

The data presented above clearly indicate a secular shift in employment in the Central Puget Sound region towards services over the past 30 years, a process of change that has tended to mirror national trends, rather than being exceptional to them. Why have these services grown rapidly? A number of arguments have been made by scholars seeking to explain this structural transformation. Let me address some of these arguments.

One view has explained the growth of the service sector as due to its historic relatively slow rate of productivity improvement (Baumol 1967), which means that as time passes and the demand for services grows relative to the demand for goods, that an even larger share of employment is associated with services production. This is often referred to as Baumol's 'cost disease', but it has been called into question recently with the emergence of the 'New Economy', in which many information technology (IT) using sectors are actually leading the expansion of productivity in the US economy (Economics Statistics Administration 2000). So, the productivity-lag argument may not be relevant to current economic history.

Another process that has been identified for the producer services is a growing demand for specialized services for a variety of reasons. These arguments include the processes of downsizing and outsourcing that involve shedding service functions internal to businesses and governments, and moving their purchase into an open market environment (Harrison 1994). However, our research indicates that this process is less important than other bases for growth in demand, including:

1 a continuing division of labour or innovation in services being offered, which are in turn purchased in the market place by an increasingly diverse set of clients;
2 co-production of services by in-house departments and service-supplier specialists where there is a division of labour or complementarity between in-house production and outsourcing;
3 a growing demand for specialized expertise that clients simply cannot produce for themselves;
4 an increasingly complex business environment that fuels the demand for specialized expertise (Beyers and Lindahl 1996).

These forces have been accompanied in the US by an enormous proliferation of business establishments in the service economy, as entrepreneurs visualize new business opportunities and start businesses to pursue them (Beyers and Lindahl 1996).

These arguments play out on the landscape in the form of demand in particular places for the production from these service industries (as well as other sectors). While my own research has given greatest emphasis to the geography of demand within the producer services, I see no reason for patterns of demand across the economy to be fundamentally different in other sectors than we have measured in the producer services (Beyers and Alvine 1985; Beyers 2000). Within the Central Puget Sound region we have considerable evidence regarding the importance of trade in services, most richly from the input-output models that have been estimated by survey data five times between 1963 and 1987 (Chase *et al.* 1993). In addition, we have data that has come from several of my research projects that have helped us understand trade in producer and cultural services (Beyers and Alvine 1985, GMA Research Corporation and Beyers 1999). Pioneering work in Vancouver has revealed similar tendencies towards an enhanced role for trade in services (Hutton and Ley 1987). While the arguments laid out above probably are relevant in explaining the shift in the structure of the economy towards a greater share of services, I am going to argue that one key force has been the expanding level of trade in services, underpinning the economic base of communities in a systematically growing manner. Figure 13.5 gives a sense of these growing levels of export orientation for a particular cohort of business establishments. This scattergram indicates a much larger number of businesses with growing export markets, than shrinking ones (Beyers 1999).

In order to address the matter of the role of services exports, I have undertaken the following exercise. Using the 1987 Washington State input-output table, I estimated the data reported in Table 13.6 (Chase *et al.* 1993). This table contains estimates of the share of sales by industry that were exported out of the Washington State economy to clients located in other parts of the US or in foreign countries, or sold to the federal government. From the standpoint of the Central Puget Sound region this is a conservative estimate of exports, because it does not include exports made from the four-county region to clients located elsewhere in Washington State. No good statistics are available on the level of intra-state export trade, but I suspect that it is significant within the services. Hence, these are conservative estimates of the share of sales garnered by non-local demand. Within the financial and services categories, I have used data from my own survey research to systematically change the share of production assumed to be moving into export markets.

The data reported in Table 13.6 were used with the employment statistics reported in Table 13.2 to estimate the level of jobs associated with export demand over the 1969–1999 time period. Figure 13.6 is a bar chart

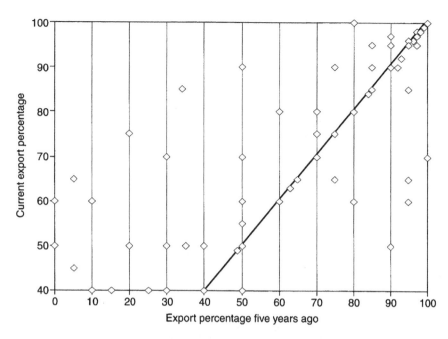

Figure 13.5 Percentage of current exports versus export percentage five years ago.

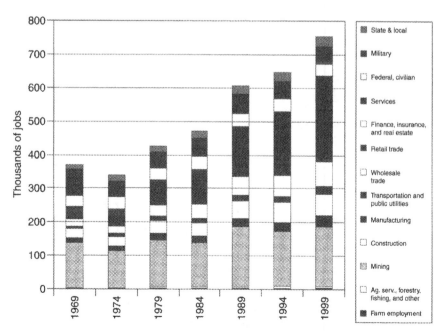

Figure 13.6 Export employment by industry in the Central Puget Sound Region.

Table 13.6 Percentage share of employment related to exports

	1969 (%)	1974 (%)	1979 (%)	1984 (%)	1989 (%)	1994 (%)	1999 (%)
Farm employment	31.1	31.1	31.1	31.1	31.1	31.1	31.1
Ag. services, forestry, fishing and other	9.9	9.9	9.9	9.9	9.9	9.9	9.9
Mining	28.1	28.1	28.1	28.1	28.1	28.1	28.1
Construction	3.7	3.7	3.7	3.7	3.7	3.7	3.7
Manufacturing	70.8	70.8	70.8	70.8	70.8	70.8	70.8
Transportation and public utilities	31.6	31.6	31.6	31.6	31.6	31.6	31.6
Wholesale trade	56.2	56.2	56.2	56.2	56.2	56.2	56.2
Retail trade	6.4	6.4	6.4	6.4	6.4	6.4	6.4
Finance, insurance and real estate	25.0	27.0	32.0	35.0	37.2	40.0	42.0
Services	25.0	28.0	30.0	31.5	33.0	35.0	38.0
Federal, civilian	75.0	75.0	75.0	75.0	75.0	75.0	75.0
Military	100.0	100.0	100.0	100.0	100.0	100.0	100.0
State and local	15.0	15.0	15.0	15.0	15.0	15.0	15.0

Source: Calculated by author from Chase, Bourque and Conway (1987).

Table 13.7 Total jobs per export job

Year	Multiplier
1969	2.49
1974	2.70
1979	2.79
1984	2.81
1989	2.79
1994	2.86
1999	2.83

Source: Estimated by the author based on data in Table 13.2 and Table 13.6.

describing the magnitude of employment assigned to exports using the shares of employment included in Table 13.5. This figure clearly indicates a growing share of the economic base of the Central Puget Sound region is lodged in services, as the length of the bars for the farm through manufacturing categories has not expanded much compared to the large rise in services. While this figure is not based on survey data for each of the years, it is based on sound data for one year (1987) and on a simulation of survey data from a strong sample of producer service establishments. Table 13.7 documents the implicit multipliers resulting from the export market estimation process just described. These are well within the range of multipliers typically measured for regional economies at a scale similar to the Central Puget Sound region.

Other perspectives on the changing role of services in the economic base of the Central Puget Sound region are found in Figures 13.7–13.9. Figure 13.8 converts the values in Figure 13.7 into shares of total export employment over time. This figure clearly depicts the growing role of services employment as a share of the economic base of the Central Puget Sound region. In Figure 13.7 there is a steadily decreasing share of exports emanating from the goods producing sectors, and a rising contribution from all services, particularly from the financial services and services (a combination of producer, health, and consumer service sectors). Figures 13.8 and 13.9 vividly illustrate the shift towards services in the economic base of the Central Puget Sound region, as the shares in the pie-graphs associated with goods production show a dramatic contraction between 1969 and 1999.

Research needs

This chapter has reported some evidence regarding employment trends in the Central Puget Sound region in relation to the themes of the Peter Wall Institute for Advanced Studies Exploratory Workshop where these results were first presented. I have addressed trends in the Central Puget

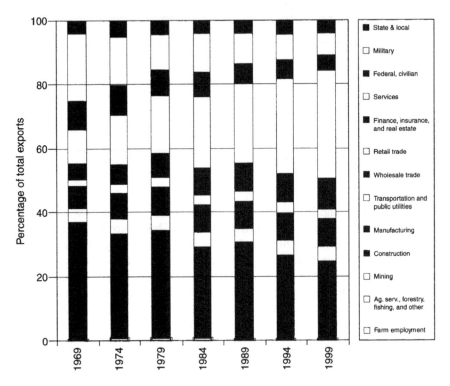

Figure 13.7 Share of export employment, Central Puget Sound Region.

Sound region in the context of the themes of services and global city formation, and economic restructuring. I have not addressed questions of the reconfiguration of urban space and form as this process of service industry development has occurred. I have also not addressed questions of occupational structural change, class formation, and policy choices of governments with regard to the development of service industries.

There has been very little comparative research on metropolitan experiences with regard to the themes just discussed. We need to work hard on getting the empirical resources that will allow us to address these key questions, and this book will provide pieces of badly-needed information in this regard. This chapter has emphasized the role of trade in services as a force driving the development of metropolitan Pacific Rim economies. Unfortunately, there is a paucity of data of this type, making it difficult to compare the role of trade in services in the Seattle metropolitan area with other major Pacific Rim cities. Moreover, even the data that was presented in this chapter in this regard was secondary in nature, and could be sharpened considerably.

We need to develop common approaches to measuring the way in

1969

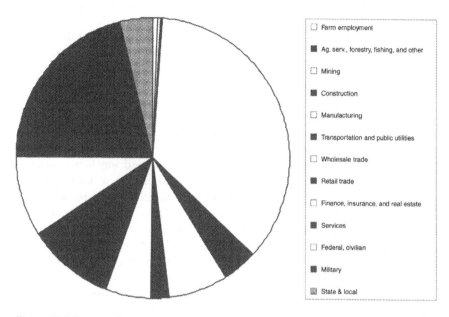

Figure 13.8 Export share of employment in 1969.

1999

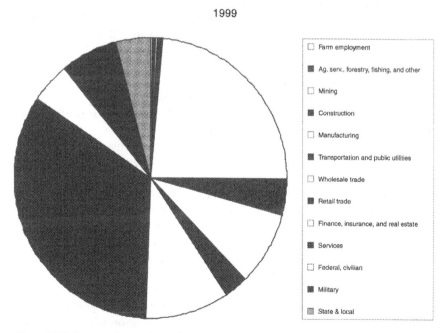

Figure 13.9 Export share of employment in 1999.

which the service economy is developing in Pacific Rim cities. This ideal is, of course, thwarted by the differences in accounting systems used by particular countries. However, future research should raise the question of where the demand is located that has fuelled the growth of the service sector in these cities. My analysis suggests that it has come about through a combination of distant clients buying services from specialized providers located in this community, and through local multiplier effects that have supported the growth of the local service sector (in large measure due to impacts of household consumption expenditures). I would hope that the chapters in this volume will stimulate more interest in researching not just the growth of services as measured by statistics like employment, but would focus on the bases of demand for these services. Having a stronger understanding of why these services are purchased, and why firms, governments and households choose to consume them, is a high priority research need. It is unlikely that government statistical agencies will soon begin to churn out such service trade or market statistics. Those of us in the scholarly community need to demonstrate these structural relationships, which may then lead agencies responsible for regional accounts to present much more detailed information on markets for all services.

The issues framing the contributions to this edited volume are numerous and complex. I believe that there are important and timely opportunities for scholars in the geographical community to sharpen our focus on these questions in the near future as urbanization in the Pacific Rim plays a more powerful role in global development trends. I have recently argued that there is a long research agenda related to service industries in the context of the development of the New Economy (Beyers 2002). Key topics that I argued need to be better researched include the following:

1   trends in the location of industrial activity;
2   trends in the location of corporate control functions;
3   a stronger recognition of the role of small enterprises including proprietorships;
4   more attention paid to shifting occupational patterns;
5   a stronger understanding of the location preferences of the gamut of service industry businesses;
6   a more solid understanding of how information technologies are shaping and reshaping the way in which services are produced and delivered;
7   greater attention to the geography of trade in services;
8   giving more emphasis to the geography of consumption, particularly the consumption of services.

In this chapter I developed some estimates of the role of trade in services from the Central Puget Sound region, but the data upon which these estimates were made are old, and likely are not of very good quality in the twenty-first century. I believe that development of accounts of this type is extremely important, and those of us in the scholarly community need to convince government agencies through our case study research of the importance of measuring such trade relationships not only for foreign trade, but also for interregional trade.

I did not explore the intraregional distribution of employment in services over time in this chapter. I could have presented information on the distribution within the four county region in more spatial detail. Agencies such as PSRC track such activity in sophisticated GIS systems, but as yet their level of sectoral detail for public use is extremely aggregate. Longitudinal GIS data stretching over the time horizon covered in this chapter simply do not exist, so that it is not possible to document the intrametropolitan evolution of service industry employment, addressing questions such as those explored by Coffey and his colleagues in Montreal (Coffey *et al.* 1996).

With the rapid evolution of information technologies (IT), there has been an outpouring of work addressing the impacts of these technologies on the organization of economic activity (Cairncross 1997; OTA 1995; Wheeler *et al.* 2000; Graham and Marvin 1996; Castells 1996). How are these ongoing developments in IT impacting the geography of production in the services? How are they impacting the face-to-face nature of work that this author has argued remains extremely important; there is no evidence the scale of personal interactions with clients and suppliers in the producer services has declined (Beyers 2000, 2002). We must conduct case study research that helps us understand the dynamic aspects of these technologies as they affect the distribution of production in the services within and among cities (and also in non-metropolitan places).

I am sure that other authors contributing to this volume have identified other key research issues. We must be successful in tackling them if we are to better understand the development of services in urban places in the Asia-Pacific region.

References

Baumol, W. J. (1967) 'Macroeconomics of unbalanced growth: the anatomy of urban crisis', *The American Economic Review*, 57, 415–26.
Beyers, W. (1999) 'Trade in services: modeling the contribution the economic base of regions', paper presented at North American Regional Science Meetings, Montreal, November 1999.
Beyers, W. (2000) 'Cyberspace or human space: wither cities in the age of telecommunications?', in Wheeler, J. O., Aoyama, Y. and Warf, B. (eds), *Cities in the Telecommunications Age: the Fracturing of Geographies*. New York: Routledge, 161–80.
Beyers, W. (2002) 'Services and the new economy: elements of a research agenda', *The Journal of Economic Geography*, 2, 1–29.
Beyers, W. and Alvine, M. (1985) 'Export services in postindustrial society', *Papers of the Regional Science Association*, 57, 33–45.

Beyers, W. and Lindahl, D. (1996) 'Explaining the demand for producer services: is cost driven externalization the major factor', *Papers in Regional Science*, 75, 351–74.

Cairncross, F. (1997) *The Death of Distance: How the Communications Revolution Will Change our Lives.* Boston: Harvard Business School Press.

Castells, M. (1996) *The Information Age: Economy, Society and Culture. Volume 1, The Rise of the Network Society.* Malden, MA: Blackwell Publishers.

Chase, R. A., Bourque, P. and Conway, R. (1993) *Washington State Input-Output 1987 Study.* Olympia, WA: Office of Financial Management.

Coffey, W., Polèse, M. and Drolet, R, (1996) 'Examining the thesis of CBD decline: evidence from the Montreal metropolitan area', *Environment and Planning A*, 28, 1795–814.

Economics Statistics Administration (2000) *Digital Economy 2000.* Washington DC: Office of Policy Development, Economics and Statistics Administration, US Department of Commerce.

GMA Research Corporation and Beyers, W. (1999) *An Economic Impact Study of Arts and Cultural Organizations in King County: 1997.* Bellevue, WA: GMA Research Corporation.

Graham, S. and Marvin, S. (1996) *Telecommunications and the City: Electronic Spaces, Urban Places.* London: Routledge.

Harrison, B. (1994) *Lean and Mean: The Changing Landscape of Corporate Power in the Age of Flexibility.* New York: Basic Books.

Hutton, T. and Ley, D. (1987) 'Location, linkages, and labor: the downtown complex of corporate activities in a medium size city, Vancouver, British Columbia', *Economic Geography*, 63, 126–41.

Office of Technology Assessment (1995) *The Technological Reshaping of Metropolitan America.* Washington, DC: US Congress, Office of Technology Assessment.

Wheeler, J. O., Aoyama, Y. and Warf, B. (eds) (2000) *Cities in the Telecommunications Age: The Fracturing of Geographies.* New York: Routledge.

14 Service industries and transformation of city-regions in globalizing China

New testing ground for theoretical reconstruction

George C. S. Lin

Urban transformation in the Asia-Pacific in the era of globalization

Despite its immense scope and diverse nature of interpretation, the extant literature of Asian urbanization has identified four common trends of urban change that characterized most, if not all, of the countries in the region undergoing globalization (McGee, 1998; Ginsburg, 1998; Yeung, 2002). First, the Asian continent has demonstrated a remarkable trend of accelerated urbanization with its urban population increasing from 235 million to 1,197 million and the level of urbanization rising from 16.8 per cent to 34.6 per cent in the period of 1950–1995 (Yeung, 1998). The continent stands out as the most populous region on earth with the largest and fastest growing urban population. Second, Asian economies have, since the 1970s, been undergoing profound structural changes in response to the intrusion of global market forces (Dixon and Drakakis-Smith, 1993; Ho, 1997; Kelly *et al.*, 2001; Friedmann, 1997). This is evident in the influx of foreign investment, the dramatic expansion of exports and, most noticeably, industrialization of nations in East and Southeast Asia. Third, new spatial forms of development are quickly taking shape in which large 'mega-urban regions', 'extended metropolitan regions', 'growth triangles', and 'transnational development corridors' dominate the economic landscapes in the region (McGee, 1991; Ginsburg, 1990; Lo and Yeung, 1996; McGee and Robinson, 1995; Forbes, 1999; Rimmer, 2002; Yeung, 2002). Finally, cities in most of the countries in Asia have been subject to profound, and at times painful, social changes characterized by the emergence of a new middle class, persistent poverty, excessively centralized governance, income inequality and social exclusion, all of which undermine the liveability and sustainability of these cities (Friedmann, 1997; Douglass, 2002; McGee, 2002; Yeung, 2002; Ho, 2002).

The rich and diverse literature on urban change in the Asia-Pacific has correctly identified new urban forms taking shape in the region and explained these forms with reference to both Asia's material conditions and discursive practices. In the theatre of global capital accumulation or the orbit of the new international division of labour, the Asia-Pacific was understood as a region that functioned as a new 'workshop to the world' specializing in manufacturing (Frobel *et al.*, 1980; Massey, 1984; Armstrong and McGee, 1985). Surprisingly, nowhere in this large body of literature can we find any serious and systematic articulation of the growth dynamics of the new services that are often located in large cities and the impacts that service industries have had on the transformation of megacities or mega-urban regions in the continent. The consequences of services growth did not seemingly garner sufficient consideration in major theoretical formulation that dealt with the up and down of the 'Asian miracles' or the rise and fall of the 'Asia-Pacific Century' (Linder, 1986; Borthwick, 1992; Palat, 1993; Dixon and Drakakis-Smith, 1993; Krugman, 1994; Lingle, 1997).

Given the fact that cities in the Asian-Pacific have now served not simply as production outlets of the manufacturing relocated from the advanced economies of the west but also as centres of services and nodes of command/control functions for multinational corporations in their global networks of decision-making and dissemination, what is the new role played by service industries in the new urban economy of the region? What is the nature of the growth of services in these cities? Many of the western advanced capitalist economies have demonstrated a progression of economic restructuring moving away from a pre-industrial society to an industrial and then a post-industrial society during which the emphasis of the economy shifted step by step from primary to secondary and then tertiary and quaternary sectors (Fisher, 1939; Clark, 1940; Bell, 1973; Daniels, 1982; Allen, 1988). Is there a similar pattern of linear progression that can be identified from the Asia-Pacific? What are the driving forces that operate behind the process of 'tertiarization'? To what extent and in what manner has the growth of services facilitated the transformation of the urban landscapes?

This chapter examines the growth of services and its impacts on metropolitan transformation in southern China which is currently one of the most rapidly expanding and 'globalizing' economic regions in the Asia-Pacific (Lin, 1997; Cartier, 2001; Hutton, 2001). I argue that this industrial-deterministic discourse of Asian urbanization may no longer be adequate to enlighten the complex patterns and processes of urban transformation, particularly metropolitan development in places such as southern China where multi-scalar forces have recently interacted to reshape a city-region toward a service economy. In the case of China, the growth of services, or, what the Chinese refer to as 'the tertiary industry' (*disan changye*), has been one of the key driving forces operating behind the dramatic expansion and transformation of large city-regions in the recent decade. The existing theory of Asian urbanization needs to be substantially updated by taking seriously on board the growth and uneven development of service activities as a new force facilitating urban change in the Asia-Pacific in the current era of globalization. Furthermore, the peculiar nature of services and the distinct process of tertiarization unfolding in the Chinese context stood in stark contrast with those found in the advanced capitalist economies. The new dynamics of tertiarization and urban change taking place in Chinese city-regions therefore represent another fertile testing ground for critical and innovative theoretical reconstruction.

Guangzhou's transformation under socialism

Situated at the confluence of the three major tributaries (west, east and north branches) of the Pearl River, Guangzhou has long been recognized as the southern gateway to China. It was one of the earliest established economic centres on the China coast ever since the 'City of Panyu' (*Panyucheng*) was built by the Qin Empire in 214BC (Xu, 1985: 169). Compared with other major cities in the country, Guangzhou is distinguished by its outward orientation and a tradition in international trade. The city is located in the southern frontier greatly distanced from the political and economic heartland in Central and North China. However, as far as ocean transportation and trade links with the European continent are concerned, Guangzhou enjoys an accessibility much more favourable than any other coastal cities, including Shanghai and Tianjin. For this and other reasons, Guangzhou was the first port city reached by the Europeans (Portuguese) in 1516 and by the Americans in 1784. When the entire China coast was closed in 1757 for defence reasons, Guangzhou was the only port city that remained open for trading with the outside world (Lin, 2002: 67). When the Treaty of Nanking was signed to end the Opium War in 1842, Guangzhou was among the first five 'Treaty Ports', along with Shanghai, Ningbo, Xiamen and Fuzhou, conceded to Britain in addition to the concession of Hong Kong. With this historical background and geographic foundation, it was not surprising to see the economy of the city highly reliant upon commerce, trade, businesses and services when the Communists took over Guangzhou on October 14 1949. The tertiary sector provided jobs for 44 per cent of the labour force whereas the primary sector accommodated 36 per cent and the secondary sector only 20 per cent (Lo, 1994: 134). In a manner similar to many other cities in southern China and Southeast Asia, Guangzhou was originally developed as a mercantile city at the outset of the socialist revolution.

The mercantile nature of the city of Guangzhou had naturally become the subject of constant assault and transformation in the era of socialism under Mao. For ideological and strategic reasons, the dominance of trading and service activities in the urban economy was unacceptable to the new regime. Ideologically, trading and service activities were considered to be 'non-productive' and exploitative in the nature that should be restrained. Strategically, industrial production, especially the production of basic industry or capital goods such as iron and steel, was identified as the key to building a great socialist enterprise, strengthening national military might and 'catching up the US and overtaking the UK' (*chaoying ganmei*). Great efforts had therefore been taken by the municipal government to transform Guangzhou from a 'city of consumption' into 'city of production'. Two major efforts had been influential enough to make their mark on the space economy of the city.

First, total fixed asset capital investment in the urban economy of Guangzhou was redistributed in such a way that the industrial sector was emphasized as the target for concentrated investment at the expense of services. When the First Five Year Plan (1953–1957) was drafted in 1952, the tertiary sector took the lion's share (74 per cent) of total fixed asset capital investment in the city (Guangzhou Statistical Bureau, 2001: 116). The secondary sector, which is mainly manufacturing, received only 25 per cent of the total capital investment. This natural and original pattern was then dramatically restructured in the following ten years in line with the socialist campaign to transform the city from consumption into production. By the year 1962, investment in the tertiary sector had substantially dropped from 74 per cent to only 33 per cent. In stark contrast, the secondary sector had emerged as the main recipient of capital investment increasing its share dramatically from 25 per cent to 62 per cent in ten years. Obviously, the new socialist regime was determined to use capital investment as an invisible yet powerful hand to facilitate the socialist transformation of the urban economy.

Second, urban planning was used as a means to materialize and 'ground' the ideological mandates of socialist urban transformation on earth. The first comprehensive master plan for the city was formulated in 1954, one year after the First Five Year Plan (1953–1957) was put in place. The goal of the city plan was to transform Guangzhou from 'a city of consumption' into 'a city of production' (Xu and Ng, 1998: 44). This comprehensive master plan as a blueprint for spatial arrangements of urban economic activities was then revised ten times during the period of 1954–1959 in response to changing political and economic circumstances domestically and internationally. Although the scale of urban planning and the approaches adopted for zoning and land use arrangements varied from time to time, a close comparison of the different revisions of the city plan identified the imperative of industrial production under socialism as a mandate common to all versions of the comprehensive master plan (Lo, 1994: 131; Gaubatz, 1999: 1512; Xu and Ng, 1998: 44).

To what extent have the efforts highlighted above helped transform the urban economy of Guangzhou from consumption into production? Figure 14.1 shows the temporal pattern of change in employment among the three economic sectors of the city. At the eve of the First Five Year

Printed by [...] Pearson [...] SEE PROFILE [...]
01/11/2024
01782629-0018

Plan in 1952, the tertiary sector functioned as the main source of employment accepting 41 per cent of the urban labour force. This was followed by the primary (32 per cent) and secondary (27 per cent) sectors. By the year 1978, both primary and tertiary sectors had dropped proportionally in the urban economy giving way to the secondary sector whose share expanded from 27 to 45 per cent. In other words, the secondary sector, originally behind both the tertiary and the primary sectors in terms of its share of the total employment of the city, now emerged as the most important source of employment leading the way in urban economic development. Such a shift in emphasis of the labour force suggested that the urban economy had responded to the socialist 'consumption to production' transformation campaign to the extent that the secondary sector had replaced services and jumped from the third to the first position. With a larger share of capital and labour devoted to industrial production, industrial output grew by 31 times from 298 million yuan in 1949 to 9.8 billion yuan in 1978 in the 1990 constant price (Guangzhou Statistical Bureau, 2001: 289).

Post-socialist changes

The death of Mao in 1976 and the subsequent demise of the radical leadership set Guangzhou on a new path of economic restructuring and urban development. The scalar reshuffling of power from the central to local government was significant enough to allow the municipal government of Guangzhou to proactively develop a market economy and interact

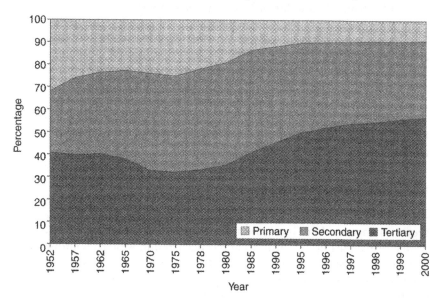

Figure 14.1 Employment structure in Guangzhou Shiqu, 1952–2000.

with forces of globalization on the basis of its inherent comparative advantages. Instead of restraining 'non-productive' consumption activities, the new policy emphasis was to satisfy the actual needs of the general population and raise the standard of living to a 'moderately well-off' level (*xiao kang shui ping*). Within this new framework, a coastal city like Guangzhou has now been considered to be the spearhead for the socialist nation to re-integrate itself with the global capitalist economy through the attraction of foreign investment and promotion of exports (Yeung and Hu, 1992; Fan, 1997). Subsequent to the establishment of four Special Economic Zones in southern China and the allowance of Guangdong and Fujian Provinces to practice 'special policies' (*teshu zhengce*) in 1979, Guangzhou, capital city of Guangdong Province, was designated as one of the 14 'open coastal cities' in 1984. With this new atmosphere favouring marketization and globalization, the 'seeds' of mercantilism deeply sown in Guangzhou prior to the Communist revolution, but forcefully suppressed by Mao's regime for nearly three decades, could now germinate, bloom and bear impressive fruits.

As shown in Figure 14.1, the urban economy of Guangzhou at the outset of the reforms in 1978 was overshadowed by the legacy of the earlier practice of transforming the city from consumption into production. The secondary sector took the lion's share of employment providing jobs to 46 per cent of the urban labour force. Since then, the share held by the secondary sector in the total employment of the city has steadily declined from 46 to 34 per cent, whereas the balance has been taken up by the tertiary sector whose share enjoyed a dramatic growth from 35 to 57 per cent during 1980 to 2000. Not only has the tertiary sector resumed its dominant position in the urban economy as it used to be prior to the Communist revolution, it has led the other two sectors by a margin even greater than what it was in 1949. If the occupational structure of the urban labour force is taken as an important indicator of sectoral transition, it is quite obvious that the urban economy of Guangzhou has undergone a significant process of tertiarization. The arbitrary distortion of the Maoist regime to the natural structure of Guangzhou's urban economy has eventually given way to the operation of forces of marketization and globalization.

Nature of services development

A proportional increase of the tertiary sector in the urban labour force seems to suggest that Guangzhou has been undergoing a process of economic restructuring similar to what has occurred in western countries, and some of the post-socialist economies. However, a close scrutiny reveals that the process of economic restructuring in the case of Guangzhou has been distinct in at least three respects. First, the tertiary sector as a new occupation of the urban labour force still lies in the low-end of the service spectrum. Table 14.1 decomposes Guangzhou's urban economy in terms of output (GDP) and employment. Among all sub-categories of the tertiary sector in the Chinese definition, the one that stands out in terms of the share in both output and employment is the group that includes wholesale, retail and catering services. The group of transportation, telecommunication and storage holds a significant share (15 per cent) in output, but it is less significant in employment (5.5 per cent). Advanced services such as finance and insurance, scientific research and polytechnic services, education, culture and entertainment all contributed a very small share of the total employment of the city. This finding is consistent with that of an earlier study conducted by Lo who showed that the commerce group in Guangzhou was one with the strongest expansion during 1978–1988 whereas the group of finance and insurance was the least significant one as far as growth in employment is concerned (Lo, 1994: 136). This pattern

Table 14.1 Urban economy of Guangzhou Shiqu: output and employment, 2000

	GDP		Employment	
	Billion Yuan	*%*	*No. of people*	*%*
Primary industry	6.55	3.03	567,841	13.41
Secondary industry	90.34	41.73	1,762,633	41.64
Manufacturing	76.73		1,495,246	35.32
Construction	13.61		267,387	6.32
Tertiary industry	119.61	55.25	1,902,966	44.95
Services for agriculture	0.17	0.08	1,313	0.03
Geological survey and water conservancy	0.22	0.10	5,701	0.13
Transportation and storage, post and telecom	33.11	15.29	233,857	5.52
Wholesale, retail and catering	23.03	10.63	723,256	17.08
Finance and insurance	15.64	7.23	50,621	1.20
Real estate	8.08	3.73	28,273	0.67
Social services	20.87	9.64	271,750	6.42
Health care, sports and social welfare	3.92	1.81	74,426	1.76
Education, culture and arts, radio, film, TV	6.06	2.80	156,309	3.69
Scientific research and polytechnic services	2.27	1.05	37,189	0.88
Government agencies/ social organizations	5.62	2.59	106,951	2.53
Others	0.63	0.29	213,320	5.04
Total	216.51	100.00	4,233,440	100.00

Source: Guangzhou Statistical Bureau (2001: 29 and 72).

stood in stark contrast with those found from the global cities or world cities in the west where advanced services have enjoyed rapid expansion as a result of the growing centralization of command and control functions in the globalization era (Sassen, 1991, 1994; Taylor *et al.*, 2002).

Second, the sectoral transition of employment did not take place in a linear, consecutive and progressive manner as in the west where the emphasis of the economy shifted one by one from the primary to the secondary, and then to the tertiary sector. A careful analysis of the temporal pattern of sectoral restructuring in the case of Guangzhou since the reforms in 1978 reveals that the decline of the primary sector in the urban labour force was accompanied by a simultaneous expansion of both the secondary and tertiary sectors throughout the 1980s (Figure 14.1). Even after 1990 when the tertiary sector surpassed the secondary sector in its contribution to the total employment of the city, the secondary sector remained an important source of employment and output generation for the urban economy. This pattern suggests that, for the initial ten years of 1980–1990 at least, the process of tertiarization did not take place after industrialization. Instead, industrialization and tertiarization occurred side by side in the case of Guangzhou. This peculiar pattern did not neatly fit the conventional wisdom of post-industrialism in which a society is believed to normally go through a linear progression from pre-industrial to industrial and then post-industrial in nature (Fisher, 1939; Clark, 1940; Bell, 1973; Daniels, 1982; Allen, 1988; Marshall and Wood, 1995; Tickell, 1999, 2001).

Finally and most significantly, the process of tertiarization in the Chinese context has been shaped and manipulated by the state for political and social considerations. Although the share of the tertiary sector in the urban labour force has grown steadily since the 1978 economic reforms, it was not until the mid-1990s that the tertiary sector surpassed the other two economic sectors and became the most important source of employment for the city (see Figure 14.1). The timing of this trend is significant. The early 1990s was a time when China started to deepen reforms of the national economy particularly the money-losing State Owned-Enterprises (SOEs). With a keen understanding that reforms or privatization of the SOEs might result in massive lay-offs and rising unemployment which could threaten social stability, the post-reform regime had to find out an outlet alternative to SOEs so that the industrial exodus could be accommodated. On June 16 1992, the Central Committee of the Chinese Communist Party and the State Council jointly made 'The Resolution to Speed Up the Development of The Tertiary Industry' and disseminated it to the entire nation. Among other things, the Resolution spelt out the definition of the tertiary industry, highlighted the 'strategic importance' of the tertiary industry as 'the main outlet to relieve the growing employment pressure faced by the nation', and listed a series of approaches that should be adopted for faster development of the tertiary industry. The growth of the tertiary industry has since then become not simply a market-driven economic affair but one on the political agenda with important implications for urban manageability and social stability. If municipal fixed asset capital investment and city planning were previously used to transform the city of Guangzhou from consumption into production in Mao's era, then they are now to be used as the instruments to promote the tertiary industry.

The impact of the new emphasis is evident when the changing composition of Guangzhou's fixed asset capital investment is analysed. The proportion of fixed asset capital investment allocated to the tertiary sector, after a slight drop from 67 per cent in 1985 to 65 per cent in 1990, increased dramatically to 76 per cent in 1995 and has kept its dominance ever since. By the year 2000, the tertiary industry had taken 85 per cent of the total fixed asset capital investment in the city. Of this investment, the bulk was found in real estate development (39 per cent), social services (19 per cent), and wholesale, retail and catering activities (Guangzhou Statistical Bureau, 2001: 125). In a similar manner, the tertiary industry occupied a prime position in city planning and land use arrangement in the most recent comprehensive plan for the terms of 1991–2000 formulated in 1994.

Foreign investment has been one of the new forces contributing to the dramatic expansion of service industries in the city. Table 14.2 lists the changing sectoral distribution of foreign direct investment in Guangzhou Municipality in the 1990s. Whereas the manufacturing sector was originally the main recipient of the bulk of foreign direct investment, its share dropped dramatically from 83 per cent to 43 per cent between 1990 and 2000. By contrast, the tertiary industry enjoyed a great proportional expansion from only 17 per cent of the total FDI to 54 per cent during the same period. In a manner similar to the allocation of domestic fixed asset capital investment, foreign direct investment appeared to be in favour of real estate development whose share rose from around 4 per cent to 41 per cent of the total FDI (Table 14.2). Although FDI accounted for only a quarter of all capital investment in the Guangzhou Municipality, it has become one of the new and powerful forces facilitating the growth of new services in the Chinese context.

Services contribution in the emerging space economy

If urban transformation is understood as a process that involves not simply internal structural change of the city but also re-positioning of the city in a broader regional or even global economy within which the city takes a part, then an evaluation of the impact of the growth of services on urban transformation has to go beyond the conventional approach that treats the city as an isolated entity and examine first the changing role played by the city in a regional context. Table 14.3 analyses the changing position of

Table 14.2 Foreign direct investment in Guangzhou municipality (*Shi*) by sector, 1990–2000 (in US dollars)

	1990		1995		2000	
	Million $	*%*	*Million $*	*%*	*Million $*	*%*
Primary industry	**0.38**	**0.14**	**6.79**	**0.30**	**22.19**	**0.71**
Secondary industry	**226.23**	**82.98**	**1,464.13**	**64.99**	**1,422.22**	**45.65**
Mining					0.14	
Manufacturing	225.89	82.86	1,265.37	56.16	1,344.34	43.15
Electricity, gas and water supply					19.32	0.62
Construction	0.34	0.12	198.76	8.82	58.42	1.88
Tertiary industry	**46.02**	**16.88**	**782.06**	**34.71**	**1,671.00**	**53.64**
Geological survey and water conservancy					0.56	0.02
Transportation and storage, post and telecom	11.31	4.15	22.53	1.00	20.37	0.65
Wholesale, retail and catering	4.70	1.72	49.12	2.18	47.59	1.53
Real estate	12.28	4.50	660.09	29.30	1,287.51	41.33
Social services					194.57	6.25
Health care, sports and social welfare			31.54	1.40	23.45	0.75
Education, culture and arts, radio, film, TV	1.61	0.59	11.67	0.52	0.08	
Scientific research and polytechnic services			1.88	0.08	2.43	0.08
Others	16.12	5.91	5.23	0.23	94.44	3.03
Total	**272.63**	**100.00**	**2,252.98**	**100.00**	**3,115.41**	**100.00**

Source: Guangzhou Statistical Bureau (2001: 471).

Guangzhou city in the Pearl River Delta regional economy after 1992 when the importance of the tertiary sector was officially recognized and promotion of services activities was initiated. It has been extensively documented that recent economic development in the Pearl River Delta since the reforms has been characterized by rapid rural industrialization, dramatic upsurge of many suburban economies, and a relative decline of the leading position held by Guangzhou in the region (Xu and Li, 1990; Sit and Yang, 1997; Lin, 1997). This observed pattern is confirmed by the data on population and GDP listed in Table 14.3. As many suburban economies grew vigorously and upgraded themselves in the region, there

Table 14.3 Guangzhou Shiqu in the Pearl River Delta region, 1992–2000

	Unit	1992	1994	1996	1998	2000
Population						
Delta	million	20.08	20.95	21.70	22.38	23.07
Guangzhou	million	3.67	3.80	3.90	3.99	4.14
Guangzhou as % of Delta	%	18.28	18.14	17.97	17.83	17.95
GDP						
Delta	billion yuan	115.93	298.36	453.39	583.31	737.86
Guangzhou	billion yuan	39.60	70.71	102.91	129.83	171.51
Guangzhou as % of Delta	%	34.16	23.70	22.70	22.26	23.24
Retail sales						
Delta	billion yuan	67.08	120.04	171.45	227.18	278.14
Guangzhou	billion yuan	16.72	35.83	56.18	75.53	93.05
Guangzhou as % of Delta	%	24.93	29.85	32.77	33.25	33.45
Utilized foreign investment						
Delta	billion USD	3.22	8.30	9.85	11.81	12.54
Guangzhou	billion USD	0.47	1.21	1.68	2.35	2.47
Guangzhou as % of Delta	%	14.60	14.58	17.06	19.90	19.70

Sources: Guangdong Statistical Bureau (2001: 626); Guangzhou Statistical Bureau (1999: 208, 230, 567 and 613); Guangzhou Statistical Bureau (2001: 29, 38, 39, 54, 401 and 470).

has been a proportional decline in terms of both population and GDP contributed by Guangzhou city. While the share of Guangzhou in the total population of the region dropped only by a small margin (from 18 to 17.9 per cent), the proportional decline of the GDP generated by Guangzhou city has been significant, a drop from 34 per cent to 23 per cent in less than ten years (Table 14.3).

Guangzhou has, however, taken on new hub functions. Over the nine years from 1992 to 2000, Guangzhou's share of the total retail sales of the delta region increased from 25 to 33 per cent. In the Chinese statistics, the output of retail sales has been the main indicator of the magnitude of trading or commercial exchange in a city or town. The share of a city in the total retail sales of a region represents to a great extent the role played by the city in trading or commercial exchange of the region. The increased contribution of Guangzhou city to the total regional retail sales suggested that the city has played an increasingly important role as the regional centre of trading and commercial exchange. Apparently, Guangzhou's upgraded position in regional trade has been one of the direct consequences of the dramatic expansion of service industries.

Another significant feature is the role played by Guangzhou city as a major recipient of foreign investment. As shown in Table 14.3, utilized foreign investment in Guangzhou city has been growing steadily in both absolute amount and its proportion of the delta region. Guangzhou's share of the total utilized foreign capital invested in the delta region expanded from 15 per cent in 1992 to nearly 20 per cent in the year 2000.

This trend has been particularly evident toward the end of the 1990s when China's accession into the World Trade Organization eventually materialized. The increase in Guangzhou's share of the total utilized foreign investment in the delta region in recent years suggested that the city has reclaimed its position as the regional centre of transnational capital and multinational corporations subsequent to both China's accession into WTO and improvements of services in the city which has made it more attractive to foreign investment. Looking at all the data listed in Table 14.3, Guangzhou city has recently resumed its functions as a regional service centre for both domestic traders and foreign investors despite its relative decline in terms of its role as a centre of industrial production.

The growth of services has also found its way to reshape urban land use and transform the urban landscape. The earlier practice of industrial-biased urban development under socialism in Mao's era had left Guangzhou city with a land use structure preoccupied by industrial production. Even in the early 1990s after a decade of market reforms, the land occupied by industrial production still represented the most important type of urban land use accounting for 27 per cent of all urban construction land. This pattern of urban land use dominated by industrial production has, since 1992, experienced profound restructuring as a result of the phenomenal development of services within the city proper. There has been a significant restructuring of urban land use characterized by a proportional decline of industrial land use and expansion of the land used for urban services and residential purposes. Whereas industrial land dropped its share of the total urban construction land from 27 to 21 per cent, the land used for urban services, recreation and public facilities expanded from 27 to 33 per cent of the urban construction land during 1992–2000. By the year 2000, urban services and facilities had exceeded industrial land use and become the most important category among all types of urban construction land.

As the land in the inner city has been converted into commercial and recreational purposes, industrial facilities have been relocated to the suburban areas, particularly in the newly annexed suburban districts of Huadu and Panyu. This process of spatial restructuring has given rise to a phenomenal urban sprawl along the Pearl River and major transportation arteries. Within the broader Guangzhou Municipality including Huadu and Panyu, non-agricultural land doubled its size, expanding from 35,000 hectares in 1988 to 70,000 hectares in 2000, whereas agricultural land shrank by nearly 47,000 hectares. Within the city, available data show that the rental of office floor space in the inner city grew from 118 thousand square metres in 1998 to 469 thousand square metres in the year 2000. The dramatic expansion of the rental of office floor space during 1999–2000 has been clearly a response to the accession of China into the World Trade Organization which has attracted many multinational corporations to set up regional headquarters in selected Chinese cities. This new growth of the rental of office floor space has favoured the Tianhe District which is a newly developed area with greater open space and transport accessibility. This district has grown by 200–238 per cent between 1999 and 2000. By comparison, the Baiyun District in the outer ring of the city has suffered from a decline in rental office space. This pattern suggested that the location of service activities has been an uneven phenomenon even within the urban district and that there has been a selective concentration of the office functions in the inner city, particularly in those areas with new and convenient urban infrastructure. The growth of services has therefore become one of the new driving forces along with industrialization that have shaped the re-formation of both urban land use and the new urban economic landscape.

Conclusion and discussion

In the dominant and pervasive discourse of globalization, the Asia-Pacific is understood as a region comprising some Newly Industrializing Economies and functioning as 'a workshop to the world' in the orbit of the new international division of labour. If there is anything that distinguished the Asia-Pacific, it is the capacity of industrial production relocated from the Atlantic core to 'the Far East' that has provided the necessary energy and power for the engine of Asian economic dynamism. Services were considered to be nothing more than a residual or by-product of the 'global shift' of manufacturing establishments fabricating the engine of Asian industrialization. With a few exceptions, the significance of the growth of services and its relationship with urban transformation have never been seriously incorporated into any theoretical formulation concerning globalization and urban change in the rapidly changing Asia-Pacific. The recent 'wake-up calls' to take services seriously should affect not only the mainstream enquiry about globalization and urban change but also the area studies specialists who are concerned with Asian urban transformation. Although important initiatives have been made to examine the growth dynamics of service industries in the Asia-Pacific, the sophisticated relationship between the growth of services and urban change in this vast and diverse region remains poorly understood. Given the complex nature of the subject and the region, any serious scholarly attempt to theorize the growth of services and urban change in the Asia-Pacific cannot skip the necessary steps of detailed investigations, comparisons, and eventually logical integration of the patterns and processes taking place in various sub-regional contexts.

This study examines the growth of services and urban transformation in one of the largest and rapidly globalizing city-regions in southern China. A historicized evaluation of the growth of the Guangzhou metropolis over the past century revealed a process of transformation from a traditional mercantile city to a city of industrial production, a path typical of what was experienced by other Chinese cities in Mao's era. Recent intrusion of the new forces of marketization and globalization has brought about a flourishing of services after decades of socialist suppression and led the metropolis to resume many of its lost urban functions in consumption and services. A distinct globalizing city-region that blends the legacy of a socialist Third World city with the new forces of marketization and globalization appears to be taking shape to characterize Guangzhou on its way of transformation toward a service economy.

In contrast with the well-documented process of tertiarization in western advanced economies, the growth of services in the Chinese situation is distinguished by its low-end or low value-added nature. Most of the services found in Guangzhou are in the popular category of wholesaling, retailing and catering. Advanced services such as finance and insurance, research and development, cultural and education have remained rather limited. The sectoral transition of Guangzhou's urban economy did not occur in a linear and progressive manner as what has been described by the thesis of post-industrialism for advanced economies in the west. Economic tertiarization in the Chinese case has taken place alongside, rather than after, the continuing process of industrialization. Furthermore, the growth of services in the Chinese context has been subject to strong state manipulation for political and social considerations. The tertiary sector has been identified by the post-reform regime as a potential outlet to accommodate the massive lay-offs from the reformed or dismantled state-owned enterprises. The dramatic expansion of services in Guangzhou in recent years has enabled the metropolis to re-position itself as a command and service centre in the rapidly industrializing and urbanizing Pearl River Delta region. Tertiarization of the urban economy has functioned as a new and powerful force to re-organize urban land use and transform the urban economic landscape.

The case of Guangzhou has raised several important theoretical questions concerning the changing dynamics of urban transformation in what Forbes and Thrift (1987) refer to as the 'socialist Third World' which has been one of the key constituent elements of the Asia-Pacific 'jig-saw puzzle'. Until recently, theoretical explanation for the peculiar pattern of urban development under socialism has been based primarily on the assumption that the imperative of rapid industrialization plays a decisive role in the formulation of the socialist urban development strategy and that industrial production functions as the most powerful force to shape the transformation of cities in the Socialist Third World (Kirkby, 1985; Forbes and Thrift, 1987; Pannell, 1990; Chan, 1992; Lin, 1994, 1998; Szelenyi, 1996; Ma, 2002). This industrial-deterministic paradigm of urban development has shed much light on the past experience of under-urbanization in many countries of the socialist Third World. It has become increasingly inadequate, however, to illuminate the new dynamics of economic and spatial transformation quickly unfolding in the cities after socialism. Industrialization remains a leading force in China's continuing urban transformation, but given the dramatic tertiarization of the urban economy highlighted in this study of Guangzhou and in other recent studies of cases elsewhere (Olds, 2001; Logan, 2002; Gong, 2002), there is a need to incorporate the growth of services and formulate a broader service-informed discourse to account for the new and divergent trajectories of urban transformation in countries of the socialist Third World undergoing globalization.

The theory of post-industrialism developed on the basis of western experience has described a linear and progressive pattern of sectoral transition in which the emphasis of the economy shifted step by step from primary to secondary and then tertiary sector and the society from pre-industrial to industrial and then post-industrial stage. This linear progression has been explained as the result of both technological advancements which facilitated cross-sectoral labour mobility and the gradual or incremental increase in personal income which created higher market demands for services. This case study of tertiarization in a major Chinese city-region suggested that the western model of post-industrialism might not hold for what has been taking place in the context of a socialist developing country in the Asia-Pacific. As the experience of Guangzhou has shown, the process of tertiarization actually took place side by side with industrialization. The dramatic expansion of services in the Chinese city-region has been driven not so much by rising personal income but more by state manipulation for non-economic considerations such as social stability. The inconsistency existing between the conventional wisdom of post-industrialism and the reality in one of the largest socialist developing countries in the Asia-Pacific testifies to the limitation of western-based theories and underscores the plea for a more inclusive and contextually sensitive theorization in the current era of intellectual globalization (Lin and Wei, 2002; Ma, 2002; Yeung and Lin, 2003).

References

Allen, J. (1988) 'Service industries: uneven development and uneven knowledge', *Area*, 20, 15–22.

Armstrong, W. and McGee, T. G. (1985) *Theatres of Accumulation: Studies in Asian and Latin American Urbanization*. New York: Methuen.

Bell, D. (1973) *The Coming of Post-Industrial Society*. London: Heinemann.

Borthwick, M. (1992) *Pacific Century: The Emergence of Modern Pacific Asia*. Boulder: Westview.

Cartier, C. (2001) *Globalizing South China*. Oxford: Blackwell.

Chan, K. W. (1992) 'Economic growth strategy and urbanization policies in China, 1949–1982', *International Journal of Urban and Regional Research*, 16(2), 275–306.

Clark, C. A. (1940) *The Conditions of Economic Progress*. London: Macmillan.

Daniels, P. W. (1982) *Service Industries: Growth and Location*. Cambridge: Cambridge University Press.

Douglass, M. (2002) 'Globalization, intercity competition and the rise of civil society: towards livable cities in Pacific Asia', *Asian Journal of Social Science*, 30(1), 129–49.

Dixon, C. and Drakakis-Smith, D. (eds) (1993) *Economic and Social Development in Pacific Asia*. New York: Routledge.

Edgington, D. W. and Hayter, R. (2000) 'Foreign direct investment and the flying geese model: Japanese electronics firms in the Asia Pacific', *Environment and Planning A*, 32, 281–304.

Fan, C. C. (1997) 'Uneven development and beyond: regional development theory in post-Mao China,' *International Journal of Urban and Regional Research*, 21(4), 620–39.

Fisher, A. G. B. (1939) 'Production, primary, secondary, and tertiary', *Economic Record*, 15, 24–38.

Forbes, D. (1999) 'Globalisation, post-colonialism and new representations of the Pacific Asian metropolis', in Olds, K., Dicken, P., Kelly, P. F., Kong, L. and Yeung, H. W. C. (eds), *Globalisation and the Asia-Pacific: Contested Territories*. London: Routledge, 238–54.

Forbes, D. and Thrift, N. (eds) (1987) *The Socialist Third World: Urban Development and Territorial Planning*. New York: Blackwell.

Friedmann, J. (1997) 'Preface: urban impacts and responses to the Asian economic crisis in the Asia Pacific region', *Asian Geographer*, 19(1&2), 1–5.

Frobel, F., Heinrichs, J. and Kreye, D. (1980) *The New International Division of Labour*. Cambridge: Cambridge University Press.

Gaubatz, P. (1999) 'China's urban transformation: patterns and processes of morphological change in Beijing, Shanghai and Guangzhou', *Urban Studies*, 36(9), 1495–521.

Ginsburg, N. (1990) *The Urban Transition: Reflections on the American and Asian Experiences*. Hong Kong: The Chinese University Press.

Ginsburg, N. (1998) 'Planning the future of the Asian city: a twenty-five year retrospective', in Yeung, Y. M. (ed.), *Urban Development in Asia: Retrospect and Prospect*, Research Monograph No. 38. Hong Kong: Hong Kong Institute of Asia-Pacific Studies, The Chinese University of Hong Kong, 3–24.

Gong, H. (2002) 'Growth of tertiary sector in China's large cities', *Asian Geographer*, 21(1&2), 85–100.

Guangdong Statistical Bureau (GDSB) (2001) *Guangdong Tongji Nianjiang (2001) (Guangdong Statistical Yearbook)*. Beijing: China Statistical Press.

Guangzhou Statistical Bureau (GZSB) (1999) *Guangzhou Wushinian (1949–1999) (Guangzhou's Fifty Years)*. Beijing: China Statistical Press.

Guangzhou Statistical Bureau (GZSB) (2001) *Guangzhou Tongji Nianjiang (2001) (Guangzhou Statistical Yearbook)*. Beijing: China Statistical Press.

Ho, K. C. (1997) 'The global economy and urban society in Pacific Asia', *International Sociology*, 12(3), 275–93.

Ho, K. C. (2002) 'Globalization and Southeast Asian urban futures', *Asian Journal of Social Science*, 30(1), 1–7.

Hutton, T. A. (2001) *Service Industries and the Transformation of Asia-Pacific City Regions*, CAS Research Papers Series No. 36. Singapore: National University of Singapore.

Kelly, P. F., Olds, K. and Yeung, H. W. C. (2001) 'Introduction: geographic perspectives on the Asian economic crisis', *Geoforum*, 32, vii–xiii.

Kirkby, R. J. R. (1985) *Urbanization in China.* New York: Columbia University Press.

Krugman, P. (1994) 'The myth of Asia's miracle', *Foreign Affairs,* 73(6), 62–78.

Lin, G. C. S. (1994) 'Changing theoretical perspectives on urbanization in Asian developing countries', *Third World Planning Review,* 16(1), 1–23.

Lin, G. C. S. (1997) *Red Capitalism in South China: Growth and Development of the Pearl River Delta.* Vancouver: University of British Columbia Press.

Lin, G. C. S. (1998) 'China's industrialization with controlled urbanization: anti-urbanism or urban-biased?', *Issues & Studies,* 34(6), 98–116.

Lin, G. C. S. (2002) 'The growth and structural change of Chinese cities: a contextual and geographic analysis', *Cities,* 19(5), 299–316.

Lin, G. G. S. and Wei, Y. H. D. (2002) 'China's restless urban landscapes 1: new challenges for theoretical reconstruction', *Environment and Planning A,* 34(9), 1535–44.

Linder, S. B. (1986) *The Pacific Century: Economic and Political Consequences of Asian-Pacific Dynamism.* Stanford: Stanford University Press.

Lingle, C. (1997) *The Rise and Decline of the Asian Century.* Barcelona, Spain: Sirocco.

Lo, C. P. (1994) 'Economic reforms and socialist city structure: a case study of Guangzhou, China', *Urban Geography,* 15(2), 128–49.

Lo, F. C. and Yeung, Y. M. (eds) (1996) *Emerging World Cities in Pacific Asia.* Tokyo: United Nations University Press.

Logan, J. R. (ed.) (2002) *The New Chinese City: Globalization and Market Reform.* Oxford: Blackwell.

Ma, L. J. C. (2002) 'Urban transformation in China, 1949–2000: a review and research agenda', *Environment and Planning A,* 33(9), 1545–69.

Marshall, J. N. and Wood, P. A. (1995) *Services and Space.* London: Longman.

Massey, D. (1984) *Spatial Division of Labour: Social Structures and the Geography of Production.* London: Macmillan.

McGee, T. G. (1991) 'The emergence of desakota regions in Asia: expanding a hypothesis', in Ginsburg, N., Koppel, B. and McGee, T. G. (eds), *The Extended Metropolis.* Honolulu: University of Hawaii Press, 3–25.

McGee, T. G. (1998) 'Five decades of urbanization in Southeast Asia: a personal encounter', in Yeung, Y. M. (ed.), *Urban Development in Asia: Retrospect and Prospect,* Research Monograph No. 38. Hong Kong: Hong Kong Institute of Asia-Pacific Studies, The Chinese University of Hong Kong, 55–91.

McGee, T. G. (2002) 'Reconstructing the Southeast Asian city in an era of volatile globalization', *Asian Journal of Social Science,* 30(1), 8–27.

McGee, T. G. and Robinson, I. M. (eds) (1995) *The Mega-Urban Regions of Southeast Asia.* Vancouver: UBC Press.

Olds, K. (2001) *Globalisation and Urban Change: Capital, Culture and Pacific Rim Mega-Projects.* Oxford: Oxford University Press.

Palat, R. A. (ed.) (1993) *Pacific-Asia and the Future of the World-System.* Westport: Greenwood.

Pannell, C. W. (1990) 'China's urban geography', *Progress in Human Geography,* 14(2), 214–36.

Rimmer, P. J. (2002) 'Overview: restructuring Chinese space in the new millennium', *Asia Pacific Viewpoint,* 43(1), 1–8.

Sassen, S. (1991) *The Global City.* Princeton: Princeton University Press.

Sassen, S. (1994) *Cities in a World Economy.* Thousand Oaks: Pine Forge Press.

Sit, V. F. S. and Yang, C. (1997) 'Foreign-investment-induced exo-urbanization in the Pearl River Delta, China', *Urban Studies*, 34, 647–77.

Szelenyi, I. (1996) 'Cities under socialism-and after', in Andrusz, G., Harloe, M. and Szelenyi, I. (eds), *Cities after Socialism: Urban and Regional Change and Conflict in Post-Socialist Societies*. Cambridge: Blackwell, 286–317.

Taylor, P. J., Catalano, G. and Walker, D. R. F. (2002) 'Measurement of the world city network', *Urban Studies*, 39(13), 2367–76.

Tickell, A. (1999) 'The geographies of services: new wine in old bottles', *Progress in Human Geography*, 23(4), 633–9.

Tickell, A. (2001) 'Progress in the geography of services ii: services, the state and the rearticulation of capitalism', *Progress in Human Geography*, 25(2), 283–92.

Xu, J. and Ng, M. K. (1998) 'Socialist urban planning in transition: the case of Guangzhou, China', *Third World Planning Review*, 20(1), 35–51.

Xu, X. Q. (1985) 'Guangzhou: China's southern gateway', in Sit, V. F. S. (ed.), *Chinese Cities: The Growth of the Metropolis since 1949*. Oxford: Oxford University Press, 167–87.

Xu, X. Q. and Li, S. M. (1990) 'China's open door policy and urbanization in the Pearl River Delta region', *International Journal of Urban and Regional Research*, 14(1), 49–69.

Yeung, H. W. C. and Lin, G. C. S. (2003) 'Theorizing economic geographies of Asia', *Economic Geography*, 79(2), 107–28.

Yeung, Y. M. (ed.) (1998) *Urban Development in Asia: Retrospect and Prospect*, Research Monograph No. 38. Hong Kong: Hong Kong Institute of Asia-Pacific Studies, The Chinese University of Hong Kong.

Yeung, Y. M. (2002) 'Globalization and Southeast Asian urbanism', *Asian Geographer*, 21(1&2), 171–86.

Yeung, Y. M. and Hu, X. W. (eds) (1992) *China's Coastal Cities*. Honolulu: University of Hawaii Press.

15 IT service industries and the transformation of Seoul

Sam Ock Park and Ji Sun Choi

Introduction

Seoul has been the capital of Korea for over 600 years since the Joseon Dynasty (the former name of Korea) was founded in 1392. The population of Seoul amounted to 10,280,523 with 25 districts (so-called 'gu') at the end of 2002. It has played a pivotal role as the key metropolitan centre of Korea, driving economic, social and cultural development.

For a long time Seoul was a representative mono-centred city in which the Central Business District (CBD) was the focus of commercial, social and civic life. However, during the last two decades, Seoul has evolved into a multi-centred city with multiple nuclei performing specialized roles as commercial, social and cultural centres. The change in spatial structure of Seoul from a mono-centred city to multi-centred city is closely related to changes in its industrial structure (Park and Nahm, 1998). The industrial decentralization policies linked to the decentralization of population in Seoul in the 1970s and 1980s had a significant effect on the decline of production activities on the one hand and the growth of service industries on the other.

The agglomeration of information technology (IT) service industries in Seoul is remarkable. The concentration of knowledge intensive firms and IT service industries has been overwhelming during the last decade, especially since the financial crisis in Korea in 1997. The Gangnam area within Seoul, located south of the Han River, has become the centre of an IT cluster based on high-tech services. The clustering of high-tech services in Gangnam has resulted in the emergence of a new core or business district as the area has been transformed into a learning region. The transformation of Seoul via the clustering of high-tech services can be regarded as a new phenomenon in the newly industrialized countries of the Asia-Pacific Rim. This chapter examines the processes through which high-tech service industries have developed and a new core has been created in Seoul, using the example of IT service industries in the Gangnam area. It also analyses the characteristics of this new core, which is often evaluated as the most successful IT service cluster in Korea.

Growth of IT service industries in Seoul and the Gangnam area

In a broad sense, the IT industry is divided into two sub-groups; IT manufacturing industry and IT service industries. The latter can be divided into telecommunications, wholesale and renting, and more narrowly defined IT services (Shin *et al.*, 2001). IT services, in a narrow definition, is composed of computer system design and consultancy (KSIC = 7210), software consultancy and supply (7220), data processing and computer facilities management services (723), database activities and online information provision services (7240), and other computer activities (7290).[1] IT service industries in a narrow sense are called 'IT service (N) industries' and are distinguished from IT service industries in a broader sense, which are called just 'IT service industries' without using the '(N)' in this chapter.[2]

Compared with the country as a whole, Seoul has specialized in IT service industries, which include IT service (N), telecommunications and IT wholesale and renting. Nationally, IT service industries account for 59.4 per cent of the total IT industry by the number of firms and 34.5 per cent of the number of employees respectively. In Seoul, however, the share of the IT service industries is much higher: 77.4 per cent by the number of firms and 67.4 per cent by the number of employees (Table 15.1). Overall, the concentration of IT service firms in Seoul is remarkable. Of the different types of IT industry, IT service (N) industries are concentrated mainly in Seoul which has 66.4 per cent and 74.6 per cent of the national IT service (N) industries by the number of firms and employees respectively. The IT industry cluster of Seoul is therefore a well-defined 'service-driven IT cluster', in which IT service firms take a leading role in the development and networking of IT industry.

The concentration of IT service industries in Seoul has accelerated during the 1990s (Figure 15.1). The Location Quotient (LQ) of IT service industry was 1.34 in 1990, suggesting that it was already specialized at the beginning of 1990s. The degree of specialization and concentration increased in 1995 (LQ = 1.95) and 1999 (LQ = 2.18). In contrast, the concentration of IT manufacturing industries in Seoul gradually declined from 1991 (LQ = 0.64) to 1995 (LQ = 0.55) and 1999 (LQ = 0.55).

The analysis of the LQs for 25 districts in Seoul reveals a clear difference in the degree of the specialization of IT service industries among the districts (Figure 15.2). There is a clear trend of concentration within the city of Seoul, where Gangnam-gu and Seocho-gu are representative districts of concentration, regardless of the scope of IT service industries.[3] The very low values of the LQs related to IT manufacturing in the two districts contrast with the extreme concentration of IT service industries.

Gangnam-gu and Seocho-gu are treated as the Gangnam area in this chapter, because the latter was separated from Gangnam-gu in 1988 even though it has similar characteristics in terms of industrial and socioeconomic

Table 15.1 Composition of IT industry in Seoul (1999)

Classification		IT manufacture	IT service (B)				IT total
			IT service (N)	Telecommunication	IT wholesale and renting	Sub-total	
Whole country	F*	10,404	5,241	2,819	7,173	15,233	25,637
		(40.6%)	(20.4%)	(11.0%)	(28.0%)	(59.4%)	(100.0%)
	E**	377,001	71,559	82,704	44,089	198,352	575,353
		(65.5%)	(12.4%)	(14.4%)	(7.7%)	(34.5%)	(100.0%)
Seoul	F*	2,523	3,482	887	4,283	8,652	11,175
		(22.6%)	(31.2%)	(7.9%)	(38.3%)	(77.4%)	(100.0%)
	E**	54,475	53,409	28,637	30,735	112,781	167,256
		(32.6%)	(31.9%)	(17.1%)	(18.4%)	(67.4%)	(100.0%)

Source: Seoul metropolitan city (2000), recalculated from Shin et al. (2001: 73).

Notes
F* = based on the number of firms;
E** = based on the number of employees.

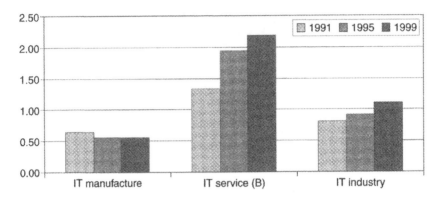

Figure 15.1 Change of LQs of IT industry based on employment in Seoul, by year.

Figure 15.2 LQs of IT industry based on employment in Seoul, by district (1999).

structure. The Gangnam area began to emerge during the 1970s and has become Seoul's new core for economic activities with numerous modern high-rise office buildings. Since the late 1980s advanced services such as software, engineering, advertising, and design services have concentrated in the Gangnam area (Park and Nahm, 1999). The concentration trend has accelerated since the financial crises in 1997 with the addition of new start-ups in high-tech and software sectors.[4]

Yeongdeungpo-gu is third in terms of the concentration of IT service industries in Seoul (see Figure 15.2). Although Yongsan-gu shows the highest value of the LQ relevant to IT service industries using the broad definition, it has a very low value for IT service (N) industries using the narrow definition. This is mainly because of the Electronic Shopping Town, located in Yongsan-gu, which is composed of innumerable small dealers and several large shopping centres specializing in computer-related equipment and IT software. However, Yongsan-gu is not considered to be a more important IT service cluster than Gangnam-gu and Seocho-gu because the higher LQ value of Yongsan-gu is related to the concentration of IT wholesale and renting services. The IT wholesale and renting industries are not regarded as technology-intensive IT service industries, which are the core of IT service industries and the focus of this study.

IT service industries in the Gangnam area have grown rapidly during the last decade (Figure 15.3). In terms of the employment, the share of Gangnam-gu relative to Seoul increased from 11.2 per cent in 1991 to 29.7 per cent in 1999. Likewise, the ratio of Seocho-gu to Seoul increased from 5.3 per cent to 16.0 per cent during the same period. In total, the share of IT service industries employment in the Gangnam area increased from 16.5 per cent in 1991 to 45.7 per cent in 1999. It was a complete contrast to the trend of decline in the traditional CBD of Seoul. The share of the traditional CBD, including Jongro-gu and Jung-gu, decreased dramatically from 26.9 per cent in 1991 to 8.4 per cent in 1999. The Yeongdeungpo-gu and Yongsan-gu areas with relatively high LQ values did not show growth in their proportions of employment during the last ten years. The share of employment in Yeongdeungpo-gu stagnated at around 12.5 per cent during the last decade. The ratio in Yongsan-gu decreased from 17.1 per cent in 1991 to 8.5 per cent in 1999.

Overall, the Gangnam area is the only place that shows rapid growth of IT service industries in Seoul in the 1990s. This has contributed to a transformation of the city's structure.

Spatial networks of strategic alliances of IT service firms in Gangnam area

The Gangnam area is now the dominant core of Seoul, replacing the traditional CBD (Jung-gu and Jongro-gu) in the central part of the city which has transformed from a city with a single core into a city with

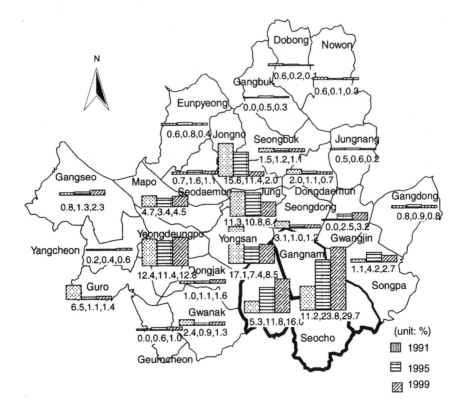

Figure 15.3 The changes of IT service industries in Seoul, by district and by year in terms of employment.

multiple cores (the traditional CBD, Yeongdeungpo-gu and Gangnam area) (Park and Nahm, 1998).

In this section, in order to understand the dominant pattern of Gangnam area, spatial networks of strategic alliances of the IT service industries have been analysed. It is clear that IT industry firms have increasingly located in Seoul since the late 1980s with the Gangnam area emerging as the core area for these IT industries since the early 1990s (Figure 15.4).

IT service industry firms that have strategic alliances with other firms are overwhelmingly concentrated in Seoul. Almost 89 per cent of the firms that have strategic alliances are located there, with conspicuous concentration of these firms in Gangnam area. Almost half of the firms that have strategic alliances are located in Gangnam area, with Yeongdeungpo-gu and the traditional CBD sharing 7 per cent and 5 per cent respectively.

The patterns of the spatial networks of the strategic alliances confirm the leading role of the Gangnam area in the IT service industry. All the

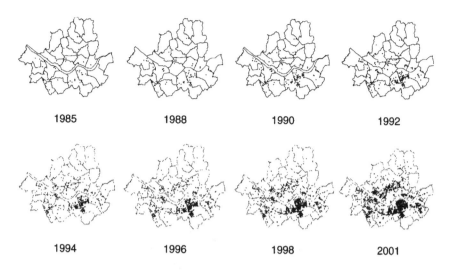

1985   1988   1990   1992

1994   1996   1998   2001

Figure 15.4 Distribution of IT industry firms in Seoul.

regions of Korea have strong networks of strategic alliances with 41 per cent of all firms that have strategic alliances having links with firms in the Gangnam area. In addition, all the districts of Seoul, except the Gangnam area and provincial regions, have more networks with Gangnam than with their own local areas.

The spatial networks of the strategic alliances of the firms in Gangnam area are shown in Figure 15.5. Among the firms that have strategic alliances with other firms in Gangnam area, 43.8 per cent are located within Gangnam area; with a further 7.3 per cent, 6.4 per cent, and 4.9 per cent of the firms in Yeongdeungpo-gu, Jung-gu, and Jongro-gu respectively. IT firms in Gangnam area have weak spatial networks with other districts (less than 3 per cent or no networks of strategic alliances). Interestingly, among the Gangnam firms that have strategic alliances, 18.4 per cent have networks of strategic alliances with foreign firms, including 8 per cent with firms in the USA and 4 per cent with firms in Japan. It is clear that within the nation, spatial networks with the traditional CBD (Jung-gu and Jongro-gu) and Yeongdeungpo-gu are comparatively strong, suggesting the networks among the three cores are important for the strategic alliances.

Among the three cores, the dominant role of the Gangnam area is remarkable. Both of the other cores, Yeongdeungpo-gu and the traditional CBD, have stronger networks with the Gangnam area than with their own local areas. For example, of the firms in Yeongdeungpo-gu that have strategic alliances with other firms, 38.5 per cent have networks that include firms in the Gangnam area, while only 13.9 per cent of firms in

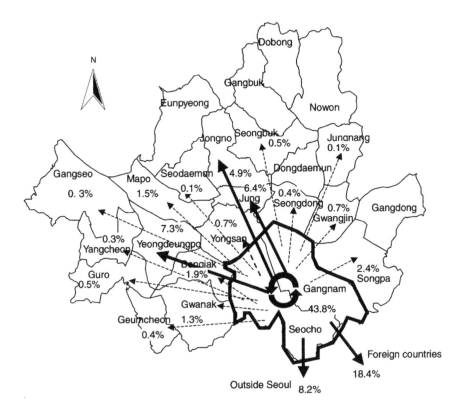

Figure 15.5 Strategic alliance of IT service firms in Gangnam area.

Yeongdeungpo-gu have networks within their own local area (Figure 15.6). About 5.7 per cent of the firms of Yeongdeungpo-gu have networks with firms in the traditional CBD. They have relatively significant networks with foreign firms, mainly with the firms in the USA (17.2 per cent). The strength of the networks with other districts in Seoul is relatively weak, as is the case in the Gangnam area.

The spatial networks of the strategic alliances of the Gangnam and the Yeongdeungpo-gu area firms suggest the following three important things. First, the Gangnam area is a dominant core for IT service industries in terms of location and networks of strategic alliances. Firms in the Gangnam area have strong local networks of strategic alliances, while firms in other areas of Seoul have strong networks with firms in the Gangnam area, much stronger than that of local networks. It is clear that the Gangnam area has emerged as a major core of IT service industries and their networking, substituting the role of the traditional CBD. Second, the spatial networks of strategic alliances between the three cores

Figure 15.6 Strategic alliance of IT service firms in Yeoungdeungpo-gu.

are stronger than those with other districts in Seoul. This suggests that the three areas are still important for high-tech services and can be regarded as multi-centres of Seoul, even though the Gangnam area substitutes the role of the traditional CBD. Third, networks with foreign firms are stronger than those with the regions elsewhere in Korea outside Seoul. The significant networks with foreign firms may represent the globalization trend of the service industries, especially in the Pacific Rim area (Daniels, 1998). The networks of strategic alliances with firms in the USA are stronger than those with Yeongdeungpo-gu. Networks with Japan and China are also considerable.

Due to the dominant role of the Gangnam area, the types of strategic alliances of the firms in the Gangnam area are further analysed. For convenience, the types of the strategic alliances are classified into those related to technology, marketing, production, capital and others. Using this classification, 82 per cent of the strategic alliances are related to technology or marketing and these were mainly formed between 2000 and 2002 (Figure

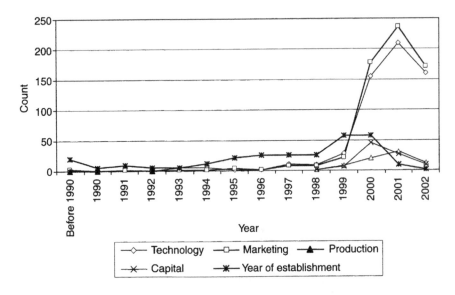

Figure 15.7 The number of strategic alliances by the beginning year of strategic alliances and the number of newly established firms by year.

15.7). The number of newly established firms in the IT service industries began to increase from 1994 and peaked in 1999 and 2000, after the financial crisis in Korea, but began to decrease after 2000. The decreasing numbers of the new firm formation in the IT sector, and the many strategic alliances developed with regard to technology and marketing in recent years suggests that IT service industries in Korea are now emphasizing cooperative networks in order to survive in a competitive world.

The Gangnam area is more important for strategic alliances based on technology and marketing rather than production and capital (Table 15.2). Interestingly, strategic alliances that incorporate foreign partners are more likely to involve production and technology. Foreign networks of the strategic alliances relating to technology are mostly linked to the advanced economies, especially the USA. On the other hand, as expected, more than half of the foreign networks with regard to production are linked to China.

Since the late 1980s advanced services such as software, engineering, advertising and design services have concentrated in the Gangnam area (Park and Nahm, 1998). This trend has accelerated since the financial crisis in 1997 with agglomeration of new start-ups in high-tech and software sectors. Many large IT firms and new high-tech start-ups have located along the Teheran Road, which crosses the Gangnam area from west to east (Shin *et al.*, 2001). 'Teheran Valley' is the new term describing the

Table 15.2 Spatial distribution of the strategic alliance of the IT service industries in the Gangnam area, by type

Type	Jung-gu/ Jongno-gu	Yeongdeungpo-gu	Seocho-gu/ Gangnam-gu	Other districts in Seoul	Outside Seoul	Foreign	Total
Technology	37 (10.5%)	22 (6.3%)	148 (42.0%)	36 (10.2%)	27 (7.7%)	82 (23.3%)	352 (100.0%)
Marketing	45 (11.6%)	35 (9.0%)	175 (45.1%)	43 (11.1%)	30 (7.7%)	60 (15.5%)	388 (100.0%)
Production	7 (11.1%)	1 (1.6%)	24 (38.1%)	7 (11.1%)	5 (7.9%)	19 (30.2%)	63 (100.0%)
Capital	6 (14.0%)	5 (11.6%)	15 (34.9%)	4 (9.3%)	5 (11.6%)	8 (18.6%)	43 (100.0%)
Others	7 (11.9%)	2 (3.4%)	33 (55.9%)	7 (11.9%)	9 (15.3%)	1 (1.7%)	59 (100.0%)

Source: Derived from the raw data materials of KRG (2003).

overwhelming concentration of IT firms, advanced service firms and other high-tech firms on the Teheran Road. Currently, more than 2,700 venture firms[5] (new start-ups and small firms in high-tech or knowledge-intensive sectors) are concentrated in the Gangnam area, which accounts for more than 20 per cent of the total venture firms in Korea. Foreign direct invested producer services are also concentrated in the Gangnam area. The concentration of new IT start-ups, high-tech firms including IT services, and R & D institutions of private firms has reinforced its position as the leading innovation centre in Korea.

Why has the Gangnam area emerged as the cluster of high-tech services and venture firms? The main reasons might be related to the supportive environments for IT firms and venture firms in the area. There are strong inter-firm networks and opportunities as more than two thirds of various advanced services such as legal, accounting, venture capital, consulting, advertising, and marketing services are concentrated in the Gangnam area. More than half of the total FDI by the producer service sector in Korea has been located in the Gangnam area (Lee, 2001). Qualified manpower is supplied from research universities in Seoul and the Gangnam area is a prestigious region for qualified employees to work and to live. Supporting infrastructure, such as IT facilities and office space, is excellent. Collective learning processes with formal and informal meetings can be regarded as a distinct culture of the Gangnam area. Although there are no leading research universities within the Gangnam area, top ranking research universities and various private R & D centres are located in Seoul. Therefore, with the exception of a leading university, the basics of a supportive environment for high-tech services and venture firms in Gangnam exist.

The Gangnam area can be regarded as a learning region in Korea with continuing processes of collective learning through intensive local networks. In order to examine the role of the Gangnam as a learning region, data from in-depth interviews and questionnaire surveys, which were conducted in early 2002, have been analysed. In-depth interviews with 100 venture firms, which were selected randomly from the 2,700 venture firms in the Gangnam area are used. Additional information from 170 venture firms, which responded to questionnaire surveys, was gathered. Based on the data derived from a total of 270 surveyed firms, sources of knowledge acquisition and spatial networks have been analysed.

The Internet is regarded as the most important source for the acquisition of codified knowledge related to product technology (Table 15.3). For the acquisition of codified knowledge on process technology, books and articles are more important than the Internet, but the Internet is still a very important source. In general, the Internet is regarded as a more important source for the acquisition of codified knowledge by IT service firms than by firms as a whole, including manufacturing firms. Related books and journals are also important sources for the acquisition of codified knowledge. Exhibitions, trade fairs and conferences were also

Table 15.3 Source of codified knowledge of product and process technology (unit: number of firms)

Sources	Books/articles	Internet	Exhibition	Newspaper/ journal/ broadcasting	Patent information	Seminar/ conference	Others	Total
Prod(S)	33 (26.6%)	51 (41.1%)	22 (17.7%)	6 (4.8%)	2 (1.6%)	7 (5.6%)	3 (2.4%)	124 (100.0%)
Prod(T)	72 (32.3%)	75 (33.6%)	42 (18.8%)	9 (4.0%)	2 (0.9%)	18 (8.1%)	5 (2.2%)	223 (100.0%)
Proc(S)	38 (35.2%)	34 (31.5%)	8 (7.4%)	4 (3.7%)	3 (2.8%)	15 (3.9%)	6 (5.6%)	108 (100.0%)
Proc(T%)	68 (35.4%)	56 (29.2%)	16 (8.3%)	5 (2.6%)	6 (3.1%)	28 (14.6%)	13 (6.8%)	192 (100.0%)

Source: Based on firm interview and questionnaire surveys.

Notes
Prod(S) = Product technology of IT service firms;
Prod(T) = product technology of all firms;
Proc(S) = process technology of IT service firms;
Proc(S) = process technology of all firms.

mentioned as a source of codified knowledge by a considerable percentage of respondent firms. It should be noted that the information infrastructure of the Gangnam area is excellent and a diversity of international conferences and exhibitions are held in the convention centre and hotels in the Gangnam area.

The sources of tacit knowledge show a completely different story. More than 40 per cent of the respondent firms regard R & D activity within firms as the most important source for the acquisition of tacit knowledge regarding product and process technology (Table 15.4). In general, IT service firms emphasize the importance of R & D for the acquisition of the tacit knowledge in product and process technology more than manufacturing and other service firms. Even though R & D activities are the most important source of tacit knowledge, more than half of the firms regard other sources as more important than R & D activities for the acquisition of the tacit knowledge about product and process technology. For the acquisition of the tacit knowledge for product technology, more than one-third of the responding firms regard the CEO's and worker's personal relations as most important. It is noticeable that the CEO's personal relations as an important source for product technology, and interactions with customer firms are relatively important for the acquisition of process technology. Overall, inter-firm relations with customers and suppliers are also important sources of tacit knowledge. About 10 per cent and 16 per cent of the respondent IT service firms said that suppliers or customers are the most important sources for the tacit knowledge with regard to product and process technologies respectively.

The results in Table 15.4 suggest that formal and informal relations with personnel and institutions are more important sources for the acquisition of the tacit knowledge for product and process technology than formal R & D activities. This result is consistent with OECD (1999) data in which R & D shares only 33.5 per cent of the total expenditure for innovation in the OECD countries.

Formal and informal meetings are important for the acquisition of tacit knowledge. Entrepreneurs in the Gangnam area have an average of two formal meetings and three informal meetings per month. There are many formal and informal meeting groups such as the E-Business Club, Software Industry Club, Network Communication Club, Venture Leaders Club, I-Partnership, etc. In these formal and informal meetings, many entrepreneurs, engineers, university professors and venture capitalists gather to share their information and knowledge.

The in-depth interview surveys confirm that the spatial networks of innovation in the Gangnam area are considerably localized (Figure 15.8). Venture firms in the Gangnam area have strong innovation and cooperative networks with other firms, entrepreneurs, private research centres, venture capitalists, professional consultants and other business services. They also have strong cooperative networks with universities

Table 15.4 Source of tacit knowledge of product and process technology

Sources	A	B	C	D	E	F	G	H	I	J	Total
Prod(S)	35 (28.2%)	58 (46.8%)	3 (2.4%)	7 (5.6%)	5 (4.0%)	7 (5.6%)	3 (2.4%)	5 (4.0%)	1 (0.8%)	0 (0%)	124 (100.0%)
Prod(T)	64 (29.4%)	88 (40.4%)	7 (3.2%)	10 (4.6%)	10 (4.6%)	18 (8.3%)	5 (2.3%)	12 (5.5%)	4 (1.8%)	0 (0%)	218 (100.0%)
Proc(S)	14 (12.4%)	53 (46.9%)	7 (6.2%)	9 (8.0%)	4 (3.5%)	14 (12.4%)	5 (4.4%)	1 (0.9%)	1 (0.9%)	5 (4.4%)	113 (100.0%)
Proc(T)	27 (13.5%)	90 (45.0%)	14 (7.0%)	16 (8.0%)	7 (3.5%)	23 (11.5%)	5 (2.5%)	8 (4.0%)	3 (1.5%)	7 (3.5%)	200 (100.0%)

Source: Based on firm interview and questionnaire surveys.

Notes
Sources of knowledge acquisition are as follows:
A = CEO's personal relation;
B = R & D;
C = Intra-firm seminar;
D = Worker's personal relation;
E: = Customers;
F: Suppliers;
G = Informal relation;
H = University/Research Institute;
I = Foreign relations;
J = Others.
Prod(S) = Product technology of IT service firms;
Prod(T) = product technology of all firms;
Proc(S) = process technology of IT service firms;
Proc(S) = process technology of all firms.

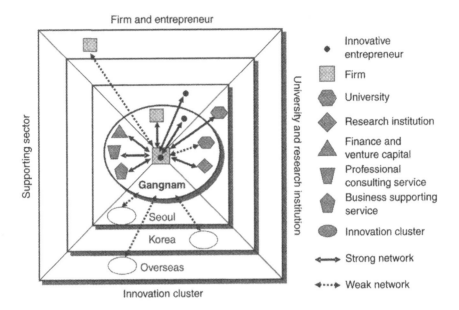

Figure 15.8 Spatial innovation networks of Gangnam area.

and entrepreneurs located in Seoul outside the Gangnam area. They have relatively weak networks with other innovation clusters outside the Gangnam area and other foreign firms. Compared to other industrial or innovation clusters in Korea, firms in the Gangnam area have relatively strong localized networks.

Firms in the Gangnam area have considerable innovation networks with foreign firms, with about 11 per cent of respondent firms saying that foreign firms had contributed to their innovation activities. The other survey shows that about 12 per cent of firms in the Guro Industrial Park located in Seoul have innovation networks with foreign firms, while only 8 per cent of firms in the Gumi Industrial Park located in southeastern part of Korea have innovation networks with foreign firms (Park, 2002).

There are some Korean ethnic networks in Silicon Valley. The Korean-American Association of Entrepreneurs has about 2,000 members, while the Korean American Professional Society, Korean American Chamber of Commerce of Silicon Valley and Silicon Valley Korean IT Forum also have a considerable number of members (Han, 2002). The Korean government supports new start-ups in the IT sector by the Korean ethnic group in Silicon Valley, such as i-Park. On the one hand, Korean engineers and entrepreneurs in Silicon Valley have connections with the Gangnam area in terms of flow of information, marketing and manpower. On the other hand, small high-tech Korean firms as well as large high-tech Korean firms such as Samsung, LG and Hyundai have invested in the Silicon Valley R & D sector, making connections between Silicon Valley and Gangnam in terms of flows of capital, manpower, knowledge and technology, and strategy. Gangnam also has international networks with innovation clusters in India and China via flows of knowledge and technology, capital, and products.

Overall, firms in the Gangnam area have a strong localized network of innovation even though the innovation networks vary along spatial scales at local, regional, national and international levels. They have significant international networks of innovation, especially with the USA, but the collective learning processes within the Gangnam area are more important for innovation than national and international networks. This suggests that localized innovation systems may be evolving.

Conclusion

The purpose of this study has been to examine the processes through which IT service industries have developed and underpinned the emergence of a new knowledge-based economy core in Seoul. The spatial distribution of the IT service industries, uneven growth of the IT industries and spatial networks of strategic alliances of the IT service firms all confirm that the Gangnam area is a newly emerging core of Seoul, substituting the former dominant role of the traditional CBD. Seoul has been transformed into a multi-core city since the 1980s. However, since the financial crisis in 1997, the role of the Gangnam area as a leading core has been strengthened.

Comparatively good local supportive environments in the Gangnam area have been a critical factor for attracting high-tech firms, including IT service firms, and developing a new core within Seoul. A high-tech venture environment, or venture habitat, has been developing as a result of the continuous interactions among spin-offs, high-tech venture firms and advanced services within the region. Formal and informal meetings contribute to the creation and transfer of tacit knowledge with regard to product and process technologies. Strategic alliances and inter-firm networks within the capital city have synergy effects on the IT service industry clustering in the Gangnam area.

The development of the Gangnam area as a main core of Seoul and a centre of IT service industry in Korea suggests that, even in the era of globalization and the Internet, regions are critical for innovation processes because organization in terms of inter-firm networks and collaboration, learning dynamics with interactive processes and territorial dimensions with regional clustering are important. Collective learning processes through formal and informal meetings, inter-firm cooperation and competition, intra-firm networks and inter-organizational collaborations are all important for local networks of innovation.

The formation of local networks of innovation and the collective learning processes are critical for the evolution of regional innovation systems in the Gangnam area. However, the significant global networks of innovation and strategic alliances of the firms in Gangnam suggest that the development of regional innovation systems is not completely related to the regional dimension and a closed regional system. Local, regional, national and global scales are all interlinked for the network of innovation. Innovation clusters are linked to each other in the global space economy beyond the regional or national boundary.

The findings of this study suggest that there are three issues to be discussed with regard to the changes of the IT services in the era of the Internet and the knowledge-based economy. They are relevant to other metropolitan areas in Korea, peripheral areas in Korea, and other large cities in the Asia-Pacific Rim areas.

First, other metropolitan areas outside the Capital region are planning to promote the development of the IT services. There has been a slight trend of decentralization of producer services from Seoul to suburban areas in the Capital region. However, the other metropolitan areas outside the Capital region have not experienced rapid growth of advanced services. Most of the industrial cities in Korea have links to Seoul for the supply of advanced services, including IT services. Other metropolitan areas, which are mostly regarded as regional centres, are struggling with attracting the advanced services in order to develop their own regional innovation systems. The development of the IT services cluster in the other metropolitan areas is not an easy task because of the dominance of Seoul, especially the cluster of IT services and their strong networks in the Gangnam area. Strategic alliances of the IT firms within the other metropolitan areas are not well developed yet. There is, however, a possibility of the emergence of clusters of advanced services in some of the other metropolitan areas outside the Capital region, such as Daejeon, in which Daeduck Science Park is located. National and local governments are planning to promote the regional innovation systems (RIS) outside the Capital region in order to promote balanced national development in the knowledge-based economy. Cluster formation of the advanced services focusing on the process of collective learning through formal and informal networks will be a critical issue for the success of the RIS in Korea.

Second, in peripheral regions, strategic alliances or cooperation between industry and local universities or research centres will be more important in the Internet era. Some cases of rural areas in Korea suggest that new paradigms in the organization of production systems and economic spaces are emerging, even in the peripheral areas, developing new ideas based on intensive local and non-local networks (Park, 2003). The processes of forming innovation networks are applicable not only to the Gangnam area, but also to developing areas of the periphery with industry-university links for using local resources and developing new products. In the peripheral areas strategic alliances and innovation networks with universities or R & D centres located in the regional centres will not be rare in the knowledge-based economy of the Internet era. Therefore, even though an actual cluster of advanced services in the rural areas cannot be easily developed, networks with regional centres will be strengthened with the development of an information society and provision of enhanced transportation networks.

Third, international networks of strategic alliances of IT services will become more important in the Asia-Pacific Rim. Inter-firm networks through strategic alliances and cooperation have significantly contributed to the development of high-tech clusters or IT service clusters in the Asia-Pacific as seen in the cases of Hsinchu Science Park in Taiwan and Bangalore in India. Ethnic networks between the USA and the Asia-Pacific have been an important process of the innovation networks and cluster of high-tech industries and services (Saxenian, 2000). Chinese networks will take a significant role in the formation of clusters and local and global networks via strategic alliances of the advanced services in the large metropolitan areas such as Shanghai, Beijing, or Singapore. Trans-Pacific networks in the Asia-Pacific, especially with Silicon Valley, have been dominant in the ethnic networks and these networks will be strengthened in the era of the Internet and the knowledge-based economy. The progress of such networks and clusters in the Asia-Pacific will have a significant impact on the reorganization of the internal structure of global cities and overall economic spaces in the Pacific Rim.

Notes

1 In this paper, IT manufacture includes Manufacture of Computers and Office Machinery (KSIC = 300), Manufacture of Insulated Wires and Cables, Including Insulated Code Sets (3130), Manufacture of Semiconductor and Other Electronic Components (321), Television and Radio Transmitters and Apparatus for Line Telephony and Line Telegraphy (322), Manufacture of Television and Radio Transmitters and Apparatus for Line Telephony and Line Telegraphy (3220), Manufacture of Television and Radio Receivers, Sound or Video Recording or Reproducing Apparatus, and Related Goods (323), Manufacture of Instruments and Appliances for Measuring, Checking, Testing, Navigating and Other Purposes, Except Industrial Process Control Equipment (3321), and Manufacture of Industrial Process Control Equipment (3322). Telecommunication is defined as Telecommunications (642) including Wired Telecommunications (6421) and Wireless Telecommunications (6422). IT wholesale and renting includes Wholesale of Computers and Non-Customized Software (51891), Wholesale of Office Appliances (51892), Wholesale of Navigating and Telecommunication Equipment (51893) and Renting of Computers and Office Equipment (7122).
2 E-commerce via ICT networks can be included in the category of IT service industries in a narrow definition. However, it was not excluded in this paper because it just began to emerge at the end of 1990s.
3 When the value of LQ equals unity, the spatial distribution of a given industry is

the same as that of the base magnitude and greater than one represent relative specialization of the given industry.

4 Along the Teheran Road, which goes across the Gangnam-gu from the west to the east, many large IT firms and new start-ups located in this area (Shin *et al.*, 2001).

5 Venture firms in Korea are defined as at least one of the following categories:

1 New start-ups invested by venture capital (venture capital: more than 20 per cent of total capital);

2 R & D intensive start-ups (R & D expenditure: more than 5 per cent of total sales);

3 New technology development start-ups (products based on patent or new technology: more than 50 per cent of total sales or export of the products based on patent or new technology is more than 25 per cent of total sales);

4 New start-ups with good technology (evaluated by Evaluation Institute).

References

Daniels, P. (1998) 'Economic development and producer services growth: the APEC experience', *Asia Pacific Viewpoint*, 39(2), 145–60.

Han, C.-H. (2002) 'Foreign entry promotion of venture firms through Korean ethnic networks', Unpublished MBA thesis, KAIST (in Korean).

Kim, J. G., Park, S. O., Bae, J.-T. and Seo, S.-M. (2002) *Venture Habitat and Long-term Development Plan for Teheran Valley*. Kangnam-gu (In Korean).

KRG (2003) *2003 White Paper of IT Market*. Seoul: 2003 Computer Associates International Inc. (CA).

Lee, B. M. (2001) 'Spatial characteristics of foreign firms' corporate networks in Korea: the case of business service', Ph.D. Dissertation, Department of Geography, Seoul National University.

Park, S. O. (2003) 'New economic spaces in the information society: networks, embeddedness, and cluster', Paper presented at the Annual Conference of the IGU Commission on the Dynamics of Economic Spaces, Vancouver, August 10–15, 2003.

Park, S. O. and Nahm, K. B. (1998) 'Spatial structure and inter-firm networks of technical and information producer services in Seoul, Korea', *Asia Pacific Viewpoint*, 39(2), 209–20.

Saxenian, A. (2000) 'Networks of immigrant entrepreneurs', in Lee, C. M., Miller, W. F., Hancock, M. G. and Rowen, H. S. (eds), *The Silicon Valley Edge*. Stanford, California: Stanford University Press.

Shin, C. H., Jeong, B. S., Kang, C. D. and Kim, R. H. (2001) *Clustering of Information Technology Industries in Seoul*. Seoul: Seoul Development Institute (in Korean).

16 Epilogue

P. W. Daniels, K. C. Ho and T. A. Hutton

In a recent article, Yeung and Lin (2003: 121) point to the rapid and dramatic transformations in the economic landscapes of Asia in recent decades as being the key impetus for a continuing interest among economic and urban geographers. This is also the central concern of this book. Service development trajectories in Asia-Pacific are not only dramatic but in many ways they are also quite different from those experienced earlier in Europe and North America. The unevenness of urban development in East and Southeast Asia means that service industries are likely to remain, at least for the foreseeable future, much more concentrated in the primate cities of these regions, compared with the relatively more dispersed distributions now typical in Europe and North America, even though certain world cities such as New York and London remain pre-eminent. As late-comers to development, many Asia-Pacific economies (especially those in Asia) have also been able to incorporate and tap into newer and more efficient information and telecommunications infrastructures in the major cities, with the result that the penetration of services, especially higher order activities such as producer services, has occurred at a much faster pace. This rapid absorption of services has also been facilitated by increasing attempts at deregulation, particularly within East Asia (notably Japan and South Korea).

The chapters in this collection demonstrate the ways in which urban economies in the Asia-Pacific are shaped by complex patterns of the service development. This makes generalization difficult but we suggest that three broad patterns emerge. First, developmental state policies have undoubtedly helped to create high technology IT and producer services clusters that are envisioned as the leading edges of national and urban economic development. Second, the continual search for new economic development niches and opportunities has also resulted in the development of new service products such as higher education and cultural products. The rapid economic growth in China over the past decade has also supported the rapid growth of lower end consumer services as well as higher value services, creating opportunities for the entry of service transnationals into major cities such as Shanghai, Guangzhou or Xi'an.

Third, the lower tiers of the Asia-Pacific urban hierarchy still largely comprise Third World cities where there is a significant informal/bazaar economy of goods and services.

The contributions to this volume also highlight a range of transformative and integrative processes at different scales in which service industries perform an important role. Regionally, air cargo services encourage internal integration of the region as well as engagement with other major economic regions in Europe and elsewhere. At the metropolitan level, producer services have transformed labour and housing markets, as well as influencing consumption practices, lifestyles and identity, while at the metropolitan-industry level transnational and local service sector actors have coalesced and developed collaborative practices. At the metropolitan-hinterland level, producer services are linking economic activities with the immediate sub-regions and providing vital knowledge and information inputs that raise the competitiveness of individual firms and city regions more generally.

Finally, it is very much the case that the publication of this volume by no means represents the conclusion to the research initiated by the small band of committed researchers that first met in Sydney almost a decade ago. The contributions provide critical insights to an enhanced appreciation of the dynamic and undeniably significant relationship between city formation and the rise of service production and consumption across the Asia-Pacific. But in many ways, they still represent the start rather than the end of the story. This does not reflect the quality of the research and the insights already available, rather it is a function of the internal economic, social, cultural and geopolitical diversity of the region combined with its evolving external relationships with the global service economy. There is therefore a great deal of scope and need for further research under the general rubric of service industries, cities, and development trajectories in the Asia-Pacific.

Some potential avenues for this additional research are offered here but we suggest that they are indicative rather than definitive in view of our own predilections for particular approaches and topics and the disciplinary contexts within which each of us is embedded. Thus, while globalization as a concept or as a process remains strongly contested there is no doubt that the integration of service industries into the economic and social fabric of cities in the Asia-Pacific is partially shaped by international trends and requirements. The cities of the region are certainly its window on, and contact point with, the international world of advanced business and professional services that have become vital players in the production process or the competitiveness of individual firms. There remains much to be understood about these synergies, the extent to which there are internal constraints (within the Asia-Pacific) on local production of advanced services, the effects of liberalization of international service transactions on the ability of Asia-Pacific service firms to 'catch-up' with producers based outside the region but represented within it (mainly in regional corporate control centres such as Hong Kong or Singapore), and the drivers needed to change the balance between internal and external supply of advanced services.

A related requirement is for more research on the ways in which the economies of the region, and more explicitly the major cities, can move to a situation in which there is less dependence on service TNCs from outside the region for the production and supply of advanced business and professional services. What are the impediments and the required mechanisms for enabling domestic suppliers of these services to acquire the quality of knowledge and human resources that will enable them to displace, or at least to work alongside, international suppliers. Is such a transition a desirable process or is the goods-services gap that is a feature of many of the economies in the region to be encouraged rather than ameliorated, on the grounds that the comparative advantage in advanced services enjoyed by regions such as Europe is sustainable provided that most of the Asia-Pacific economies maintain a comparative advantage in goods production.

It may, however, be unrealistic to envision a role for Asia-Pacific economies and cities that is less service intensive than North America and Europe. The most advanced economies of the region have experienced a substantial 'hollowing out' of basic manufacturing capacity and labour (Japan, Hong Kong, Singapore; areas of Pacific North America and Australasia), exalting the significance of service industries at the regional and national levels. More broadly, the imperatives of capitalism dictate a constant search for lower cost factors of production which has initially been reflected in the transfer of goods production into the region but which more recently has also included a growing demand for outsourced service work. The rapid expansion of outsourced computer services and software production to India and the relocation of call centre jobs from Europe and North America to various countries in the Asia-Pacific merits closer examination in terms of the true scale of the process, the motivations of those transferring the jobs, the determinants of locations for investment, and the economic, social, cultural and physical impacts on the cities that are recipients of these developments. There is much debate about the significance of service work outsourcing for both the home and recipient countries but there seems no doubt that it offers significant potential for some Asia-Pacific economies, and certainly many of the cities use service industries as contributors to their development trajectories in ways that would not have been envisaged even ten years ago. Such investments involve raising the quality and capacity of the physical and economic infrastructure of the participating locations and this may trigger other trickle down/multiplier effects that make those locations attractive to further rounds of service-related investment. These trends and processes will likely comprise important elements of the evolving research agenda on services and new development trajectories within the Asia-Pacific.

Reference

Yeung, H. W. C. and Lin, G. C. S. (2003) 'Theorizing economic geographies of Asia', *Economic Geography*, 92(2), 107–28.

Index